ENERGY MANAGEMENT PRINCIPLES

ENERGY MANAGEMENT PRINCIPLES

Applications, Benefits, Savings

Second Edition

CRAIG B. SMITH
Dockside Consultants, Inc.

KELLY E. PARMENTER
Applied Energy Group

ELSEVIER

AMSTERDAM • BOSTON • HEIDELBERG • LONDON • NEW YORK • OXFORD
PARIS • SAN DIEGO • SAN FRANCISCO • SINGAPORE • SYDNEY • TOKYO

Elsevier
Radarweg 29, PO Box 211, 1000 AE Amsterdam, Netherlands
The Boulevard, Langford Lane, Kidlington, Oxford OX5 1GB, UK
225 Wyman Street, Waltham, MA 02451, USA

Notices
Knowledge and best practice in this field are constantly changing. As new research and
experience broaden our understanding, changes in research methods, professional practices,
or medical treatment may become necessary.

Practitioners and researchers must always rely on their own experience and knowledge
in evaluating and using any information, methods, compounds, or experiments described
herein. In using such information or methods they should be mindful of their own safety
and the safety of others, including parties for whom they have a professional responsibility.

To the fullest extent of the law, neither the Publisher nor the authors, contributors,
or editors, assume any liability for any injury and/or damage to persons or property as a
matter of products liability, negligence or otherwise, or from any use or operation of
any methods, products, instructions, or ideas contained in the material herein.

ISBN: 978-0-12-802506-2

British Library Cataloguing-in-Publication Data
A catalogue record for this book is available from the British Library.

Library of Congress Cataloging-in-Publication Data
A catalog record for this book is available from the Library of Congress.

For Information on all Elsevier Publishing publications
visit our website at http://store.elsevier.com/

Typeset by MPS Limited, Chennai, India
www.adi-mps.com

Printed and bound in the US

DEDICATION

To Nancy J. Smith and Verne, Cory, and Tenaya Parmenter
for love and unwavering support.

CONTENTS

LIST OF FIGURES

LIST OF TABLES

FOREWORD TO THE FIRST EDITION

It is now common wisdom that energy supply is a crucial need of any society. Analysis of the historical influence of energy availability and cost on the development of all societies indicates that the effects are substantial and diverse. The variety of energy resources, forms, and end-uses is so great that there exists a broad spectrum with regard to societal needs, efficiency of use, importance to public welfare, and flexibility to change. As with national food consumption patterns, we have the lean and the fat, good nutrition and malnutrition, the efficient and the wasteful, the cheap and the costly. What is clear from energy systems studies is that the quality of life, material welfare, health, employment, and income are demonstrably dependent on ample energy availability and low cost.

It now appears that there will be no return to the cheap abundant energy of the past. With the bulk of the world's population avidly seeking to increase their well-being, mostly through energy-dependent technology, future energy supply and its effective use is becoming a crucial issue worldwide. Compounding this problem are the undesirable byproducts of expensive energy use resulting from the mining of fuels, their transportation, and the discharge of waste into the biosphere.

The inherent finiteness of our world, and those of our ultimate resources, makes it obvious that indiscriminate use of energy cannot continue indefinitely. Even the use of renewable energy resources has its environmental problems. There'll always be a continuing pressure for reduced energy demand and for conservation to use our existing and future resources more efficiently. Today, there exists no realistically conceivable set of circumstances that would eliminate the desirability of carefully managing our energy uses.

However, in order to fully grasp the potential for energy husbandry, we must understand the more professional aspects which underlie it. In *Energy Management Principles*, Craig Smith provides us with the theoretical basis for conservation, as well as the avenues for its application. All too often we find conservation literature either too technical in its approach or else so general that it fails to develop any commonality of approach. In this book, Smith has managed to strike a delicate balance between theory and practice. Furthermore, by avoiding the political and ideological aspects of conservation, he provides us with a factual and coherent piece of analysis.

Extensive opportunities exist for the more efficient use of energy in all sectors of our society. By providing us with an analysis of our inefficiencies, the author shows the potential for the reduction. Using this approach, with reference to specific case examples, the author has successfully extracted a set of general principles of energy management.

This book is designed for the serious student who wants to learn and understand the general principles and methods of energy management. I think you will find it a very informative and useful tool for developing your own ideas about energy's role in society and the need for energy management program to deal effectively with the many energy-related problems we face today and will face in the future.

Chauncey Starr
Vice Chairman
Electric Power Research Institute
Palo Alto, CA, USA

Author's note: Dr. Chauncey Starr, 1900–2000, was a pioneer in many aspects of the energy field. He began his career working on the Manhattan Project, and then later headed up Atomics International, a company that made many pioneering developments in nuclear power, including the first nuclear systems used in space. He served as Dean of Engineering at UCLA, and later was responsible for creating the Electric Power Research Institute, and was its leader for a number of years. He remained active in the energy field up until the time of his death, remaining an innovative thinker and active participant into his 90s. His Foreword, written in 1980, remains accurate and applies today.

PREFACE

From the Preface to the First Edition:

Since 1970, I have been examining the opportunities for more efficient energy use. In 1972, one of my projects was to help with the planning of a NATO Science Committee conference on the subject, which was held at Les Arcs, France. I recall hearing an announcement of the Arab-Israeli war on the plane as my wife and I were flying to this conference in October 1973.

Events have changed the world since 1973. It is now clear that new approaches must be developed, new strategies must emerge, as regards energy use. I remain convinced that a resource of major dimensions is those actions that can be taken by end users themselves. Basically, they have untapped energy sources in their homes, offices, and factories—the energy wastes of past years.

Now that prices are higher and supplies more limited, it makes economic sense to collect these wastes—formerly uneconomic to reclaim—and put them to work.

That is what this book is about.

Craig B. Smith

Although written three decades ago, the words appearing above still apply. Since 1973, we have experienced additional oil shocks, dire warnings of natural gas shortages, the demise of nuclear energy as a commercial power source, and the emergence of wind power and solar power as growing but still limited sources of renewable energy. There has been a wide range of improvements in energy use efficiency across all sectors of the economy. In the transportation sector, automobiles, trucks, and even aircraft have made significant improvements in fuel efficiency. Common household appliances such as refrigerators have also made great gains in energy efficiency. There have been significant efficiency gains in electric motor drives for industry and much improved space conditioning systems and controls for buildings. However, there is still a large untapped potential for greater gains.

During the last three decades, I've continued to track the improvements made in the energy sector, most recently in collaboration with Kelly Parmenter, Ph.D. Over this period of time in the United States and in other countries worldwide, the energy used to produce one dollar of gross national product has dropped dramatically. At the same time, we have witnessed a steady increase in the price of crude oil, vehicle fuels, and electricity. While it is clear that global production of petroleum will peak and eventually decline, the warnings about imminent shortages of natural gas proved to be temporarily incorrect. However, there is little doubt that natural gas, like all fossil fuels, is finite in supply.

As long as a half a century ago, some scientists expressed concerns about the prospects of increased CO_2 concentrations in the atmosphere affecting global climate. While controversy rages about whether the climate changes we are experiencing are natural or man-made, there is little doubt but that they are occurring. In addition to effects on reefs, ocean food supplies, and altered weather patterns of drought and storms, we face inundation of coastal areas due to melting of the polar ice caps.

Resolution of these problems requires international cooperation and collaboration of unprecedented scope. Nations must put aside nationalistic interests and work together for the common good. Current efforts to do this on an international scale have met with great resistance, with the United States failing to take a leadership position. We each hope that things will change for the better in the future, but in the meanwhile each one of us can assist in reducing this threat to our wonderful planet by making certain that our own personal energy use is as efficient as possible.

Craig B. Smith
Balboa, CA, USA
Kelly E. Parmenter
Santa Ynez, CA, USA

ACKNOWLEDGMENTS

The authors gratefully acknowledge individuals who have assisted in the preparation of this book. In particular, Mr. Kim Koch, President, Farpointe Construction Company, kindly reviewed current building code issues, gave a tour of job sites, and provided several photographs. Mr. Keith Garrison, General Manager, GBF Enterprises, provided a factory tour, photographs, and commentary on new manufacturing technologies. We thank John Lockwood for the LED landscape photo in Figure 9.2. Other photographs by the authors unless otherwise noted. John Murphy shared data on energy management projects in a manufacturing plant. We are grateful to Cheryl Prather and Laurel Husain at Punahou School, Honolulu, for information concerning Punahou's most recent efforts to develop a sustainable campus.

We appreciate our reviewers. Mr. Joe Prijyanonda, Applied Energy Group, reviewed Chapter 8. Professor Kuppu Iyengar, University of New Mexico, reviewed Chapter 12. Professor Emeritus Gerald J. Thuesen, Georgia Institute of Technology, reviewed Chapter 13. We are very grateful to Charlotte McKernan and Christopher Murphy who did a great job converting sketches and faded Xerox copies into legible illustrations. We thank our past and current colleagues from ANCO Engineers, AECOM, EPRI, Global Energy Partners, EnerNOC, and Applied Energy Group/Ameresco for all we have learned from them. We extend our sincere appreciation to all of our loyal clients who have afforded us opportunities to share and continue to develop energy management practices. We also commend the Association of Energy Engineers for the work it does to promote and educate energy engineers. Finally, this book would not have been possible without the encouragement and support of the staff at Elsevier, especially Natasha Welford, editorial project manager, Engineering, and Lisa Reading, senior acquisition editor, Energy.

CHAPTER 1

Introduction

INTRODUCTION

Energy is essential to life and survival. Reduced to bare essentials, stripped of thermodynamics, economics, and politics, this is how we must view it.

Energy may well be the item for which historians remember the last half of the twentieth century, as it marked the beginning of a new era of change, an era of possibly greater fundamental significance than the Industrial Revolution. For several centuries mankind grew lazy, lulled into complacency by the ease with which multitudes could be fed, housed, and transported using the abundant supplies of low-cost energy that were readily available.

Then, in less than a decade (1973—1981) the bubble that had taken 114 years to swell (since Drake's first well in 1859) finally burst. Long unheeded warnings took on a prophetic aspect as fuel shortages and rising costs nearly paralyzed industrial economies and literally shocked the world into an inflationary period that lasted years.

It is remarkable that our lives could be so affected by one perturbation to the world economy. Figure 1.1 shows what this perturbation was— initially, a tenfold increase in crude oil prices in less than a decade, followed by two decades of relatively constant prices as efficiency measures were invoked worldwide to curtail demand. Then, at the beginning of the new millennium, prices skyrocketed again, more than tripling in 8 years. Note that Figure 1.1 shows the *average* annual oil prices, so the spikes and dips are smoothed out. Following 2008 the global recession brought about a drop in demand, causing the price to plummet, but in 4 short years it returned to hover near US$100 per barrel. Next, as U.S. dramatically moved to become a net exporter of oil, the OPEC countries, principally Saudi Arabia, began flooding world markets with oil, causing a precipitous plunge in the average price. Over a period of a few months it dropped from US$95.85 per barrel in September 2014 to US$40 per barrel in August, 2015.

One thing is certain—low oil prices undercut the incentive for higher cost renewable energy and for electric or hybrid vehicles. Low prices also detract from efforts to reduce greenhouse gas emissions. A more draconian

Energy Management Principles.
DOI: http://dx.doi.org/10.1016/B978-0-12-802506-2.00001-X

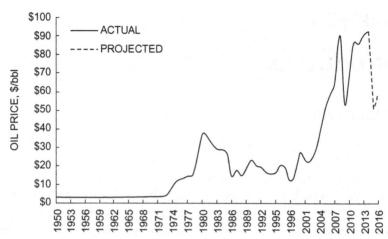

Figure 1.1 Average historic oil prices.

objective is to push higher cost shale oil, tar sands, or offshore oil producers out of the market, and even drive them to bankruptcy.[1] A certain sign of this is a rapid increase in the number of idle drilling rigs, as has occurred in early 2015. On the one hand this leads to a loss of jobs; on the other hand, cheap fuel reduces transportation and manufacturing costs, so it is not without some short-term economic benefit.

Of course, in reality the problem is much more complex, involving not only oil prices but also the uneven geographical distribution of energy resources, the exponential growth of populations and fuel consumption, the desires of poorer nations throughout the world, political and national security considerations, and long-term environmental effects.[2]

Tragically, the finiteness of energy resources can be a cause for moving the world into war. Resources of all types are essential to war, and in themselves can be causes for the rise and fall of nations. Twenty-five centuries ago, Greece denuded its forests building ships to continue the Peloponnesian wars; in 1940, Germany seized the Rumanian oil fields at Ploesti when it could no longer import petroleum due to the British blockade; a year later Japan attacked the U.S. Pacific fleet at Pearl Harbor in order to gain access to oil and mineral resources in the South Pacific. In the Six Day War (1967), the Israelis captured the Egyptian oilfields in

[1] Bell, Ruth Greenspan, and Roddenbeck, Max, (2015) "The Cheap Oil Strategy," p.A13 *Los Angeles Times*.

[2] This Chapter is based on Smith, Craig B. (1981), Chapter 1, pp.1—10, *Energy Management Principles: Applications, Benefits, Savings*. Pergamon Press: Oxford.

Sinai, while in its 1980 attack on Iran, Iraq went after the large Abadan refinery complex and other strategic points in Iran's oil-producing western province of Khuzestan. Later, Iraq's invasion of Kuwait in 1990 and its threatening of the vast oil resources of Saudi Arabia triggered the first Gulf War and indirectly went on to cause a huge turmoil in the Middle East that has continued for more than two decades at an enormous cost in lives and money.

Efficient energy use, therefore, not only increases one's independence of external energy supplies, but also helps diffuse a potentially unstable international situation. Energy independence has been touted as a goal by several U.S. presidents beginning with Jimmy Carter. The same is true of other industrialized countries. However this goal has proven to be more elusive than first thought.

RESPONDING TO A CRISIS

In 1973, the Community Concourse (six city-owned buildings in San Diego, California) used 21 million kWh of electricity per year at a cost of $270,000. By the end of 1975, the cost increased by 22% to $330,000 annually due to dramatic increases in electricity rates, even though stringent energy management measures had been instituted immediately following the oil embargo in October 1973. These measures, which included an employee awareness campaign, adjustment of lighting levels by delamping, changes to thermostat set-points, and revised operating procedures in the building HVAC systems among other actions, resulted in a savings of roughly 8 million kWh per year or 37% relative to the 1973 level. Without the energy management program, the cost of operating this facility in 1976 would have doubled to approximately $520,000 per year, to be paid by local taxpayers. This example describes what happened in six large municipal buildings. There are thousands of buildings throughout the U.S. and other countries for which similar stories may be told.

Meanwhile, farther to the north, citizens in Seattle were asked to approve participation in a nuclear power plant project. The project was under consideration because additional low-cost hydroelectricity capacity was no longer available.

After extensive investigation in 1976, Seattle decided not to participate in the new power project. Instead, the city proposed to undertake an energy management program and use the savings gained by more efficient energy use to offset future power needs. This bold proposal—not without the possibility of some severe economic penalties if Seattle's optimism was overstated—hypothesized that nearly half (230 MW) of predicted future growth needed by 1990 could be met by an energy management program. The program included formation of a city office of conservation, residential insulation retrofit, new construction standards, appliance standards, energy use disclosure reports, heat pump projects, and energy management research and development. The program was a success. In 2008, Seattle launched another innovative energy management program with the title "Building a World-Class *Conservation* Power Plant."

Europe and the United Kingdom launched programs similar to those in the U.S. to address the energy crisis. They established new speed limits, curtailed use of automobiles on Sunday, imposed space heating temperature limits, and invoked new lighting standards. Even once the supply shortages were no longer a concern, Europe was left with a tenfold increase in oil prices compared to a few years earlier. This had a severe effect on European economies.

Two years after the 1973 embargo, Arizona moved to ban all new hookups of natural gas. Other states began reviewing energy supplies and uses. New Mexico proposed a tax on energy exported out of the state. Three years later, the California Public Utility Commission established priorities for natural gas use; it was prohibited as a fuel in generating plants. Over the next several years natural gas was to be phased out in industry; first as a boiler fuel, then for all process heat applications for which a substitute fuel—usually oil—could be found.

The impact varied from firm to firm. In a large manufacturing plant, the potential loss of gas-fired boiler capacity led to an investigation of heat recovery possibilities. It appeared possible to reclaim heat dissipated by several 4,600 hp air compressors; before, the heat was extracted by interstage coolers and discharged to the atmosphere from a cooling tower. (See Chapter 11).

In a smaller plant that manufactured agricultural antibiotics, the crisis meant that no natural gas was available to fuel a drying oven needed to expand the plant's capacity. Looming in the future was the possibility of fuel curtailment, resulting in a shutdown of the plant's boiler and existing drying ovens (Text Box 1.1).

Jumping back to the 1970s, Los Angeles passed an emergency ordinance following the oil embargo when it became apparent the city did not have sufficient fuel to meet all needs. Commercial users were asked to reduce electricity use by 20%, industry by 10% and residential

TEXT BOX 1.1

Jumping ahead three and a half decades to 2015, we can add an ironic footnote to the international concerns regarding natural gas shortages. History has reversed itself; today there are abundant supplies of natural gas, to the point that it is now being widely used as a utility power plant fuel, being cheaper than fuel oil and creating fewer environmental emissions.

consumers by 10%. The City set up an energy management program for its own facilities. (See Chapter 4).

One Southern California family installed florescent lighting, better insulation, and additional switches for lighting, as well as changed thermostat settings, and operated appliances more efficiently. As a result, annual electricity use for a family of four went from 6,859 kWh per year ($156/year cost) in 1972 to 3,868 kWh per year in 1974. By then, rising prices had brought the cost back up to $141/year; in 1975 the cost was the same as 1972, even though the usage had dropped to about 56% of pre-embargo level. Yet, without the *energy management* efforts extended by this family, they would have incurred a sharp increase—perhaps a doubling—of utility costs. A little more than 40 years have passed, yet we vividly recall these experiences, as it was our home.

We have drawn each of the examples discussed above from our own experiences. The examples have one thing in common: they illustrate the response that was taken all over the world as people encountered rapidly escalating energy prices. Over the succeeding decades, the cumulative results were remarkable. As we will show, national energy use in many countries declined, while gross industrial output increased. These examples illustrate the practice and benefit of energy management.

PURPOSE OF THIS BOOK

When energy problems caused by rapidly increasing demand in the face of dwindling fuel supply first became apparent in the early 1960s, the immediate response was to seek new supplies and alternative fuels. Later, consideration was given to the end-user as a means of conserving fuel and capital: by improving end-use efficiency, supply problems were automatically eased. The oil embargo of 1973 gave an additional stimulus to users—in both industrialized nations and in Third World countries—to make the most effective use of fuels and energy.

Approaching energy problems from the user's end, rather than the supply end, introduces new challenges. First, the number of users is much greater than the number of suppliers, thus complicating the problem. Second, communication with users is difficult due to their number and diversity. Third, the full range of end-use technologies is not readily dealt with by legislative or regulatory controls, also due to diversity. Fourth, the technological sophistication of end-users varies widely, as do their capital resources, limiting the technical improvements that are feasible. Finally, the nearly infinite variety of uses invokes the need for a great many different technologies, materials, and equipment.

In concert with improving energy use efficiency, the substitution of renewable energy forms for fossil fuels also can help reduce greenhouse gas emissions. The subject of skepticism for many years, solar and wind generation have been expanding rapidly around the world. In our state of California, the main utilities have signed contracts that will make more than one-third of the electricity produced in the state come from renewable sources.[3] The governor has proposed a goal of 50% renewable energy by 2030. In 2006, California enacted a comprehensive law to reduce greenhouse gases.[4] This law, the first in the U.S., requires the state to reduce greenhouse gas emissions to 1990 levels by the year 2020. This is being accomplished by regulation, economic incentives, advanced technologies, and by cap and trade and other innovative programs. In 2015, California Governor Jerry Brown issued an Executive order to further reduce greenhouse emissions in the state to 40% below 1990 levels by the year 2030, paralleling the goals of the European Union.[5]

On the positive side, changes made by end-users can have an immediate (minutes) or short-term (months) impact on energy use and demand, compared to 5—10 years needed to add new energy supply capacity. The previous availability of energy, coupled with its low cost, resulted in situations in which there was little incentive for more efficient energy use. Both of these conditions have changed dramatically over the past few decades, along with greater awareness for environmental protection to avoid climate change. Now financial and other drivers to manage energy

[3] Olsen, David, and Hochschild, David (2015) "Clean Energy's Revenge," p.A17, March 12, *Los Angeles Times.*
[4] California Assembly Bill AB-32, "The California Global Warming Solutions Act of 2006."
[5] Megerian, Chris, and Finnegan, Michael, (2015) "Brown Sets Further Cuts on Emissions," April 29. p.A1, *Los Angeles Times.*

effectively are broadly available. Finally, even though the diversity of end-use technology is considerable, there are certain basic approaches or "general principles" that apply in a wide variety of applications.

The purpose of this book is to set forth these basic principles, provide examples, and supply a general methodology and the tools to implement it to manage energy use cost-effectively. In an effort to stress the practical, we provide examples throughout, such as those in this chapter.

DEFINING ENERGY MANAGEMENT

The energy industry uses many terms to describe different ways for using energy more effectively. The terms include energy management, demand-side management, energy efficiency, energy conservation, fuel switching, load management, and demand response, to name the most common. Table 1.1 shows common energy industry terms and the specific actions associated with them for managing energy more effectively. The subsections below define each term in more detail.

Table 1.1 Energy management terminology

Term	Behavioral changes	O&M procedures	Energy efficient equipment	Process improvements	Fuel conservation	Energy recovery	Temporary load reductions	Permanent load reductions	Distributed energy resources
Energy management aspects potentially encompassed by term									
Energy Management	●	●	●	●	●	●	●	●	●
Demand-Side Management	●	●	●	●	●	●	●	●	●
Energy Efficiency	●	●	●	●	●	●		●	○
Fuel Switching			○	○	●	○	○	●	○
Load Management	●	●	○	○	○	○	○	●	○
Demand Response	●	●					●		○

● Primary aspect.
○ Secondary aspect.

Energy Management and Demand-Side Management

The authors prefer the term *energy management* because it encompasses all aspects of managing energy, ranging from behavioral changes and better operation and maintenance practices to energy efficient equipment retrofits and process improvements to fuel conservation and energy recovery to temporary and permanent peak demand reductions and even to distributed energy resources.

Demand-side management is another term that encompasses all aspects of managing energy, but it is generally associated specifically with utility programs aimed at improving energy use at customer sites, the "demand" side of the meter. Perhaps the most widely accepted definition of demand-side management is the following: "Demand-side management is the planning, implementation, and monitoring of those utility activities designed to influence customer use of electricity in ways that will produce desired changes in the utility's load shape, that is, changes in the time pattern and magnitude of a utility's load. Utility programs falling under the umbrella of demand-side management include: load management, new uses, strategic conservation, electrification, customer generation, and adjustments in market share."[6,7] This definition can extend to suppliers of all forms of energy, not just electric utilities.

Energy management practices by the user can relate to reduction in and better control of energy use (kilowatt hours of electricity or joules [Btu] of fuels) through more efficient systems and automated controls. Or, the user can take measures to reduce needed capacity to obtain more favorable energy pricing. To illustrate the second point, it is conceivable for a homeowner to switch off the electric water heater or increase thermostat set-points for the air conditioning system simply to save money on kilowatt hour or demand charges. These actions might be done at the homeowner's convenience—for example during the evening hours when demand for hot water is low or during the weekday when family members are at work or school. Or the utility might provide encouragement by giving a favorable electricity rate during the evening. Thus, by providing appropriate storage capacity controls, the user could meet hot water requirements by off-peak electricity use or the air conditioning controls could automatically decrease the temperature set-points

---

[6] Gellings, C.W., and K.E. Parmenter. (2007). "Demand-Side Management," in *Handbook of Energy Efficiency and Renewable Energy*, edited by F. Kreith and D.Y. Goswami, CRC Press: New York, NY. (New edition expected in 2015.)

[7] Gellings, C. W. (1984–1988). *Demand-Side Management: Volumes 1-5*. EPRI: Palo Alto, CA.

to begin cooling the house with off-peak electricity shortly before occupants return home. It is possible to have a situation where the strategy would be the same regardless of who implemented it, but the ends might be quite different depending on whether the action was taken by the utility or by the customer.

Energy Efficiency

The term *energy efficiency* refers to using an alternative method, process, or piece of equipment to produce a given outcome (a product or service, for example) with less energy. Implementation of these types of change hinges largely on the availability of technology and economic justification.

Fuel Switching and Distributed Energy Resources

Fuel switching is substituting one fuel or energy source for another. This may or may not lead to net energy savings, but can shift the requirement from one fuel to another, saving scarce resources, or reducing fuel costs.

In their most general sense, distributed energy resources (as opposed to large central power plants) include technologies for distributed generation (non-renewable and renewable), combined heat and power, energy storage, and power quality. The main efficiency advantage is the reduction of transmission and distribution losses. In addition capital costs may be lower. Distributed energy resources can be applied at the utility-scale where they feed into the distribution system, or they can be applied at the local (building) level. The focus here is building-level distributed energy resources since they can be considered a demand-side energy management alternative.

Load Management and Demand Response

Load management refers to the supply end of the system and encompasses those activities taken by utilities to manipulate the load seen by their generating systems to achieve the most favorable and economic operating condition. Generally, the utilities' major concerns will be to improve their load factor and reduce peak demand wherever possible since a high peak demand requires maintaining generating capacity that is infrequently used and is expensive both in terms of initial capital cost and operation.

Demand response is a subset of load management that refers to actions that temporarily reduce load in response to price or other signals from the utility. It is often differentiated from other load management strategies, like thermal energy storage or energy efficiency improvements that yield permanent load reductions. An example would be to have electricity customers reduce their consumption at critical times or in response to market prices. Demand response can be driven by one of two methods—incentive-based (direct load control, interruptible/curtailable rates, demand bidding/buyback programs, etc.), or time-based rates (time-of-use rates, critical-peak pricing, and real-time pricing).

Terms to Avoid

We avoid the terms *energy conservation* and *energy consumption* because they are technically inaccurate. According to the first law of thermodynamics, energy is always conserved, so this usage is a contradiction in terms. What is significant in most practical applications is preserving the *quality* of an energy form. In a strict thermodynamic sense this invokes consideration of the second law of thermodynamics as well as the first law. Likewise, energy is not consumed, although fuels are, and also the quality of an energy form is consumed, or more precisely, *degraded*. Again, referring to thermodynamic terminology we can measure the quality of an energy source in terms of its *available work*. Available work is consumed in the process of utilizing energy and thus conservation of available work is a meaningful goal of energy management policies.

Energy Management Units

Most nations of the world have approved the International System of Units (SI), although complete adoption has not occurred in certain countries (most of them English-speaking). SI units are used throughout this book, with non-SI units in common use sometimes shown in parentheses for clarity. The units for energy and power are the joule and the watt:

Energy, heat, work: Joule (J) = 1 newton · meter = 1 watt · second
Power: Watt (W) = 1 joule/second

These units are small for practical purposes so we use Gigajoules (10^9 J) or Megawatts (10^6 W) for large quantities and Megajoules (MJ) or kilowatts (kW) for most energy management applications. For convenience, we often use kilowatt · hour (equal to 3.6 MJ) when referring specifically to electrical energy.

Table 1.2 lists a few basic conversion factors. For a more complete listing, refer to the appendices.

Table 1.2 Basic conversion factors

Multiply	By	To Obtain
Btu	1.055×10^3	Joule
Calorie	4.190	Joule
Foot pound force	1.356	Joule
Btu/hour	0.2933	Watt
Horsepower	7.46×10^2	Watt

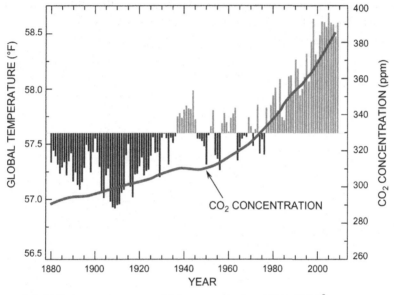

Figure 1.2 Global temperatures and CO_2 concentration 1880–2014.[8]

CONCLUSIONS

Around the world, in industrial and non-industrial nations alike, there is a heightened awareness of the central role played by energy in the economy, food supply, and national productivity. The other side of the coin is growing awareness that emissions from internal combustion engines and power plants contribute to global warming. The average earth surface temperature was higher in 2014 than ever before, continuing a steady rise that began in 1980. The 10 hottest years have occurred since 1998. As shown in Figure 1.2, accompanying this change we have seen a steady

[8] U.S National Oceanic and Atmospheric Administration, NOAA National Climatic Data Center. https://www.ncdc.noaa.gov/indicators/ Accessed April 20, 2015.

increase in atmospheric CO_2. There is also a continuing rise in the average sea level, and a drastic reduction in the polar ice caps, with the Arctic region warming and its ice shrinking. In Antarctica, the situation is more complex, with land ice shrinking but sea ice increasing.[9]

The world is at a historic balance point where large emerging economies in Asia and South America are experiencing a rapid growth in demand for energy. How this demand is met has far-reaching consequences. Energy management promises to be of increasing importance in enabling humankind to meet the challenges of the future: providing employment, food, and security for future generations, without despoiling the "blue planet."

[9] Mohan, Geoffrey (2015) "Heat Hits New High," p. A1, *Los Angeles Times*. (Mohan is reporting on a recent report by scientists at the U.S. National Aeronautics and Space Administration and National Oceanic and Atmospheric Administration.) See also: NOAA National Climatic Data Center, State of the Climate: Global Analysis for March 2015, published online April 2015, retrieved on April 20, 2015 from http://www.ncdc.noaa.gov/sotc/global/. In addition to NASA and NOAA, meteorological agencies in Japan and the UK report the same conclusions.

CHAPTER 2

Catalysts for Energy Management

INTRODUCTION

The supply constraints resulting from the oil embargos of the 1970s were a significant catalyst for new energy policies and programs aimed at improving energy management worldwide. There were widespread efforts to reduce wasteful energy use and to develop energy efficient technologies, processes, and improved methods of supplying and managing energy. The efforts yielded substantial technology advancements and programmatic advances, effectively reducing the energy intensity of equipment and processes and slowing the global per capita rate of increase in annual energy use. These improvements continued to have momentum into the 1980s. However, in the 1990s, energy management improvements slowed for various reasons, perhaps due in part to a degree of complacency setting in. The energy management policy and programs simply were not as effective as their predecessors. In the 2000s, new (and renewed) concerns started to take hold again.[1] Today, these concerns are still at the forefront of the energy industry's collective consciousness. They contribute to a need, perhaps even an urgency, for better energy management as a means to effectively deal with future shortfalls in supply, rising costs, and the specter of global warming.

As in the 1970 and '80s the drivers revolve around the increasing worldwide demand for energy and uncertainty in the availability of energy resources to supply that demand. This time around they include greater concerns over adverse environmental impacts, in particular, greenhouse gas emissions. Of course, the costs and regulatory implications associated with energy availability and environmental impacts also constitute important drivers. In this chapter, we describe these key drivers for energy management programs and discuss the many ways energy management improvements benefit end-users, energy companies, nations, and society in general.

---

[1] Gellings, C.W., K.E. Parmenter, P. Hurtado, C. Arzbaecher. (2007). *A Renewed Mandate for Energy Efficiency: Discussion Paper*. Prepared for Attendees of the End-Use Electric Energy Efficiency Workshop: Serving Customer Power and Energy Needs in a Carbon-Constrained World, Madrid, Spain, Nov. 26–27, 2007. Palo Alto, CA: EPRI.

Energy Management Principles.
DOI: http://dx.doi.org/10.1016/B978-0-12-802506-2.00002-1

GROWING DEMAND FOR ENERGY[2]

What has recent history shown us? Overall, some remarkable changes have occurred in the 35 years since publication of the first edition of *Energy Management Principles*. At that time in 1980, the world was still recovering from the second oil shock, when the Shah of Iran was over-thrown in 1979 and Iran cut oil production. Meanwhile, oil production in the U.S. had plateaued and was headed downward, making the country a net importer of oil. Today, 35 years later, the U.S. is a net exporter of oil, proven world reserves have increased by 60% in the last two decades, but unrest in the Middle East first pushed the price of oil to US$100 per barrel, only to have it plunge to US$50 per barrel as OPEC attempted to force other producers out of the market.

In 1980, there were dire concerns about natural gas shortages; those fears dissipated with new finds and a 50% increase in proven reserves between 1993 and 2013, despite the fact that consumption increased by nearly 30% over the same period. By 2013, the price per GJ (MBtu) of natural gas had dropped to about one-fifth that of oil.

During the same period, the price of coal see-sawed as environmental concerns and the recession impacted European and U.S. demand. In the Asia-Pacific region, production and consumption of coal doubled in the 10-year period 2003−2013, while North American consumption declined by 20%. China today accounts for 50% of global coal consumption and has a 4% annual growth rate.

Nuclear power, once considered an "inexhaustible energy resource" has fallen out of favor due to public concerns over safety and radioactive waste management. The Chernobyl disaster in 1986 and the Three-Mile Island accident both resulted from a loss of cooling, causing fuel melt-down and release of radioactive fission products. The consequences were quite different, however. In Chernobyl, the reactor lacked the contain-ment building that is mandatory in modern Western nuclear plants. Consequently radioactivity was spread over broad area, causing the evacuation of residents and closing an area within a 20 mile radius of the plant. This area is still closed and the residents have not returned.

At the Three-Mile Island accident in 1979, the containment building kept radioactivity from escaping and the impact of the accident outside of

---

[2] Data source unless otherwise noted: BP Statistical Review of World Energy, June 2014, bp.com/statisticalreview.

the plant boundary was negligible. As a precautionary measure, residents living within a 20 mile radius of the plant were evacuated. They returned to their former homes within 3 weeks.

The 2011 Tōhoku earthquake and tsunami, the largest ever recorded to hit Japan, caused a meltdown in three reactors at the Fukushima nuclear power plant. The area within a 12 mile radius of the plant was evacuated. There have been no reported radiation deaths. The tsunami resulted in about 18,500 people dead or missing, most by drowning. This toll is far greater than the known deaths from radiation due to Three-Mile Island, Chernobyl, and Fukushima combined.

The World Health Organization, in a 2005 statement titled *Chernobyl: The True Scale of the Accident,* reported that a team of 100 experts in the field estimated that up to 4000 people might ultimately die as a result of radiation exposure from the Chernobyl incident. While radiation spread over broad area of Europe, the resulting levels were quite low in most places. Initial studies indicate that long-term exposure in Japan will be far less than experienced at Chernobyl, and radiation-induced deaths will be difficult to detect from those due to other causes.

There are great misconceptions regarding radioactive waste management, the first being that spent fuel "must be safeguarded for millions of years." One hundred years is closer to the truth. Also, while the U.S. has spent billions in developing storage facilities that are still not accepted for use, France successfully manages its radioactive waste with little fanfare in a densely populated country slightly smaller in area than California and Nevada combined. France operates 58 nuclear power plants and generates 75% of its electricity from nuclear energy.

As a consequence of public perceptions of radiation risk, worldwide production of nuclear power peaked in 2010 and has been declining since then, except in the case of China and India. After Fukushima, some countries decided to close their nuclear power plants or not build new ones.

Hydropower increased by 40% in the decade 2003−2013, with one-half of the new hydro capacity originating in China. The supply of other renewable energy forms (solar, wind, geothermal, and biomass) increased over 400% in the decade, but still amounts to only 2% of total global energy usage, as shown in Table 2.1.[3] Summing up, primary world energy usage remains dominated by fossil fuels.

[3] Ibid., p. 41.

Table 2.1 World primary energy use, by Source, 2013

Fuel	Million tonnes oil equivalent (10^6 TOE)	10^9 GJ	Share of primary energy use (%)
Oil	4,185	188	33
Natural gas	3,020	136	24
Coal	3,827	172	30
Nuclear	563	25	4
Hydro	856	38	7
Renewables	279	13	2
Total	12,730	572	100

Source: BP Report.[2]

Total global energy use in 1975 was 200×10^9 GJ versus an estimated 572×10^9 GJ in 2013, an increase of 286%.[4]

Electricity is an energy form of growing importance worldwide. Table 2.2 shows growth trends for electricity. We note that an increasing share of energy use takes place in the form of electricity. The electricity growth rate is greater than other energy forms, but slowed considerably following the 2008 global recession.

Historically, electricity production has grown about 28% faster (365% overall vs 286%) compared to all other energy forms during the four decades from 1975 to 2013. In the U.S., nearly 40% of all energy consumed is used to generate electricity. Of this total, nearly two-thirds is generated from coal and natural gas, with one-third coming from nuclear generation, hydropower, wind, solar, and biomass (2013 data). The big change here is the decline in coal and oil consumption, offset by an increase in natural gas and wind generation. In the case of coal and natural gas, it requires anywhere from 2.5 to 3.3 units of fuel to produce one unit of energy ultimately delivered as electricity. Consequently, the efficient use of electricity by end-users has a multiplied effect on savings for the national economy and a corresponding reduction in power plant atmospheric emissions.

Electricity is used in a variety of ways: for space conditioning, lighting, motive power, refrigeration, process heat, and for electrolysis. Chapters 8 and 9 describe applications of electricity to HVAC and lighting. Electricity use in transportation and industrial processes is discussed in Chapters 10 and 11.

---

[4] See Smith, Craig B. (1981), *Energy Management Principles* 1st ed., p 16 for 1975 data, and Table 2.1 for 2013 data.

Table 2.2 Approximate world electricity production

Year	U.S.	Rest of world	Total in 10^6 MWh	Total in 10^9 GJ
1961	878	1,575	2,453	8.8
1963	1,011	1,838	2,849	10.3
1965	1,157	2,183	3,340	12.0
1970	1,638	3,263	4,901	17.6
1973	1,947	4,095	6,042	21.8
1975	2,003	4,436	6,439	23.2
Average annual % increase	6.0%	7.7%	7.1%	
2008	4,119	15,042	19,161	69.0
2009	3,950	15,112	19,062	68.6
2010	4,125	16,129	20,254	72.9
2011	4,100	16,980	21,080	75.9
Average annual % increase	0%	3.2%	2.5%	

Sources: For years 1961-75, Smith, Craig B. (1981), *Energy Management Principles* 1st ed., p. 17; for years 2008–2011, U.S. Energy Information Administration, "International Energy Statistics," can be found at: http://www.eia.gov/cfapps/ipdbproject/IEDIndex3.cfmhttp://www.eia.gov/cfapps/ipdbproject/IEDIndex3.cfm.

The rapid increase in electricity production is due to the popularity of this energy form, which is preferred because of its flexibility, ease of transmission and control, and versatility of fuels that can be used to produce it. In the last several decades, myriad new applications have emerged for electricity, including now ubiquitous personal computers and related equipment, portable electronic devices, electric vehicles, and a long list of electric appliances, from ultrasonic toothbrushes to heat pump water heaters. In addition, there has been a surprising and unplanned increase in "vampire" electricity usage, all those glowing neon or LED indicator lights from cell phone chargers, cable TV boxes and dozens of other low power devices that operate 24 h/day, 7 days/week.

Energy Use by Sector

Roughly 2/3 of all energy use occurs in industrial and transportation sectors. The balance is used in the residential and commercial sectors, with a small portion committed to nonenergy uses, such as feedstocks for industrial processes (see Table 2.3).

Table 2.3 Energy use by sector, 2011 (10^9 GJ/year)

Area	Industry	Transport	Residential	Commercial	Total
World	281	109	97	65	552
	50.9%	19.7%	17.6%	11.9%	100%
U.S.	32.3	28.6	22.8	19.0	102.7
	31.4%	27.9%	22.2%	18.5%	100%

Sources: World Data: U.S. Energy Information Administration, "International Energy Outlook 2013."
U.S. Data: U.S. Energy Information Administration/Annual Energy Review 2011, DOE/EIA-0384
(2011) September 2012.

SUPPLY CONSTRAINTS

History has shown that it is the *availability* of accessible and affordable energy that is critical for communities to flourish. Civilizations have risen and fallen for a multiplicity of causes, but energy availability may be surmised to be one of the major reasons.

During millions of years of human evolution, from the Stone Age to the industrial revolution, humankind depended primarily on solar energy for direct heat, food, and fuel. Civilizations flourished first where the availability of natural resources, water, fertile land, and food made it convenient, and later where the geography provided security and access to trade. When agricultural productivity decreased, or when the forests were stripped, the community moved on to another location or declined. Changing climatic condition caused human and other life forms to migrate, and often created deserts in its wake.

With the development of trade, transportation, and the first crude pumps and engines, establishment of permanent cities became feasible. Now, instead of being susceptible to climate changes and depending only on energy and food resources of the immediate vicinity, these necessities could be brought in from afar, creating a much broader resource base to support the community. This was accompanied by the discovery of cheap, transportable, and readily available fossil fuels, which made a dramatic difference in human energy use and is linked to the origins of the Industrial Revolution.

Today's industrialized communities rely heavily on both domestic and foreign energy supplies, food, and other goods to thrive. We are accustomed to having exotic tropical fruits airlifted one-fourth of the way across the earth from Hawaii to brighten the breakfast of New Yorkers. We also expect the lights to turn on when we flip a switch and we expect to be able to purchase gasoline any day of the week.

Many of us take energy availability for granted. However, the energy resources we have come to depend on, both locally and internationally, are most certainly limited. Examples of these very real supply constraints range from the 1973 oil embargo when gasoline purchases were restricted, to the rolling electricity blackouts in California during the early 2000s when homes and businesses experienced disruptions to their electricity supply. Ensuring that we have access to energy when we need it is an important driver for energy management activities at the facility and community level.

World Fossil Fuel Resources

There are literally hundreds of books dealing with various estimates of world energy resources. Debate surges back and forth, with pessimists concluding that fossil fuels are nearly exhausted and optimists arguing that there are undiscovered resources many times greater than the amounts that have been used to date. The one point about which there has been some agreement is that fossil fuel resources are finite and sometime in the next century or so will be exhausted to the point where further uses as fuels is uneconomic.

This point was made convincingly by Hubbert in a classic study in which he calculated an upper limit on fossil fuel formation using various methods for estimating the maximum possible extent of coal, oil, and natural gas resources.[5] Combining this with historical rates of extraction, he formulated graphs such as those shown in Figure 2.1. Hubbert's work provided a useful technique for estimating a lifetime of remaining fossil fuels, even though his numbers are no longer current.

In reality, it is unlikely that fossil fuel resources will ever be exhausted. Instead, the cost of extracting or recovering resources having increasingly dilute concentrations will become prohibitive to the point where they will no longer be economical fuels.

McKelvey states:

The era of abundant, low-cost fossil fuels which has supported large increases in world population and material wealth over the past two centuries is approaching an end as exponential growth and demand proceeds to deplete remaining readily recoverable resources.[6]

---

[5] Hubbert, M. K. (1962) *Energy Resources*, Publication No. 1000-D, National Academy of Sciences, Washington, DC.

[6] McKelvey, V. E. (1977) "World Energy—The Resource Picture," pp. 11−18 in Fazzolare, R. A. and Smith, C. B., eds., *Energy Use Management: Proceedings of the International Conference*, Vol 1, New York: Pergamon Press.

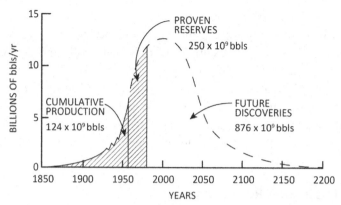

Figure 2.1 Ultimate world production of crude oil.

Table 2.4 Estimated proven reserves of fossil fuels

Fuel	Quantity	Unit	Approximate energy content (10^9 GJ)	Share of proven reserves (%)
Oil	1,688	10^9 bbl	10,339	29
Natural gas	6,558	10^{12} cubic feet	6,925	19
Coal	891,531	10^6 metric ton	18,529	52
Total			35,793	100

Source: BP Report.[2]

Table 2.4 summarizes a current estimate of *proven* reserves of fossil fuels prepared by British Petroleum.[7] Coal accounts for approximately 52% of proven reserves, followed by oil at 29% and natural gas at 19%.

Table 2.1 showed that fossil fuel resources currently supply more than 85% of the worldwide demand, which was 572×10^9 GJ/year in 2013. Based on this recent level of world energy use, the e*stimated proven reserves* in Table 2.4 constitute a pool that could last for slightly more than 60 years. However, energy use in much of the world has been increasing at an annual rate of 2–4% due to population growth and greater industrialization. If this rate of increase continues, all fossil fuel resources would be exhausted sometime in the next century. That is, assuming fossil fuel energy use increases at the *current* rate of growth, the current estimates of *proven* oil and natural gas reserves will last fifty-plus years, and coal, over 100 years.

[7] Op. cit., BP Statistical Review.

Another significant issue with fossil fuel resources is where they are located. Petroleum and natural gas have high concentrations in the Middle East and Eastern Europe. Political instability is a great concern for countries that rely on imports, both from an economic and national security perspective.

There are also several unknowns to consider with respect to fossil fuel reserves. The first is, "What is the magnitude of the *unproven* fossil fuel resources?" Past history has shown that exploration continues to yield new gas and oil fields, albeit with greater recovery costs. Technological advances also can increase reserves. In Hubbert's analysis, cited above, he postulated future oil discoveries would be 3.5 times as great as the proven reserves known at that time. The U.S. Energy Information Administration estimated in 2009 that U.S. *unproven* resources of oil and natural gas are 7—9 times greater than proven reserves.[8] If those numbers are typical for global sources, fossil fuels would hypothetically last for several centuries. Of course this would only apply if the cost of recovery was within an economic range.

The next unknown is exponential growth. Herein lies another argument. How much more growth can be tolerated? Zero growth is intolerable, since this would imply no increase for the developing nations, or at best an increase at the expense of the industrial nations. Certainly a slowing of growth rate in the developed nations is possible, accompanied by the inevitable increases in energy use that must occur in the rest of the world. Still, it is difficult to imagine that this will *not* lead to an increased demand for energy over the next 25—50 years. To deny this is to deny the irresistible forces of liberty and nationalism which are in motion in the world today. So, growth rates in energy use are likely to increase.

The final, and perhaps most critical unknown, is "What will be the effect of dumping additional megatons of CO_2 from combustion into the atmosphere?" This debate is beyond the scope of this book. Suffice it to say that we believe this extremely serious problem is not receiving sufficient attention by governments worldwide. It is this concern that drives our dedication to the efficient use of energy as one mitigating measure that the entire world can accept.

---

[8] U.S. Energy Information Administration. (Sept. 2012). *Annual Energy Review.* Table 4.1 "Technically Recoverable Crude Oil and Natural Gas Resource Estimates, 2009." Washington, DC: U.S. Energy Information Administration.

Other Energy Resources

There are three other major classes of energy resources in addition to fossil fuels:

- Nuclear fuels (uranium and thorium).
- Unconventional hydrocarbon fuels (tar sands, shale oil, gas entrapped in sandstone and shale, and geopressurized zones).
- Alternate energy sources, which include hydropower, solar energy, wind power, biofuels, geothermal energy, and ocean energy.

Scientists estimate worldwide nuclear fuel reserves, if used in conventional light water reactors, potentially have an energy content of the same order of magnitude as oil and gas reserves. Thus nuclear energy use in conventional reactors does not change the global picture greatly, although it could be an important economic modifier for certain countries if widely substituted for imported oil. Another potential advantage of nuclear energy is the beneficial aspect of eliminating CO_2 released to the atmosphere as the result of combustion. However, as described previously, ongoing public concerns about nuclear power plant safety and radioactive waste management are significant barriers to more widespread use of conventional nuclear power.

Nuclear energy becomes potentially even more significant if the technical and social issues surrounding the use of alternative forms of nuclear energy are solved. For one, if breeder reactors were to become available, the future energy supply horizon would be extended perhaps another 500 years, even with increased rates of energy use. However, the practicality of this option is still in question and public concern over diversion of nuclear fuels for use as terrorist weapons is a formidable barrier. Another alternative form—fusion energy—despite years of research and huge investments, so far remains a feasible source only on the sun and, of course, in nuclear weapons.

Large amounts of energy are potentially available in the unconventional hydrocarbon resources. Early efforts to develop economical production did not meet with much success. However with the increased price of oil, renewed efforts at extracting shale oil and recovering oil and gas from depleted reserves have shown success. Remarkably, in a few short years, the U.S. has gone from being a net importer of oil to a net exporter, in part driven by hydraulic fracturing of shale ("fracking") and other methods of resource recovery. Hydrogen may also prove to be an energy resource in the future, but producing, transporting, and storing it currently present significant economic, safety, and technical challenges.

Alternative energy sources such as hydropower and renewable resources were essential in the past and will be important in the future. Hydropower, in particular, accounts for a considerable share of worldwide energy use (7% in 2013), albeit this share is still small compared with fossil fuel use. Renewables such as solar, wind power, ocean energy, biofuels, and geothermal energy represent a small but growing share of energy use (collectively 2% in 2013).

As noted previously, solar energy has historically been the major human energy source. Huge amounts of it reach the earth; roughly 10 times the fossil resources shown in Table 2.4, or 3000×10^{12} GJ, are incident *every year* on the land and sea. Much of this is reradiated and therefore is unavailable for human use. Some is captured in the form of wind and wave power, and about 0.1% is stored annually by the biosphere. Renewables will be an important future source in any event; however, it will be of greatest interest if economical means can be developed to use solar and other alternate energy forms as substitutes for fossil fuels. This is technically and economically much more difficult to accomplish. Indeed, it is true, as McKelvey stated, that while solar energy may be free, it isn't cheap.[9] Nevertheless, there has been progress in replacing energy use for space heating, water heating, and low-temperature process heat with solar thermal systems. Solar photovoltaic systems for generating electricity have shown steady increase in efficiency and in cost reduction. Wind power installations have expanded around the world. Still, the applications have been limited, and the equipment relatively expensive.

The finite supplies of conventional fuels and the technical and economic challenges of alternative energy sources make supply constraints a significant driver for energy management.

ENVIRONMENTAL IMPACTS

The environment is directly and indirectly affected by energy production and use. The impacts are too numerous to mention here, but examples include the following:

- Extraction of coal in mines causes a variety of health effects in miners, including pneumoconiosis (black lung).
- Oil spills from tankers damage marine and coastal environments.

[9] Op. cit. McKelvey (1977).

- Release of sulfur dioxide and nitrogen oxides into the atmosphere from power plants causes acid rain, which is harmful to forests and pollutes streams and lakes.
- Greenhouse gas emissions from fossil fuel combustion are linked to climate change.
- Hydropower dams and turbines have negative impacts on wildlife.
- Severe damage to nuclear power plants can lead to radioactive contamination.

These few examples show that most, if not all, of the energy resources we depend on have environmental hazards, with some more destructive than others. Because of the relatively large share of fossil fuels feeding our energy supply today, combustion of fossil fuels is responsible for the most deleterious environmental effects. Mitigating these impacts by reducing demand for fossil fuels and seeking advancements in more environmentally sensitive alternative energy forms is therefore another important driver for energy management. This driver may originate from an individual concerned with protecting the environment, or from a corporate or government policy.

POLICY AND REGULATORY MATTERS

Numerous government and corporate actions are underway worldwide to affect patterns of energy use. Policies and programs exist at the international, national, state, local, and corporate levels, with leadership from government organizations, energy companies, regional coalitions, and heads of corporations. These actions address energy management improvements to the economy as a whole as well as in individual economic sectors, including energy supply and delivery, industry, residential and commercial buildings, and transportation. Table 2.5 lists examples of some of the types of policy and program instruments that have proven successful in improving energy management.[10]

The move toward a strong energy efficiency push on the part of public utility commissions and other regulatory agencies represents a radical departure from the preoil embargo days. Utilities have a definite need for

---

[10] For a more detailed discussion of policies and programs, see Gellings, C.W., K.E. Parmenter, P. Hurtado, C. Arzbaecher. (2007). *A Renewed Mandate for Energy Efficiency: Discussion Paper.* Prepared for Attendees of the End-Use Electric Energy Efficiency Workshop: Serving Customer Power and Energy Needs in a Carbon-Constrained World, Madrid, Spain, Nov. 26–27, 2007. Palo Alto, CA: EPRI. Chapter 4.

Table 2.5 Examples of policies and programs for energy management

General	Energy supply and delivery	Industry	Buildings	Transport
Subsidies for research and development	Minimum standards of efficiency for fossil fuel-fired power generation	Strategic energy management	Stronger building codes and appliance standards	Improved fuel economy
Public goods charge to fund energy efficiency programs	Reduced fossil fuel subsidies	Energy efficiency standards, including standards for advanced motors, boilers, pumps, compressors, etc.	Labeling and certification programs	Mandatory fuel efficiency standards
Tax incentives	Carbon taxes or charges for fossil fuels	Process-specific energy efficient technologies	Advanced lighting initiatives	Advanced vehicle design and new technologies
Incentives for private sector investment	Innovative rate structures	Energy management systems	Leadership in procurement of energy efficient buildings	Taxes on vehicles, fuel, parking, etc.
Leadership in procurement of energy efficient equipment, vehicles, and facilities	Decoupling of profits from sales to encourage energy efficiency	Negotiated improvement targets	Energy efficiency obligations and quotas	Mass transit improvements
Public information and education to increase awareness	Mandatory demand-side management programs (including energy efficiency and demand response measures)	Incentives	Energy performance contracting	Transportation infrastructure planning
Trade allies	Meeting larger portion of future demand with energy efficiency	Research initiatives	Voluntary and negotiated agreements	Telecommuting
		Benchmarking	Tax incentives	Mode switching
			Subsidies, grants, loans	Leadership in procurement of energy efficient vehicles
			Carbon tax	
			Education and information	

(Continued)

Table 2.5 (Continued)

General	Energy supply and delivery	Industry	Buildings	Transport
	Greater use of combined heat and power Improved infrastructure (including smart grid) Reduced natural gas flaring Tradable certificates for energy savings		Mandatory audit and energy management requirement Detailed billing	

Sources:
1. *Realizing the Potential of Energy Efficiency, Targets, Policies, and Measures for G8 Countries*, United Nations Foundation, Washington, DC: 2007.
2. *Climate Change 2007: Mitigation, Contribution of Working Group III to the Fourth Assessment Report of the Intergovernmental Panel on Climate Change*, B. Metz, O. R. Davidson, P. R. Bosch, R. Dave, L. A. Meyer (eds), Cambridge University Press, Cambridge, United Kingdom and New York, NY, U.S.: 2007.
3. *Energy Use in the New Millennium: Trends in IEA Countries*, International Energy Agency, Paris, France: 2007.

energy management technology that will enable them to serve their customers, providing both power and information concerning its efficient use. The end-user needs to have critical energy requirements met without interruption at a reasonable cost. Today, utility load growth does not lead to lower costs.

Carbon Policy

Worldwide there are many new regulations, policies, and actions aimed at reducing greenhouse gas emissions from fossil fuel combustion and other factors that contribute to global warming. Some countries are further along than others, with the U.S. relatively slow to adopt policies. In the Copenhagen Accord of the United Nations Framework Convention on Climate Change in December of 2011, the Obama administration put forward a target to lower greenhouse gas emissions in the U.S. to 2005 levels by 2020.[11] The U.S. EPA has also begun to establish a framework under the Clean Air Act for regulating greenhouse gas emissions that includes rules requiring the largest industrial emitters to submit annual greenhouse gas emissions data and mandates that all sources emitting greenhouse gases above a certain threshold obtain permits.[12] Most recently, on August 3, 2015 the Obama administration and U.S. EPA announced finalization of the Clean Power Plan, which is a rule to reduce carbon emissions from power plants.

However, several state carbon polices are ahead of federal regulations. For example, California is implementing a cap-and-trade program to ensure the state achieves its target of reducing emission levels to 1990 values by 2020. The entities covered in the cap-and-trade program consist of power plants, transportation fuels, refineries, cement manufacturing plants, and large industrial combustion sources.[13] Cap-and-trade programs and the use of allowances and carbon offsets as compliance instruments are an innovative and somewhat controversial policy to reduce greenhouse gas emissions.

In addition to government policies, interesting incentive programs have emerged for carbon reductions; an example is contributing money

---

[11] United Nations Framework Convention on Climate Change, Appendix I—*Quantified economy-wide emissions targets for 2020*, http://unfccc.int/meetings/copenhagen_dec_2009/items/5264.php.

[12] Arzbaecher, C.E., and K.E. Parmenter. (2014). "Carbon Policy Impact on Industrial Facilities," *Strategic Planning for Energy and the Environment*, 34(1), 11−39.

[13] California Environmental Protection Agency, Air Resources Board, http://www.arb.ca.gov.

for reforestation to offset air passenger miles. Corporate carbon policies are also on the rise, (i) because corporations want to improve public image and (ii) as a result of pressure from customers and shareholders. Indeed, numerous companies have been asked by shareholders to adopt emission reduction targets, to publish sustainability reports, or to tie executive compensation to sustainability goals.[14]

ENERGY COSTS

The issues associated with all of the drivers discussed so far have implications on energy costs. Growing worldwide demand for energy amidst constraints in the availability of low-cost energy resources pushes energy costs upwards, particularly in locations with limited supplies. Even in areas where fuels and electricity are inexpensive, it is apparent that the efficient and judicious use of resources will benefit the user, his or her nation, and humankind in general. It also benefits energy companies, since better energy management delays capital expenditures for new power plants and associated infrastructure.

Saving money on energy use is usually the principal driver for end-users to pursue energy management actions. It is also one of the main benefits that users derive from energy management programs. Increases in energy prices affect household budgets, operating costs in commercial and industrial facilities, and ultimately the cost of manufactured goods, transportation, essential human services, and agriculture. For residential customers, lower energy costs mean more disposable income. For industrial and commercial users, energy management can mean cost savings, better utilization of existing capital equipment, or plant expansion (by using existing equipment more efficiently) without new capital expenditures (Text Box 2.1).

For energy utilities, properly conducted energy management programs can postpone the addition of new generating capacity at a time when sites are scarce, capital costs are high, and permitting is challenging, both from societal and environmental perspectives. Energy management is a highly cost-effective energy resource that is yet to be fully tapped.

---

[14] Interfaith Center Corporate Responsibility, *2013 Shareholder Resolutions /Environmental Performance by Sector*, http://www.iccr.org/shareholder/trucost/index.php.

TEXT BOX 2.1

When the cost of fuels, irrigation, fertilizer, pesticides, transportation, and food processing increase, the impact is felt in the grocery stores where we buy food. Feedlot operations are particularly energy intensive. Energy audits indicate that feedlot beef requires input of 20–30 energy units for each energy unit of edible meat produced. Much of today's high-yield agriculture is energy intensive; as fuel prices increase, so do food costs.

BENEFITS OF ENERGY MANAGEMENT

There is an inextricable link between the drivers of energy management and its benefits. It is impossible to have a discussion of one without the other. Energy management has widespread benefits to end-users, energy companies, nations, and society at large, many of which tie directly back to some of the key drivers—growing energy use and supply constraints, environmental impacts, and policies and regulations—and the increasing energy costs accompanying these issues. Invoking energy management actions and programs can address and mitigate these driving forces while delivering a wide variety of additional benefits, including nonenergy benefits.

Nonenergy Benefits

Also referred to as nonenergy impacts, ancillary benefits, co-benefits, and benefits "beyond the meter," nonenergy benefits represent an important but often overlooked value proposition of energy management programs. These benefits vary depending on the beneficiary, whether it be the end-user (residential, commercial, or industrial), utilities, or society in general (see Table 2.6 for specific examples).[15]

Residential users potentially realize benefits ranging from improved comfort *(better weatherization reduces drafts and keeps interior environments at comfortable temperature)* to increased property value *(energy efficient improvements can equate to lower operating costs and better resale value)*. Commercial users may experience fewer sick days and greater productivity *(improving lighting levels, temperature and humidity, and air quality lead to a healthier*

---

[15] Parmenter, K.E. (2013). "Who Gets the Benefit from Non-Energy Benefits?" Presented at Association of Energy Services Professionals Spring Conference. Dallas, TX. May 1, 2013.

Table 2.6 Nonenergy benefits of energy management programs

	End-users			Society	
Residential	Commercial	Industrial	Utilities	National	International
Improved health and safety	Fewer sick days	Increased production	Reduction in costs associated with a customer's inability to pay bills	Job creation/decrease in unemployment	Greenhouse gas emission reductions
Increased comfort	Greater productivity	Improved product quality	Delayed construction of new plants	Increase in tax revenues	Natural resource preservation
Water savings	Reduced liability	Reduced liability	Lower infrastructure costs	Increase in economic output	Decreased water usage
Greater disposable income	Lower operation and maintenance costs	Lower operation and maintenance costs	Greater flexibility and reduced costs in meeting demand	Higher asset values	Improved air quality
Increased property value	Compliance with codes	Compliance with environmental regulations	Compliance with state and federal regulations	Energy security/national security	Achieving sustainable development
Greater convenience and/or ease of operation	Improved occupant comfort	Improved process control	Environmental stewardship	Improved environment	Energy price moderation
Noise reduction	Water savings	Water savings		Reduced public spending on energy	
	Environmental stewardship	Competitiveness		Improved health/lower healthcare costs	
				Reduced poverty/lower welfare costs	

working environment) and lower operation and maintenance costs *(some efficient technologies such as LED lighting last longer and require less maintenance).* Energy management in industry can result in improved product quality *(better tasting food, enhanced material properties),* competitiveness *(superior products for lower costs),* and water savings *(repairing steam leaks eliminates wasted water).*

Utilities benefit from a reduction in billing costs related to collecting from customers who cannot afford to pay high energy bills, greater flexibility and reduced costs in meeting demand, and compliance with state and federal regulations. An effective energy management program is an important element in regulatory acceptance of utility forecasts, rate requests, and permit applications for new generating capacity. In the words of one utility executive, "an unexpected value of our energy management programs has been the improvement in our credibility that has resulted."

Societal benefits range from socioeconomic to environmental to national security. For example, improved health and safety lowers healthcare costs; reduced poverty decreases welfare costs; and reduced greenhouse gas emissions that result from more efficient use of fossil fuels has huge implications for human welfare.

A LOOK BACK

When the first edition of this book was published in 1980, world energy use had grown at an average annual rate of 4.4% during the period from 1961 to 1975. During the same time period, energy use in the U.S. had grown at an average annual rate of 3.3%, reaching 75.1×10^9 GJ (71.2 quads) in 1975. Projections at that time foresaw energy use in the U.S. reaching 106×10^9 GJ (100 quads) by 1992; instead, it only reached 90×10^9 GJ (85 quads). The elusive mark of 106×10^9 GJ was reached in 2004 and again in 2005, but by 2013 energy use had dropped back to 103×10^9 GJ (97.8 quads).

Over the last two decades, 1995−2013, primary energy use in the U.S. increased from 96.0×10^9 GJ (91.0 quads) to 103×10^9 GJ (97.8 quads), or by a *total* of 7%. Meanwhile, Gross Domestic Product (GDP) grew from US$$10.16 \times 10^{12}$ to US$$15.76 \times 10^{12}$, an increase of 55% during the same period. (Dollar values are "chained" to 2009.) What this signifies is that the input of energy per dollar of GDP declined by over 30% between 1995 and 2013. Some of this is no doubt due to the changing industry base in the U.S., but overall it represents a truly remarkable transition in energy efficiency.

What if this hadn't happened? First, the energy component of the cost of goods would be 30% higher, affecting consumer spending and international competitiveness. Second, had a 3% annual growth rate of total energy use continued, the U.S. alone now would be using over 200×10^9 GJ (189 quads) of energy per year, a crippling amount from an economic and environmental perspective.[16] During these years, the number of residential and commercial energy consumers has grown, as has the various ways they use energy. We are indeed fortunate that remarkable improvements in the energy efficiency of refrigerators, air conditioning units, lighting systems and other devices have offset the growth in demand. Besides huge economic savings to consumers, greater end use efficiency has meant less combustion, less greenhouse gases, and less environmental pollution. The era of energy management is a huge success story for humankind.

CONCLUSIONS

There are compelling reasons for efficient use of energy on a global scale. Besides the economic incentive (which in most countries is a sufficient cause in itself) the global resource picture is clouded. Readily available and low-cost fossil fuels that had been the mainstay of the past century are becoming increasingly costly and scarce. The processes required for mining, processing, transporting, and using these fuels will be subject to increasing environmental concerns as global usage doubles and then at least doubles again and possibly again in the future.

Nor can we expect major miracles in the form of new fuel sources. Experience has shown that new oil finds, such as those in the North Sea, Alaska, and West Africa, are not magical solutions to the problem. Instead, they are part of the fulfillment of the prophecy of "undiscovered recoverable reserves." In fact, we must expect new major finds of this type every year or the global situation will be worse than described above.

---

[16] An additional 100×10^9 GJ of energy use would be roughly equivalent to dumping an additional 6000 million metric tons of CO_2 into the atmosphere, or double the current amount.

Energy management, in addition to having important short-term economic benefits, buys time for the future. It provides time to make a transition to readily available fuels. It eases the transition to alternate fuels. Finally, it is the only near-term (2—5 years) way to get more "mileage" from existing fuels and energy sources. This is much quicker than the time needed to bring new wells and mines into production or the time required to build new power plants. Energy management also has many nonenergy benefits than are all too often overlooked when valuing energy management proposals.

CHAPTER 3

General Principles of Energy Management

INTRODUCTION

Life was a reward for successful energy management by prehistoric man. Limited by what could be hunted, dug from the ground, or carried on the back or as body fat, our ancestors had little latitude in dealing with the vicissitudes of nature. Predators, droughts, disease, and natural disasters took their toll, and the bands' elders were lucky to reach 40 years of age when times were good. For primitive peoples—even for those living today—energy management requires balancing the roughly 2000 to 3000 kcal per day (approximately 3.0 to 4.6 GJ per year or 2.9 to 4.3 MBtu/year) work and metabolic energy expenditure with an equivalent food intake. Peak demand also was, and still is, important; even with adequate energy supply, an excessive power requirement (such as subzero weather) could lead to failure of the human system. Thus, the concept of energy management is not new to human affairs, but has been an essential aspect of human survival for centuries.

The industrial revolution changed this situation by allowing humans to draw upon a greater diversity of nature's stored sources of energy—first firewood, then fossil fuels, and eventually nuclear energy. Humans have also learned to use energy from the sun, wind, and water much more effectively than our ancestors did, and we have developed increasingly innovative ways to capture energy from geothermal resources, agricultural crops, and biological waste. While these developments have been significant only in the last few hundred years—really, in the last century—of man's million year-plus history, they represent a revolutionary accomplishment of enormous significance to human and other life forms.

Today, most citizens of our energy-intensive cities manage or mismanage the equivalent of 100 times as much energy as our early ancestors—much more than is needed for bare survival. In the face of diminishing natural resources, a worldwide increase in demand for energy, and the threat of global climate change, efficient and judicious use by all people will be needed if costs are to remain reasonable and environmental

Energy Management Principles.
DOI: http://dx.doi.org/10.1016/B978-0-12-802506-2.00003-3
35

impacts manageable. Energy management can contribute to this goal in the home, on the farm, and in factories and cities.

APPROACHES

At the most elementary level, energy management may be thought of as *task energy use*, that is, the provision of as much energy as is needed, when it is needed, where it is needed, and with the quality required. Since there is often limited flexibility in the timing and locational aspects of task energy use, the primary areas of focus for energy management are to maximize utilization of energy *quality* and minimize utilization of energy *quantity*.

Appropriate "cascading" of energy use from high quality to progressively lower quality forms ensures maximum utilization of energy quality for each task at hand. Therefore, implicit in energy cascading is meeting energy requirements with "waste" heat, or recovered energy dissipated from another process whenever lower quality energy is acceptable for the task. This concept is in itself a fundamental aspect of energy management, since "not every Joule (or Btu) is created equally." Why not? Because the energy content of a swimming pool of lukewarm water is roughly the same as a liter of gasoline. However, while the energy content is the same, the useful "work" that can be accomplished by a liter of gasoline is *much greater* than the useful "work" that can be done by a swimming pool of lukewarm water! By the same token, when people use high grade fuels such as oil or natural gas to generate hot water or low-temperature process steam, in a strict thermodynamic sense they are misusing the potential value of the fuel, even though this practice is quite commonplace. Extending this concept to the use of electric resistance elements for water heating shows the potential for even greater misuse of high quality energy, since only about 1 unit of electricity is generated for every 3 units of fuel combusted in a conventional coal-fired power plant. An example of the best use would be to employ a topping cycle such as a gas turbine to extract work initially and then use the low-quality "waste" heat for process steam production or water heating.

There are three basic approaches for minimizing the *quantity* of energy used:

- **Reduce use by downsizing or eliminating an end-use** (or shifting usage to a different time period, which effectively redefines the "when it is needed" portion of task energy use) due to a self-imposed change or because of regulatory or economic pressures. In this approach, unless the end-user has been operating systems superfluously, the user generally makes a sacrifice to reduce energy quantity. That is, they no longer realize the same level of "functionality" they previously obtained from their end-use systems. Except in circumstances where the financial benefits far outweigh the sacrifices [e.g., participation in a utility's demand response program may be a very attractive option with minimal detriment to end-use functionality (see Chapter 5)] or in cases of needless energy use (e.g., process equipment in use when no longer needed, unused space with 24-h per day illumination, or unoccupied rooms being heated and cooled), this approach should usually be viewed as a last resort.

- **Increase efficiency** with better operation and maintenance procedures, more efficient equipment, advanced controls, improved processes, or different material inputs. The aim of this approach is to achieve the same level (or, ideally, a much improved level) of functionality from the end-use systems while minimizing energy quantity and lowering lifecycle costs. Increasing efficiency is a primary focus of this book.

- **Substitute energy form** with another in less demand or with a more appropriately matched energy quality for the task in question (this is also referred to as "fuel switching"). Preferably, the replacement energy form would serve both purposes: reduce demand for the constrained or expensive energy form *and* provide the suitable quality of energy. However, in some cases, the most cost-effective energy substitute may not reduce overall energy quantity, but it would still reduce the required quantity of the original energy source.

GENERAL PRINCIPLES

Although there is a very great diversity in energy end-use technology, there are certain basic approaches or general principles that apply to a wide range of applications. Identification of fundamental principles for energy management is an attractive concept because it suggests an initial approach to the problem. The principles alone will not improve energy use efficiency, but they can provide a rational basis for developing more specific technological responses.

In Table 3.1 we summarize some general principles that are applicable to a wide variety of situations. The table also provides an approximate, highly qualitative assessment of relative costs, implementation time, complexity, and benefit based on our experience. The following discussion helps clarify how these principles could be applied to a typical energy user.

Table 3.1 General principles of energy management

Principle	Relative cost	Relative time to implement	Relative complexity		Relative benefit (Typical)
1. Review of historical data	Low	1 year	Low		5–10%
2. Energy audits (review of current practices)	Low	1 year	Low		5–10%
3. Operation and maintenance ("housekeeping")	Low	1 year	Low		5–15%
4. Analysis of energy use (engineering analysis, building simulation, system modeling, availability studies)	Low to moderate	1–2 years	Moderate to high		10–20%
5. Economic evaluation (cost/benefit, rate of return, life-cycle costing)	Low	1 year	Low		5–15%
6. More efficient equipment	Moderate to high	years	Moderate to high		10–30%
7. More efficient processes	Moderate to high	years	Moderate to high		10–30%
8. Energy containment (heat recovery, waste reduction)	Moderate to high	years	Moderate to high		10–50%
9. Material economy (scrap recovery, salvage, recycle)	Low	1–2 years	Low to high		10–50%
10. Substitute material	Low to moderate	1 year	Low		10–20%
11. Material quality (purity and properties)	Low	1 year	Low		5–10%
12. Aggregation of energy uses	Moderate to high	years	Moderate to high		20–50%
13. Cascade of energy uses	Moderate to high	years	Moderate to high		20–50%
14. Alternative energy sources (substitute fuel or energy form)	Moderate to high	years	Moderate to high		10–30%
15. Energy conversion	Moderate to high	years	Moderate		10–30%
16. Energy storage	Moderate to high	years	Moderate to high		10–30%

Review Historical Data

The first principle is to *review historical energy use*. It helps establish typical seasonal, monthly, and even daily energy use patterns and facilitates identification of anomalies such as unexpected spikes or dips in usage, energy use during non-business periods, or even gradual energy increases over time that may signal degradation of equipment. Sometimes seasonal variations or scheduling discontinuities are present but unrecognized; the review process brings them to light and may suggest ways of combining operations, reducing demand charges, or otherwise affecting savings. For example, a plant may experience a surge of manufacturing during a certain season, yet maintain space conditioning all year-round. Often the question "why do we do this?" and the answer "that's the way we've always done it" flag an area for immediate savings. Chapter 4 discusses historical review in greater detail.

Energy Audits

Historical energy use data are never sufficient, however, since they provide the total picture but not the details. It may be necessary to collect other types of data to better understand the factors driving energy use which might include weather, production, or occupancy. In addition, *energy audits* are the means for investigating energy use by specific processes and machines, and provide insight into inefficient operations. Chapter 6 provides a comprehensive discussion of building and site energy audits.

Operation and Maintenance

Improving *operation and maintenance* in the plant will generally save energy. Well-lubricated equipment has reduced frictional losses. Cleaned light fixtures transmit more light. Changing filters reduces pressure drop. Repairing steam leaks prevents waste of high quality energy. Operation and maintenance practices are applicable to all types of end uses. Chapters 8, 9, 11, and 12 describe common measures for HVAC, lighting, process, and building envelop systems.

Analysis

Analysis goes hand-in-hand with the energy audit to determine how efficient equipment is, to establish what happens if a parameter changes (reduce flow by 50%), or to simulate operation (computer models of building or process energy use). Chapter 7 presents general techniques for energy analysis.

Economic Evaluation

Economic evaluation is an essential tool of energy management. New equipment, processes, or options must be studied to determine costs and returns. The analysis must include operating costs, investment tax credits, taxes, depreciation, and the cost of capital for a realistic picture, particularly when considering escalation of energy prices. Chapter 13 discusses various methods of cost effectiveness analysis.

More Efficient Equipment

More efficient equipment can often be substituted to fulfill the same function (e.g., LED or high output T5 fluorescent lamps rather than T12 or even T8 fluorescent lamps for area lighting, or premium efficiency motors instead of standard or high efficiency motors). Most types of industrial, commercial, and residential equipment are now rated or labeled in terms of their efficiency; there are wide variations among different manufacturers depending on size, quality, capacity, and initial cost, but there are many online resources for comparing the different technologies. Chapters 8, 9, and 11 provide several examples of high efficiency HVAC, lighting, and process equipment.

More Efficient Processes

More efficient processes can often be substituted without detrimental effect and often yield improved product quality. A classic example is a continuous steel rolling mill, which uses a continuous process to produce steel products, avoiding energy loss involved in cooling and reheating in batch production. Another example is powder metallurgy rather than machining to reduce process energy; still another is a dry papermaking process which reduces energy expended to remove water from the finished product. Inert atmosphere ovens can reduce energy used for drying solvent-based paints, compared to ultraviolet bake ovens. Membrane separation in food processing can result in better tasting products than heat treatment technologies.[1] Drying with microwave or radio-frequency radiation increase drying rates and minimizes surface drying and cracking relative to conventional drying processes. For example, an energy management team conducted a study to find a replacement for gas-fired drying oven

---

[1] Parmenter, K., C. Sopher. (2006). *Membrane Separation in the Food Industry,* Lafayette, CA: Agriculture Production and Food Processing Technology Application Service, Global Energy Partners.

used in the processing of agricultural feed additives. The study tested a microwave oven, electric resistance heating, and a solar oven. The relative drying time using these three technologies was in the approximate ratio of 1:10:100. Not only was the microwave process the fastest, but it reduced waste heat losses and improved product quality. See Chapter 11 for additional examples of efficient processes.

Energy Containment

Energy containment seeks to confine energy, reduce losses, and recover energy. Examples include repair of steam and compressed air leaks, better insulation on boilers or piping, air sealing of building envelopes, and installation of heat exchangers or power recovery devices. For example, the flue gases from boilers and furnaces and other systems that depend on combustion provide excellent opportunities for heat recovery. Depending on flue gas temperatures, the exhaust heat can be used to raise steam or to preheat the air to the boiler. Figure 3.1 shows an example of such a system, where an ammonia reformer heater is designed to conserve fuel by using a steam generator and air preheater to recover heat from the stack gas.[2]

There is overlap between this energy management principle and the operation and maintenance principle discussed above (Principle 3) and the energy cascading principle discussed below (Principle 13).

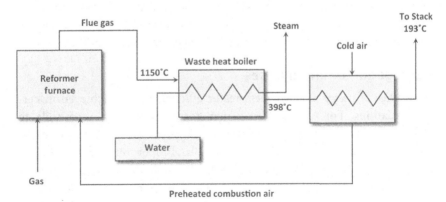

Figure 3.1 Heat recovery using an air preheater.

[2] Smith, C. B., ed. (1978). *Efficient Electricity Use*, 2nd Ed, p 60. New York: Pergamon Press.

Material Economy

Material economy implies recovery of scrap, reduction of waste, and "design for salvage." The powder metallurgy example cited above also illustrates this principle. Product design that permits salvage or recovery of reusable parts, motors, and components is another example. Structures, in fact, can be designed for reuse and relocation.

Substitute Materials

Substitute materials can sometimes be used to advantage. For example, in low-temperature applications, low melting point alloys can substitute high-temperature materials. A material that is easier to machine, or that involves less energy to manufacture, can replace an energy-intensive material. Water-based paints can be used without baking in certain applications. An emerging technology for primary aluminum production that uses wetted cathodes and inert anodes instead of carbon anodes promises to reduce energy use and lower greenhouse gas emission while increasing productivity and lowering costs over the conventional Hall-Héroult process; these savings are due largely to the fact that current carbon anodes are consumed during the process whereas the inert anodes do not corrode or release carbon dioxide emissions.[3]

Material Quality Selection

Material quality selection is extremely important, since unnecessary quality almost always means higher costs and often means greater energy use. For example, is distilled water needed, or is deionized sufficient? Purity of chemicals and process streams has an important impact on energy expense; trace impurities may not be important for many applications.

Aggregation of Energy Uses

Aggregation of energy uses permits greater efficiency to be achieved in certain situations. For example, in a manufacturing plant it is possible to physically locate certain process steps in adjacent areas to minimize the energy use for transportation of materials. Proper time sequencing of operations can also reduce energy use, for example, by using temperatures generated by one step of the process to provide preheating needed by another step.

---

[3] Industrial Technologies Program. (2011). *Ultrahigh-Efficiency Aluminum Production Cells.* Washington D.C.: Energy Efficiency and Renewable Energy, U.S. Department of Energy.

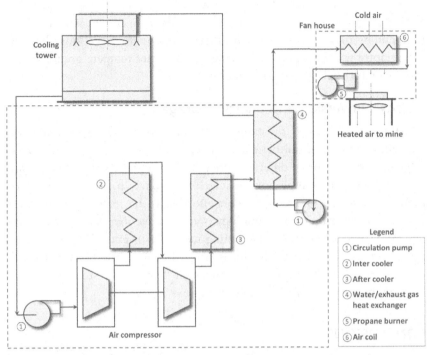

Figure 3.2 Flow diagram of mine air heating and compressor cooling cycle.

Cascade of Energy Uses

Heat recovery is an example of *cascading energy use*, whereby high tempera-ture heat is used for one purpose and the waste heat from that process applied to another process step, and so on. There are many sources of waste heat in commercial and industrial facilities. Figure 3.1 showed an example of recov-ering heat from a gas–fired reformer furnace. Energy in the form of heat is also available at a variety of noncombustion sources such as electric motors, crushing and grinding operations, air compressors, and air thickening and drying processes. These units require cooling in order to maintain proper operation. The heat from these systems can be collected and transferred to some appropriate use such as space heating. An example of this type of heat recovery is shown in Figure 3.2.[4] All the energy supplied to the motor in electrical form is ultimately transformed into heat and nearly all of it is avail-able to heat buildings or for domestic water or mine air heating.

[4] Smith, C. B., ed. (1978). *Efficient Electricity Use*, 2nd Ed, p 63. New York: Pergamon Press.

As the temperature of waste heat decreases, the opportunities for applying it to other processes diminish; however, in some cases industrial heat pumps may be a viable and efficient option for accepting the low temperature waste heat and delivering it at a higher temperature for applications requiring higher quality energy.[5]

Energy Conversion and Energy Storage

Careful consideration of the energy source and form can lead to improved efficiency, environmental benefits and costs savings. Consider whether an *alternative energy* source, different *energy conversion* process, or *energy storage* is applicable. For example, onsite solar photovoltaic panels for electricity production are becoming more cost effective as the technology matures. In addition, solar thermal technology is an effective means for water heating. Also, thermal energy storage using ice banks or eutectic salts is a useful means to shift cooling loads to off-peak periods and battery technology is rapidly advancing, which will make onsite storage of electric energy increasingly viable in the future.

CONCLUSIONS

The three basic approaches and 16 general principles for energy management we present here provide a starting point for initiating energy management efforts in any activity. Applying these concepts can take place at several levels. Modification and retrofit can be applied to existing equipment and facilities and involve either operating budget or capital dollars, depending on project size and complexity. Many utility companies offer incentives to lower the costs of energy efficiency improvements. In new facilities, plant designers have numerous opportunities to improve process efficiency, often without increasing capital costs, simply by planning that takes into account the anticipated costs and availability of energy resources. In building construction, builders can specify the most efficient equipment and materials economically justified. When energy costs are negligible, the initial cost of the project is often the important consideration. Now, with increasing operating costs to be expected, a higher initial cost may be justified if it saves money over the project's lifetime.

---

[5] Parmenter, K., E. Fouche, R. Ehrhard. (2007). *Tech Review: Industrial Heat Pumps for Waste Heat Recovery.* Lafayette, CA: Industrial Technology Application Service, Global Energy Partners.

CHAPTER 4

Planning For Energy Management

INTRODUCTION

The stimulus to start an energy management program must come from somewhere. It could originate from a variety of potential sources, including a concerned individual who has noticed excessive compressed air leaks in the production area, a facility maintenance manager discouraged by the increasing time requirement for repairing old equipment, a company president who is suddenly made aware of rising energy costs, a corporate (or government) mandate to reduce carbon footprint, a utility account manager who notifies the company of an opportunity for great incentives, or the more extreme case of a local utility announcing it is going to curtail the factory's fuel supply.

Reducing energy costs or complying with regulations of one sort or another are usually the motivation. However, even companies that do not face high energy costs find that an energy management program pays for itself by eliminating waste and reducing costs; it may also offer the company a marketing advantage or improved public image because they can potentially tout themselves as a green business. For example, in a group of California hospitals, the Hospital Association correctly recognized that an energy management program could reduce operating costs. Perhaps more importantly, the association realized that such a program would be visible evidence that the hospitals were attempting to control costs, and therefore had important *political* implications, even though energy costs were small fraction of total operating costs. In many cases, there are several simultaneous motivating factors for establishing an energy management program due to the myriad drivers and benefits, such as those outlined in Chapter 2.

Now suppose that the time is appropriate to initiate a program, regardless of whether the stimulus was passed down from top management or was passed up the line from the operations end of the firm.

Energy Management Principles.
DOI: http://dx.doi.org/10.1016/B978-0-12-802506-2.00004-5
45

Where does one begin? An energy management program can be organized in many ways, but we suggest organizing it in three primary phases:
1. Initiation and planning.
2. Audit and analysis.
3. Implementation and continuous assessment.

Table 4.1 outlines the planning steps necessary to establish the program. Being proactive and following this systematic process, rather than

Table 4.1 Planning an energy management program

Initiation and planning phase

1. Commitment by management to an energy management program.
2. Assignment of an energy manager.
3. Creation of an energy management committee of major plant and department representatives.

Audit and analysis phase

1. Review of historical patterns of fuel and energy use, production, weather, occupancy, operating hours, and other relevant variables.
2. Facility walk-through survey.
3. Preliminary analyses, review of drawings, data sheets, equipment specifications.
4. Development of energy audit plans.
5. Energy audit covering (i) processes and (ii) facilities and equipment.
6. Calculation of projected annual energy use based on audit results and expected weather, operation, and/or production.
7. Comparison with historical energy records.
8. Analysis and simulation (engineering calculations, heat and mass balances, theoretical efficiency calculations, computer analysis and simulation) to evaluate energy management options.
9. Economic analysis of selected energy management options (lifecycle costs, rate of return, benefit-cost ratio).

Implementation and continuous assessment phase

1. Establishment of energy effectiveness goals for the organization and individual plants.
2. Determination of capital investment requirements and priorities.
3. Implementation of projects.
4. Promotion of continuing awareness and involvement of personnel.
5. Formation of measurement and verification procedures. Installation of monitoring and recording instruments as required.
6. Institution of reporting procedures ("energy tracking" charts) for managers and publicize results.
7. Provision for periodic reviews and evaluation of overall energy management program.

just reactively implementing projects when energy efficiency problems can no longer be ignored, greatly increases the likelihood of on-going success and continuous energy improvement.

INITIATION AND PLANNING PHASE

Importance of Management Commitment

Regardless of the motivation for the program, it will not succeed without a commitment from the firm's top management. For this reason, Table 4.1 lists this as a first step in the initiation and planning phase. Management must be convinced of two key things, first, the need, and secondly, the potential economic returns that will result from investing time and money in the program.

Obtaining management commitment often requires the presentation of facts, figures, and costs concerning current energy usage, along with estimates for the future and projected savings. Therefore, it may be necessary for the person responsible for encouraging program development to do some degree of historical review prior to the audit and analysis phase to help sell the concept to management, unless, of course, management is the stimulus for the program.

Energy Champions

Once management commits to the program, the next step is to name one individual the energy manager. The energy manager may be a member of the engineering staff in a large firm, or a maintenance supervisor, electrician, or foreman. The energy manager's core responsibilities are to ensure the energy management program is accepted by staff and operates effectively. This is no easy task without the support of management and other energy champions.

Therefore, the energy manager's first step might be to formulate an energy management committee with representatives from each key department or division using energy, depending on the size and complexity of the firm. A representative from the accounting department would be another good addition. Next, the energy manager should explain to the department heads and line supervisors the need for the program, taking into consideration the economic and other motivating factors driving the program.

Collectively, the committee's main responsibilities will be to ensure the program has reasonable targets and that goals are successfully met from the "ground up." Therefore, the committee should take steps to

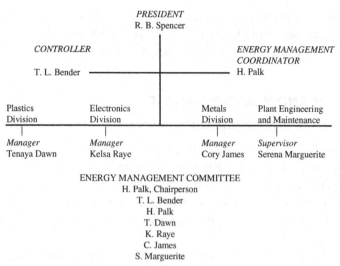

Figure 4.1 Energy management organization chart.

inform all personnel—from office staff to the production line—of the need, emphasizing that efforts will be placed on reducing waste and improving productivity and profitability. The energy manager and committee could even devise an incentive system whereby personnel are awarded for identifying energy management improvements.

Figure 4.1 shows an example of an energy management organization chart for an industry with three principal divisions. The president established an energy management committee consisting of an energy manager (appointed by the president) and representatives of each of the three manufacturing divisions, plant engineering and maintenance, and the central power plant. The purpose of this committee is to coordinate plans, bring in new ideas and perspectives, and to ensure that actions taken in one part of the plant do not have an unfavorable effect on another part.

A similar approach can be taken by a city. For example, following the 1973 oil embargo, Los Angeles experienced serious shortages of fuel oil and was forced to implement a mandatory program of electricity cutbacks in the residential, commercial, and industrial sectors. As the city struggled with the problems caused by these changes, the mayor created an interdepartmental energy conservation committee (Figure 4.2). This committee met periodically, reviewed or proposed new rules and regulations,

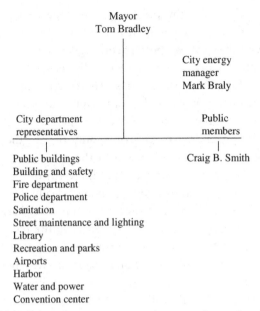

Figure 4.2 Interdepartmental energy conservation committee, city of Los Angeles.

initiated a system of energy reporting for the various city departments, initiated energy audits in public buildings, and provided liaison for a series of other energy management initiatives.

Addressing Institutional Barriers

There are often instances where efficient energy use is discouraged by other factors. The energy manager should be aware of these barriers and should understand how to deal with them when confronted. They fall in several broad categories:

- **Economic:** Rate of return is too low, or lower than alternative investments; capital is not available; unwillingness to make short-term investments for long-term returns. This is one of the most prevalent challenges faced by energy managers and energy champions. However, there are funding opportunities that can help address this barrier. Most utility companies—gas and electric—offer a wide variety of programs that provide services or financial incentives to encourage efficient energy use. Exploring utility program opportunities is an important first step for energy managers. (see Chapter 5 for more information on utility rates and programs). Additionally, some energy service companies (ESCOs) offer financing support or energy performance contracts to firms whereby the ESCO pays the capital costs and the firm repays the debt out of money saved on their energy bills; these contracts typically involve performance guarantees so that the firm only pays if the energy savings were actually realized.

- **Ownership:** Unwillingness to make investments in leased buildings or equipment. A classic example is a tenant in a leased building refusing to make investments to improve the inefficient air-conditioning system on the grounds that "it would only benefit the owner." This decision could be the correct one. On the other hand, if the investment would pay back in less time than the term of the lease, if might be justified by the operational savings alone. Ideally the landlord could be enticed to participate and provide some cost-sharing, especially since the energy upgrades would help attract future tenants.

- **Tradition, precedent:** "This is the way we've always done it"; "we'd rather invest in expanded production capacity"; "we'd have to hire new maintenance personnel or train existing staff on how to use the more sophisticated systems and controls"; "it's easier to patch problems as they occur instead of taking the time to make the case to management for new equipment." These are all real issues, but some of the easier issues to overcome with a little education and training. Even the desire to invest in expanded production capacity can be addressed at least partly by efficiency improvements that inherently increase productivity.

- **Priorities:** Energy is a low cost item, a small part of value added in manufacturing, or not a core business focus; "we have to worry about more important things." We have heard these arguments from many of the businesses we have talked to over the years. The staff is extremely busy and so focused on the core business function that they simply do not have time to explore energy management improvements. In this case, the impetus for a program may actually come, at least in part, from an external party such as a utility account representative. Some utilities, such as BC Hydro, even have programs that provide energy managers and energy management training for their customers.

The energy manager will have to explore each situation on a case-by-case basis to find appropriate solutions.

AUDIT AND ANALYSIS PHASE

After the program initiation and planning phase, the audit and analysis phase begins. This phase consists of a detailed review of historical data, energy audits, identification of energy management opportunities, energy analysis, and economic evaluation. It involves determining where and how energy is being used and identifying opportunities for using energy more effectively.

Historical Review

First consider the methods and objectives of the historical review. Data for the historical analysis can be compiled from utility bills, facility records of operating schedules and shifts, equipment inventories, production statistics, or any other available source of data. The objective is to understand both near- and long-term trends in energy usage. For example, what is the reference base, or *baseline*, of energy use that the energy management program will attempt to modify? Also, what are the past patterns of energy use and what do they signify for the energy management program?

Insight into the following types of trends can be useful to the energy manager:

- Is historical energy use increasing or decreasing? (Consider the past 2−5 years.)
- Are there seasonal variations in energy use? (Summer or winter peaks?)
- How complete is the database? (Energy use for the whole plant, for each division, etc.)
- What have been past trends in energy costs? (10% annual escalation or what?)
- Are there temporal variations in energy use? (Off-shift versus on-shift; weekend versus weekday, etc.)

Example. Historical energy use in a small manufacturing plant. (See Figure 4.3.) In reviewing the historical energy use data, we note that there is a summer electrical peak and a winter gas peak. One might ask: "Are these related to process energy use or space conditioning?" The summer peak could be caused by greater plant output or longer hours of operation. Or, it could be due to the use of air conditioning. This suggests an area for future investigation. Another questions: even during the summer months, there is gas usage of 10,000 therms per month. Why is that?

In many cases, it is useful to relate historical energy use to variables that affect usage. There are four variables that commonly influence energy use patterns:

1. Weather conditions.
2. Operating hours or production schedules.
3. Building occupancy.
4. Some measure of productivity.

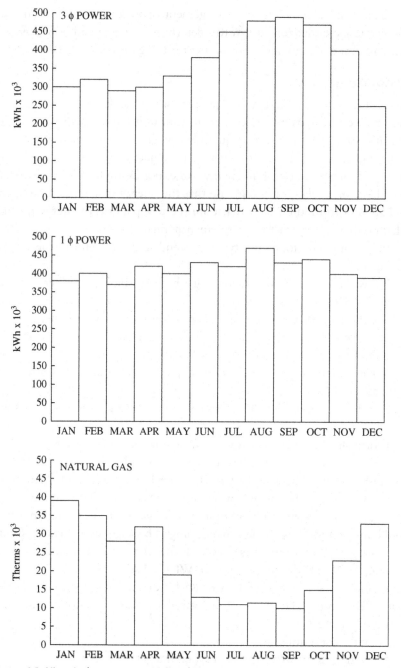

Figure 4.3 Historical energy use, XYZ industries.

There could also be other relevant variables driving energy use, including the nature of the raw materials used to manufacture the products. For example, lower quality materials may be more energy intensive to process.

In some climates certain types of construction tend to follow ambient temperatures. If this is the case, energy use for space conditioning will correlate with the degree days of heating and cooling.[1] The relationship between energy use and degree days can be readily estimated with regression analysis. Where such an analysis is appropriate, comparison of energy usage and weather data for the past 2—5 years can give very valuable insight into the space conditioning component of energy use. For example, if the results show a consistent correlation across the years, then the systems are probably functioning properly. However, changes in weather-adjusted energy use patterns can signify problems in the equipment or its operation.

Energy use also tends to trend with operating hours and production schedules, so usage should usually be lower when businesses are closed or not fully staffed. Access to interval energy data—that is, data collected at hourly, 15 min, or even smaller intervals—allows analysis of daily or weekly trends, in addition to the seasonal and yearly trends available from monthly energy bills. Reviewing energy data at a more disaggregated level can reveal problems such as lights, space conditioning, or other equipment operating when they are intended to be off. This could indicate, for example, a programming error in the facility's energy management and control system, which is an easy fix. It could also be due to employee behavior, which can be addressed with education and training.

Often it is useful to relate historical energy use to some measure of productivity. This not only gives information useful in assessing the operation of a particular plant or process, but can also provide comparative data between two or more similar facilities for benchmarking purposes. (Unless they are situated in locations having similar climates, be careful of comparisons that do not take into account the effect of weather on energy usage.)

---

[1] The number of cooling degree days (CDDs) is a measure of how much higher the outside air temperature was relative to a base temperature, and for how many days during the period (single day, month, year). The number of heating degree days (HDDs) is a measure of how much lower the outside air temperature was relative to a base temperature, and for how many days during the period.

Such *energy indices* can be found in a variety of businesses. Typical indices are energy content in MJ/kg for metals, chemicals, and cement, or fuel consumption per passenger or ton-kilometer, for transportation systems. Other indices evolve to meet specific needs (e.g., energy use per meal served in a restaurant; energy use per guest day in a hotel; annual energy use per patient day or per bed in a hospital; energy use per widget produced in a factory).

Sometimes energy indices reveal trends not apparent from the straight historical data. Figure 4.4 shows total energy use and energy use per employee for a large government research facility that launched an energy management program. The broken line shows total use, which decreased slightly up to 1977 and then increased in 1978. The solid line shows energy use per employee. Interestingly enough, during the 5-year period shown, the number of research programs underway in this facility grew significantly, leading to increased employment. When the effect of increased employment is included, the payoff from energy management program is more impressive.

For a meaningful comparison of trends, it is important to make sure the timing of the production data aligns with the timing of energy data when developing the energy indices. When there is a mismatch in the time periods, it is difficult to get an accurate correlation. Also, for facilities that produce more than one product, it may be necessary to develop a different energy index for each product unless they have similar energy

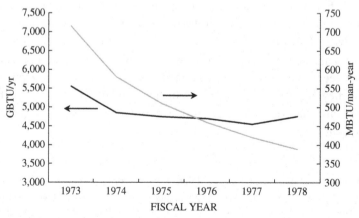

Figure 4.4 Energy use per capita.

intensities. (A recent paper we contributed to describes baseline development in industrial facilities in greater detail.)[2]

Besides revealing certain patterns and trends in energy use, historical review provides the basis for comparison with the energy audit and other subsequent work. It is a critical step for establishing the initial baseline at the process or facility level. However, the historical review in itself never provides sufficient information for formulating an effective energy management program. Few facilities will have sufficient metering instrumentation to provide a detailed breakdown of energy use at the system or equipment level throughout the plant.

Energy Audit

In the energy audit, the auditor or audit team collects detailed information for each piece of equipment, lighting systems, Heating, ventilating, and air conditioning (HVAC) systems, and processes, and sometimes information on the building construction. The energy manager and energy committee then use results of the audit to delineate major areas of energy use and to formulate the next steps in the energy management plan. The audit can be done on a process-by-process basis or on a building or facility basis, depending on the scope defined during the planning and initiation stage. Auditors may include members of the firm's maintenance or technical staff or outside energy specialists could be brought in to conduct the audit. Ideally, the audit team will have a combination of experienced energy engineers who know how to identify issues and opportunities and facility personnel who are intimately familiar with the facility's systems and operations.

To obtain detailed system data, the auditor may choose to measure loads and equipment operating hours (in hours per day, week, or month) using metering equipment, or he or she may use equipment nameplate specifications and knowledge of the typical loads and operating hours to estimate energy and demand. If weather-sensitive loads like space conditioning equipment represent the largest end-use in the facility, the auditor may decide to develop a model of the building and then simulate energy usage based on different weather scenarios. After accounting for and calculating energy use for all of the major loads, auditors can compare the

---

[2] Gilless, C., P. Hurtado, R. Milward, K. Parmenter, M. Perakis, V. Vesma, G. Wikler, C. Williamson. (2013). *Energy Baseline Methodologies for Industrial Facilities.* Prepared for Northwest Energy Efficiency Alliance. Walnut Creek, CA: EnerNOC Utility Solutions.

findings with historical records. Agreement does not have to be perfect—and it most likely will not be—but significant discrepancies should be investigated to determine the source and to verify that major items have not been overlooked or usage overestimated.

Since the audit includes inspection and analysis of all equipment and systems within the scope of the program plan, it is one of the major ways to identify and flag options for more efficient energy use. Because energy audits are so useful, we have dedicated an entire chapter to them; see Chapter 6 for more detail.

Energy and Economic Analyses

The next step is to investigate more thoroughly the options discovered during the audit along with any other potential opportunities under consideration. This investigation involves energy and economic analyses for each energy management opportunity. The analysis results help the energy management committee define goals and select promising projects to implement based on the organization's priorities. Energy and economic analyses of potential projects represent the most technically challenging portion of the energy management program and, thus, account for the majority of this book. Chapters 7–13 cover these topics in detail. Throughout those chapters we identify additional sources of information to assist new energy management practitioners.

IMPLEMENTATION AND CONTINUOUS ASSESSMENT PHASE

The final phase in the energy management program is really an on-going process. It comprises establishing energy usage goals; prioritizing and implementing projects; defining measurement, verification, and reporting procedures; promoting on-going awareness and involvement of personnel; and continually assessing program goals and achievements.

Establishing Goals

After the audit and analysis phase, the energy management committee has all the necessary information for establishing meaningful energy management goals and realistic energy usage targets at the system, process, building, plant, or organization level. In the case of large organizations or multinational corporations, some goals may actually have been set prior to initiating the energy management program.

Prioritizing and Implementing Projects

Prioritizing and implementing projects identified thus far is one of the most critical aspect of the entire program, since taking action to realize improvements is the central goal of an energy management effort. Project ranking will depend somewhat on the specific priorities of the organization, such as expected economic return, meeting regulations, carbon footprint, fuel availability, production requirements, etc.

Obviously, one requirement of the implementation phase is that the organization or firm be prepared to make the investments necessary to begin saving energy. It is generally useful to categorize the energy management opportunities identified into three groups:

- Operations and maintenance ("housekeeping") options.
- Retrofit and modification options.
- New design or major construction options.

These groups call for an increasing scale of capital investment, ranging from zero to minimal for housekeeping changes, to extensive for options requiring new construction.

Inform, Train, and Motivate Personnel

The implementation and continuous assessment phase also includes actions to inform, train, and motivate personnel so that the organization fosters a strong sense of involvement and ownership of the energy management program by everyone from the factory worker or office employee to the maintenance personnel and all the way up to top management. This point deserves emphasis since it is ultimately human beings that are entrusted with these marvels of engineering that are supposed to save all this energy and money. Experience indicates that more efficient equipment and improved processes are only "half the battle." Obviously, it makes little difference how efficient the plant and equipment are if any of the following are true:

- Operating personnel do not understand the need for efficiency.
- They do not believe in the need.
- They do not know how to operate their new, improved equipment.

The human element is vital, and is all-to-often ignored.

Measure, Verify, and Report Performance

A very important element of this phase is to take measurements, monitor equipment, and verify that systems are operating as expected and energy use and performance targets and goals are being met. These actions reflect the fundamental management concept that people are only able to operate effectively if two conditions are in play:

- They know what they are supposed to accomplish.
- They receive feedback that tells them how well they are doing.

For long-term success and to prevent inefficient habits from returning, it is essential for this assessment to take place on an on-going basis.

Continuous Program Assessment

Finally, the program must succeed. It must be reviewed periodically to determine its strengths and weaknesses. It should be flexible, capable of responding to changing economic conditions (energy prices, cost of goods and services), new regulations (equipment, building, environmental), corporate mandates (energy indices, carbon footprint, other priorities), and to evolving program needs (a new process or building is added, an old one is shut down). Continuous assessment also permits a review of the success of implemented projects and provides a basis for revaluating other projects that failed to pass the first screening during the original implementation plan. Chapter 14 describes the implementation and continuous assessment phase in more detail.

CONCLUSIONS

An effective energy management program must begin with management commitment. The next step is to evolve a plan for subsequent actions. A review of historical patterns of energy use provides the foundation for energy audits and further engineering studies and analysis. Early in the program, suitable criteria must be established for evaluating possible energy management projects. Training, personal awareness, and information programs are vital. The success of any program depends as much on human motivation as it does on technology.

CHAPTER 5

Understanding Utility Rates and Programs

INTRODUCTION

High energy costs are a major driver for energy management activities. Understanding utility pricing options and the factors that influence rates is an important first step in setting up an energy management program. Energy companies are also a useful source for information on energy management opportunities and for incentives for energy management projects.

There is such a wide variation in utility rate structures and utility programs that it is impossible to describe all of them within the limitations of this book. The approach we recommend is to contact your utility representative and ask for an explanation of each charge on your bill and also ask if a more favorable rate schedule is available. The utility representative will be able to recommend energy efficiency, demand response, or other energy management programs suitable for your building or facility. Most utilities list their various rate schedules and programs on their websites.

ELECTRICITY RATES

Typically four components make up the charges on a monthly electric bill. They are energy, demand, "adjustments," and taxes and fees.

Energy Charges

Energy charges are based on the number of kilowatt hours (kWh) used per billing cycle. Depending on the rate structure, the price per kilowatt hour may be fixed (flat rate), tiered (block rate), variable (time-based rate), or interruptible. Rates can also be adjusted seasonally, for example in summer when air-conditioning loads are high, or during the winter when fuel costs may be higher. In addition, there are many variations across the rate schedules offered by a given utility depending on the type and size of the customer.

Fixed or flat rates based on kilowatt hours only (without demand charges) are a less common type of pricing. To illustrate a flat rate, a

Energy Management Principles.
DOI: http://dx.doi.org/10.1016/B978-0-12-802506-2.00005-7

typical residential customer in Oxford, U.S. Kingdom pays 12.62 p/kWh (about US$0.20/kWh) for a fixed period of time (in this example, until 2015).

A very common form of tiered pricing is an *inverted block* rate where the price per kilowatt hour increases with each higher "block" of usage. For example, a typical residential customer in Southern California currently pays US$0.15/kWh for the first block of 0−386 kilowatt hours per month, US$0.19/kWh for the next block of 387−501 kilowatt hours, US $0.28/kWh for the next block of 502−771 kilowatt hours, and US$0.32/ kWh for 772 or more kilowatt hours per month. There are also "declining block" rates. While the basic principle is the same as the inverted block rate, the price per kilowatt hour decreases with each higher block of usage.

The energy rates for large commercial and industrial customers are often lower than for residential users. For example, typical energy prices across the U.S. are roughly US$0.08−0.14/kWh for commercial customers and US$0.04−0.09/kWh for industrial customers, depending on location. For rate schedules with a demand component, the energy charges are usually low—on the order of a couple cents per kilowatt hour—since the cost of providing the electricity is captured in the demand component.

With the growing "Smart Grid" infrastructure, interval data (data collected at hourly, 15 min, or even smaller intervals) is also becoming more prevalent, making time-based pricing another option for many. In time-based pricing, energy charges are time-dependent and often seasonally dependent. These types of rates more closely follow wholesale prices of electricity, which vary depending on the balance (or *imbalance*) of electricity supply feeding the grid relative to demand.

Examples of time-based pricing include *time-of-day* rates (often referred to as *time-of-use* or TOU rates), critical peak pricing (CPP), critical peak rebates (CPR), and real time pricing (RTP).

- TOU rates impose higher charges during peak hours when a utility's generating costs are highest, and offer reduced rates during off-peak hours including at night and on weekends when the demand is low.
- There are also other variations to TOU rates, including CPP which imposes very high costs for a limited number of critical events during the year. Customers who curtail loads during those critical events are rewarded with lower off-peak rates.
- A similar type of time-based rate is CPR where customers receive a rebate for reducing loads during critical peak hours.

- In RTP, energy costs vary hour by hour and season to season depending on wholesale market prices. The utility informs the customer a day ahead or in "real time" about pricing and the customer can plan or make last minute energy management decisions to reduce usage during high cost hours. For example, this is potentially desirable to a medium-sized industrial or commercial customer with the flexibility to alter operations to leverage lower cost time periods. On typical days the customer benefits from the lower relative rates offered to them, especially during "shoulder" and off-peak hours. A possible downside, however, is that rates can hit extremes. Though unlikely, is it conceivable that the rate at 5:00 PM on an extremely hot summer weekday might be US$4.00/kWh because of severe supply constraints, while at night the rate might drop to US$0.04/kWh!

Interruptible rates are another form of electricity pricing that helps balance supply and demand on the electricity grid. In this type of pricing, customers are given short notice and asked to drastically reduce or cease use of loads in response to capacity constraints on the grid. In exchange, the customers receive a reduction in normal rates or some type of financial rebate. If the customer fails to interrupt loads during the prescribed period, they may face steep financial penalties.

- In addition to the basic energy charge per kilowatt hour, some utilities will add a "fuel adjustment charge" in situations where the fuel used to generate electricity is rapidly escalating in cost. This charge is also per kilowatt hours used.

Demand Charges

The second component in the electricity bill is the demand charge, based on the generating capacity, measured in kilowatts, the utility must have to service the customer's load. To meet its customers' requirements, the utility needs to have available sufficient generating capacity (operating and standby), substations, and transmission and distribution systems to meet the needs of all of its customers. The demand charge can be determined in several different ways—based on average demand during a fixed time interval, or based on the peak demand recorded during a specified period.

Some utilities include a "demand ratchet" in their contracts. The purpose of this is to penalize customers who require an unusually high demand for short period of time. The ratchet clause causes them to have to continue paying a higher demand charge for an additional period of time, in some cases up to one year.

Power factor may also influence demand charges. Reactive power, measured in kVARs, is power that is not available for use by customers because it is stored in the magnetic fields of motors, transformers, and certain other electromagnetic devices. (Refer to Chapter 7 for additional details on electric load analysis). A classic example is electric motors that are lightly loaded. For maximum efficiency, electric motors should be sized so they operated at or near rated capacity. Otherwise, their *power factor* will be low.

Power factor *pf* is calculated using Equation 5.1 and is always less than 1.0. Low power factor means that the reactive load is high.

$$pf = real\ power(kW)/apparent\ power(kVA) = kW/(kW + kVARs) \quad [5.1]$$

While reactive power is not useful to customers, utilities must design and build their systems with sufficient to capacity to supply both the actual kW load and the reactive load. Typically pricing is based on the assumption that the power factor is 0.9 or higher. If the power factor is lower than 0.9, additional charges will be assessed to recover the added expense.

Typical demand charges are US$1−25/kW per month.

Rate Adjustments

The next category can include a wide variety of additional charges (and sometimes credits) the utilities impose. These may be called fixed charges, customer charges, standing charges or other names. They include the cost of bill preparation, meter reading and other utility expenses. They may be due to regulatory agency requirements or because they are required by local authorities. Examples include a surcharge to support energy efficiency rebate programs, subsidies for low income customers, system upgrades, emergency funds to repair storm damage, nuclear power plant decommissioning reserve, and myriad other charges or credits. Some of the adjustments are fixed and others are calculated on a per total kilowatt hour used basis and added into the effective rate being billed to the customer.

In the residential flat rate example above for Oxford, U.S. Kingdom, there is a standing charge of 26p/day (about US$0.42/day or US$12.8/month) to recover fixed costs. Also, in the RTP example above, there is a fixed "customer" charge of US$194/month.

Taxes and Fees

Local jurisdictions (city and County) may add taxes or fees to the total utility bill.

ENERGY MANAGEMENT OPPORTUNITIES FOR ELECTRICITY CHARGES

Obviously the first step is to understand each component of the bill. To do this, one should look at a complete 12-month billing cycle. It is also helpful to examine daily kilowatt hour and kilowatt usage on an hourly basis for a typical workday and a weekend. Hourly energy and demand information is readily available from the utility for customers with interval meters ("smart" meters). The data can also be obtained by installing temporary metering or sub-metering or even by reading and recording the utility's electric meter every hour over a representative day or days. With this information several actions are possible to lower energy costs with minimal or no investment of capital.

- If energy charges are excessive, see if some operations can be moved to times when rates are lower. This change would only yield savings if the facility is already on, or decides to shift to, a time-based rate schedule.
- If peak demand charges are excessive, stagger equipment operation to reduce the monthly peak demand, or if the facility is on time-based rates, reschedule or turn equipment off during peak hours.
- If there is a demand ratchet clause in your contract, avoid short-term demand peaks (or get a different contract)!
- If power factor is less than 0.9, take steps to reduce kVARs. Capacitors can be installed in parallel with offending equipment and, with proper engineering design, capacitors can increase power factor.

NATURAL GAS RATES

Like electricity, the charges for natural gas include a number of components. There is the cost of the fuel itself, storage and distribution costs, customer charges, and factors for seasonality. A block schedule is typical for natural gas pricing.

In Southern California, there is an increasing (*inverted*) block rate for residential customers. For the first block (up to 14 therms/month or

1.4 MBtu/month) during the summer months, the rate is US$8.92/MBtu. This block increases to 5.0 MBtu or more during the winter months, depending on the customer's climate zone. Above block 1 (called "Baseline" use), the non-baseline rate jumps to US$11.52/MBtu. In addition there is a customer charge of US$0.164/day and additional taxes and regulatory fees. Combining all charges, these rates amount to US$10−15/MBtu. The utility has announced a 5.5% increase in residential rates for 2015−2016. *Note*: one sometimes encounters the unit MMBtu in gas rate schedules, where M is 1000. In this book "M" always has the SI meaning of 10^6.

Gas customers in Kingston, Ontario have a *declining block* pricing structure. The rate includes the fuel price, transportation and storage, and local distribution costs.

- Block 1   1st 1500 m^3 CA$ 0.367/m^3
- Block 2   next 3500 m^3 CA$ 0.338/m^3
- Block 3   next 70,000 m^3 CA$ 0.317/m^3
- Block 4   all over 75,000 m^3 CA$ 0.304/m^3

In addition, there is a residential service charge of CA$ 21.00 per month. These gas prices range from US$8−9/MBtu.

Commercial and industrial customers in Missouri also have a declining block rate. They pay US$0.31 per hundred cubic foot (Ccf) for the first 7000 Ccf and US$0.20/Ccf over 7000 Ccf. Since 1 Ccf equals 0.1025 MBtu, these prices correspond to a range of US$2−3/MBtu. The monthly service charge for these customers is US$28.83.

COAL AND FUEL OIL RATES

Two other important fuels are coal and fuel oil, used for electricity generation, transportation, and heating. For the purposes of this book, the primary uses of interest are residential and industrial space heating and industrial boiler fuel and process heat.

The prices of coal and fuel oil vary by location and season. In the eastern U.S., fuel oil currently costs from US$4.00−4.50/gallon, with an average of US$3.36/gallon, or about US$25/MBtu. This price fluctuates from winter to summer and with the international price of crude oil. In Norfolk, United Kingdom, the price is 50p/liter, or about US$3.06/gallon, to which must be added delivery charges and taxes. Coal in the eastern U.S. is priced at US$180−335/ton, which averages US $9−17/MBtu. A recent UK price was £230/mt or about US$17/MBtu, in both cases with additional charges for delivery and taxes.

ENERGY MANAGEMENT OPPORTUNITIES FOR OTHER FUELS

The foregoing discussion shows that delivery costs and seasonality of supply and demand impact the prices of natural gas, fuel oil and coal. In addition, oil is subject to price variations caused by international tensions and coal is under greater scrutiny and regulation due to environmental concerns. The first and most obvious energy management opportunity is fuel switching to take advantage of the price differential. Some firms have dual-fuel equipment, so they can switch to the fuel with the best price at a given time. Another option is to plan and adjust operations to reduce fuel needs during times when prices are typically higher, for example, in winter time.

UTILITY PROGRAMS

Utility programs are an excellent source of financial and technical support for energy management activities. Energy management programs offered by utilities are often referred to as *demand-side management* (DSM) programs. They may be called *demand response* (DR) programs if specifically aimed at customer peak load reductions in response to pricing or other signals from the utility.

There are numerous programs in operation, providing many different types of services and financial incentives. Utilities often design their programs to target specific markets and end-use systems. The goals of the programs may be overall energy savings, peak load reductions, energy education, or support for low income customers. Many programs are aimed at *resource acquisition*, whereby the utility supports energy management activities as a means to secure additional energy resources for balancing supply and demand. Other programs aim to transform the market (*market transformation*) by taking actions to accelerate adoption of energy efficient technologies and practices in given markets. Figure 5.1 illustrates the relationship between targeted markets, example program types, and potential program delivery channels.

Many utilities have strict program rules and guidelines that must be followed to receive the program benefits. Thus, it is very important for energy managers to understand and follow the correct procedures when participating in a program to ensure the facility ultimately receives the incentives the energy manager has built into the cost-effectiveness calculations.

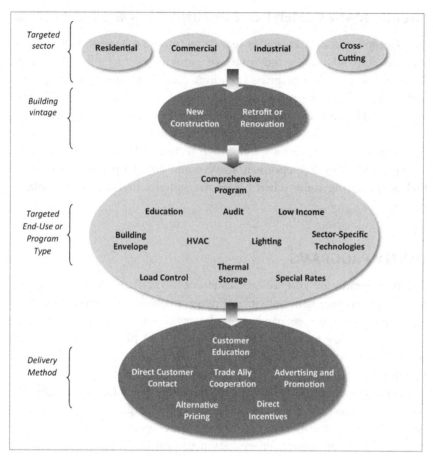

Figure 5.1 Examples of utility programs.[1]

We strongly recommend that energy managers contact their utility representative to learn about program offerings and eligibility requirements. In the U.S., another good resource for identifying utility as well as other state programs is the Database of State Incentives for Renewables & Efficiency (DSIRE™).[2]

---

[1] Source: Fig 5.1: Adapted from "Demand-Side Management," Gellings, C.W., and K.E. Parmenter, p. 539 in (2007) *Handbook of Energy Efficiency and Renewable Energy*, edited by F. Kreith and D.Y. Goswami, New York: CRC Press.

[2] Database of State Incentives for Renewables & Efficiency (DSIRE™) website, http://www.dsireusa.org.

CONCLUSIONS

Understanding utility rates and exploring utility program opportunities are important first steps for energy managers when planning an energy management program. The availability of certain services, such as energy audits, or pricing structures from the utility may help dictate the early direction of the energy management program. In addition, rebates and other incentives for targeted end-use equipment such as lighting, motors, air compressors, etc., may help the energy management committee prioritize measures.

CHAPTER 6

Building and Site Energy Audits

INTRODUCTION

An energy audit, like most other engineering activities, benefits from professional judgment and experience. Audits can provide important guidelines to the energy manager and insight in the major areas of energy use, if properly performed. An audit can even lead to immediate savings in some cases, simply by making people aware of how much energy is being used. For example, it is not uncommon for audits to reveal lights that remain on 24 h a day, controls that no longer function as intended, leaks and misuse of compressed air, or fans that operate around the clock.

More important, however, is insight that an audit provides into how and where major quantities of energy are being used and possibly misused. The energy manager can then draw upon this information to guide the formulation of the energy management program and establish where to place priorities.

Suppose a cursory walk-through of a warehouse revealed more than one hundred 100-watt lamp fixtures, a few small appliances, and a 130-hp (100 kW) fan motor. At first glance it is clear the auditor should not devote much of his or her resources to "counting light bulbs," since these would account for less than 10% of the installed load. It would be more useful to focus immediately on the fan, where 10% improvement in efficiency would lead to savings equal to the entire lighting load. The energy management plan could then include a lighting upgrade during a future phase if deemed cost-effective.

No textbook can provide answers for every situation. Our point here is that the auditor must develop the ability to stand back and look at the overall situation and to avoid getting mired down in the details.

GENERAL METHODOLOGY

The general methodology of an energy audit consists of assessing all building and site energy and utility systems to determine and document the following:
- Energy entering the site or building.
- Energy generated on the site or in the building.

Energy Management Principles.
DOI: http://dx.doi.org/10.1016/B978-0-12-802506-2.00006-9

- Energy distributed on the site or in the building.
- Energy used on the site or in the building.
- Energy leaving the site or building.
- Site-wide waste energy collected, treated, and/or discharged on-site.
- Site-wide waste energy recirculated or otherwise reused on-site.

Typical energy and utility systems include electricity, fuel oil, natural gas, LPG, coal, steam, chilled water, hot water, compressed air, and potable water supply. In some cases, sites may use other energy forms such as solar, wind, or various types of biofuel.

Waste streams and waste energy considered include the following:

- Wastewater.
- Solid wastes—dry combustible (paper, sawdust etc.), wet combustible (garbage), dry noncombustible (coal ash), and wet noncombustible (sewage sludge).
- Liquid combustible waste (waste oil).

In general, a facility will consist of a site occupied by one or more buildings, as sketched in Figure 6.1. Energy enters the site in the form of purchased fuels or utilities, and is transformed, converted, and distributed among the various buildings on the site. Within the buildings, people and systems perform operations that lead to useful products or services. The systems enabling the operations are called *energy end-uses.*

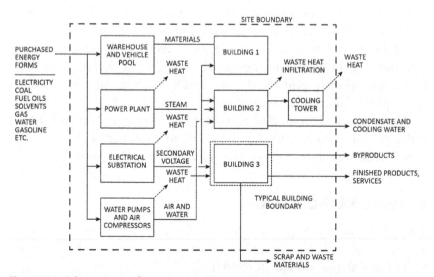

Figure 6.1 Schematic site for energy audit concept development.

Outputs from the site include the useful products or services, any by-products, waste (that may or may not contain energy in some form), and wasted or dissipated energy.

Now suppose we draw a hypothetical boundary around the site, as indicated in Figure 6.1. (Imagine this boundary is like a site fence or wall.) Equation 6.1 summarizes the energy balance at the site boundary:

$$E_i = \sum_{i=1}^{n} \left\{ E_{bi} + E_p + E_L - E_r \right\} \qquad [6.1]$$

where:

E_i = sum of input energy forms (primary voltage electricity, fuels, etc.) less any energy in storage

E_{bi} = sum of energy used in building number i (end-uses plus losses)

E_p = energy leaving site (in products, transfers, or other forms not included in E_b)

E_L = energy losses from site, excluding the losses from buildings (waste heat from cooling towers, steam line losses, transformer losses, condensate, etc.)

E_r = energy recovered; i.e., heat recovery, etc.

Equation 6.2 shows a similar energy balance equation for E_b for the ith building, which sums all the energy entering and leaving a hypothetical building boundary (say the building envelope):

$$E_{bi} = E_{ui} + E_{CLi} \qquad [6.2]$$

where:

E_{ui} = sum of energy end uses within building i (heating, cooling, lighting, process, office, domestic)

E_{CLi} = sum of energy conversion losses for building i (motor losses, transformer losses, etc.)

There are several advantages to using the energy balance approach described by Equations 6.1 and 6.2. First, the site energy input is usually available from records (utility bills, fuel purchase invoices, etc.). Thus, it is easy to obtain the total energy input. As described in Chapter 4, the energy manager can sum these historical data (along with total cost) and then use them to check against the results of the site audit. For example, the total of the building audit results plus any site losses or outputs minus any recovered energy should equal the site input. If a preliminary check shows this not to be true, it could indicate that a building has been overlooked, losses underestimated, or certain processes have not been included. This provides a useful check on the audit process.

A very important feature of the site audit is that it may reveal energy management opportunities that would not be apparent from building audits alone. For example, in an audit of a large government research facility, auditors found that a significant volume of water was pumped under high pressure from one area to another. The site audit flagged this practice and indicated the need for further investigation. Auditors subsequently determined that the reason was to provide a pressure suitable for fire protection in a small outlying building. The energy efficient alternative proposed was to use a small booster pump to provide the required pressure, permitting a reduction in the pipeline pressure and substantial energy savings. Building audits alone would not have uncovered this opportunity.

In the limited case of a single building, the site audit may be unnecessary, since the only "site" energy use may simply be the exterior lighting area.

Because of the importance attached to site and building audits as part of an energy management program, the sections below describe approaches for each in greater detail. In addition, Table 6.1 summarizes general principles for site and building audits, and Figure 6.2 is a flowchart of the overall process for entering and analyzing building and site audit data.

Table 6.1 Principles of site and building audits

Important Activities

1. Coordinate with operating management (solicit assistance of people working at the facility).
2. Obtain and review historical data prior to survey.
3. Conduct preliminary walk-through of facility.
4. Plan the energy audit survey (who does what, where does survey team go, when does team meet, etc.).
5. Conduct the energy audit survey, following the plan as a guide and using the proper facility energy audit forms.
6. After the survey, review the forms to assure completeness, readability, and reasonableness of values.
7. Recheck suspicious entries before leaving the site.

Important Survey Items

1. *Lighting*: check for inefficient lighting technologies and control strategies, excessive levels, unnecessary lighting in halls, stairwells, unused areas, storage areas, and parking lots.
2. *HVAC*: check thermostat settings (too high or too low), filter maintenance and other system maintenance performed at

proper intervals, controls, system capacity, equipment efficiency, overall operation.

3. *Process areas*: check total capacity measured against needed capacity, heat losses and vapor losses, equipment use schedules, equipment efficiency, controls.
4. *Furnace and ovens*: check total capacity measured against needed capacity, idling temperatures, process temperatures, air to fuel ratio, heat transfer and insulating surfaces, need for constant operation.
5. *Plant air systems*: check for leaks and maintenance procedures, misuse of compressed air, compressor sizing, sequencing, heat recovery, controls.
6. *Boilers and steam lines*: check for efficiency of burner settings, proper pressures and temperatures, steam leaks, faulty traps, lack of insulation, condensate return, opportunities for heat recovery.
7. *Numerically and hydraulically controlled machines*: check need for full operating pressure to maintain hydraulic fluid flows and temperatures.
8. *Electrical and other special building equipment*: check need for continuous operation, demand control, power factor, etc.
9. *Water*: check pumping capacity requirements, pump efficiency, head losses, water minimization, water treatment and reuse.
10. *Material transport*: check for more direct routes, less energy intensive modes, and operating requirements.
11. *General:* verify need for all energy-using equipment.

SITE AUDITS

Figure 6.3 illustrates a site audit methodology. The first step is to identify all the possible site energy forms, and then obtain historical data on their annual use and cost. It is usually convenient to convert them to a consistent set of units (MJ or GJ per month or year) although the physical units (liters, kilograms, cubic meters) could also be used (see Appendices for conversion factors).

The second step is to obtain or prepare simplified site plans that show the location of each building, substations, transmission lines, steam lines, water lines, etc. This information helps establish how and where the various energy forms are used at the site. Some specific suggestions follow.

Electricity

Electricity generated from power plants is stepped up to very high voltages (hundreds of kilovolts (kV)) for long distance transmission. Distribution substations then use transformers to lower the voltage to a medium range to carry the electricity to transformers near customers.

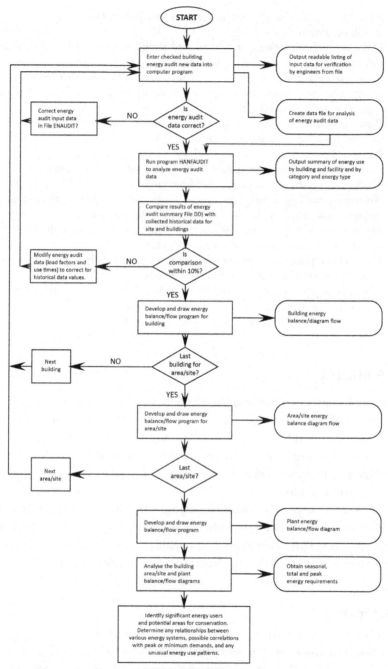

Figure 6.2 Flowchart of building and site audit data appraisal process.

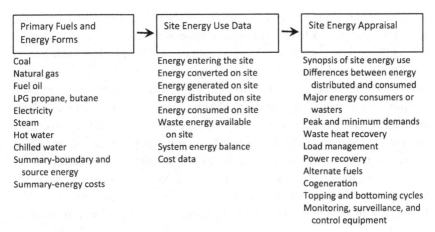

Primary Fuels and Energy Forms	Site Energy Use Data	Site Energy Appraisal
Coal	Energy entering the site	Synopsis of site energy use
Natural gas	Energy converted on site	Differences between energy
Fuel oil	Energy generated on site	distributed and consumed
LPG propane, butane	Energy distributed on site	Major energy consumers or
Electricity	Energy consumed on site	wasters
Steam	Waste energy available	Peak and minimum demands
Hot water	on site	Waste heat recovery
Chilled water	System energy balance	Load management
Summary-boundary and	Cost data	Power recovery
source energy		Alternate fuels
Summary-energy costs		Cogeneration
		Topping and bottoming cycles
		Monitoring, surveillance, and
		control equipment

Figure 6.3 Site audit methodology.

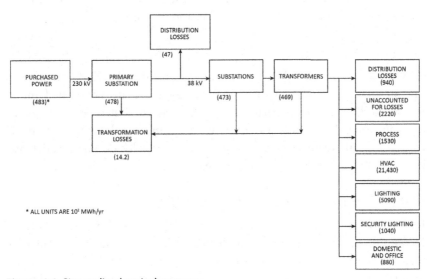

Figure 6.4 Site audit: electrical systems.

This medium voltage power is considered the primary distribution level and has voltages in the range of a few to several dozen kV depending on the distribution system. At the customer site, the electricity is transformed again to secondary voltages (e.g., 480 V/240 V in the U.S., 380 V/220 V in Europe, etc.) for powering end-use systems. Figure 6.4 shows an example of how the voltage transforms from the transmission system to the primary distribution system and finally to the secondary distribution systems serving end-users, illustrating the energy losses incurred along the way. This example is from a large government research facility.

For large sites, where customers connect at the primary distribution level, incoming electrical energy will equal the sum of end uses (lighting, power, heating) plus transformation and distribution losses associated with stepping the voltage down and distributing the electricity to the buildings at the site. If a master meter and sub-meters are available, an energy auditor can identify the losses directly. If not, they can be measured or estimated from manufacturer's data (for transformers) and calculated for distribution lines. As a rough guide, transformer losses are typically less than 1% and distribution losses will typically be less than 5%. Chapter 7 discusses methods for calculating electrical loads and losses in more detail.

For most sites, customers connect to the secondary distribution lines, so the transmission and distribution losses incur on the utility's side of the meter and do not need to be included in the site energy balance.

Steam

Steam may be generated in central boilers and distributed throughout a site, or may be generated locally. Normally there are boiler fuel input records and steam meters that permit direct assessment of steam generation efficiency. If not, boiler efficiency can be measured. Other sources of energy loss in steam systems are line losses, trap losses, and condensate losses. Figure 6.5 show a typical site steam audit.

Water, Compressed Air, and Other Utilities

Except for water, it is not common to find meters for these utilities. However, the approach for site audits of these energy forms is similar to that described above. Use meter data if available. If not, install temporary metering to measure water or air flow under typical operating conditions or make engineering estimates. The important aspects are to (i) account for all energy forms and (ii) determine the losses. Figure 6.6 shows a typical site audit for a water distribution system serving a large area.

Summary

After completing individual site audits, the next step is to total each energy form and compare with the historical records. This comparison provides a final check on the site audit accuracy. A summary sheet also

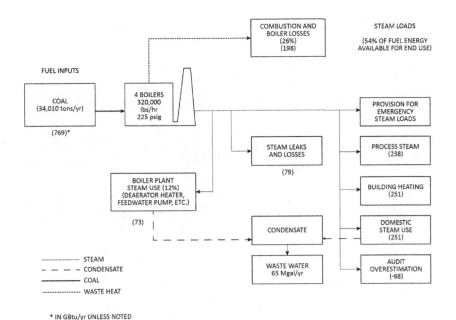

Figure 6.5 Site audit: steam.

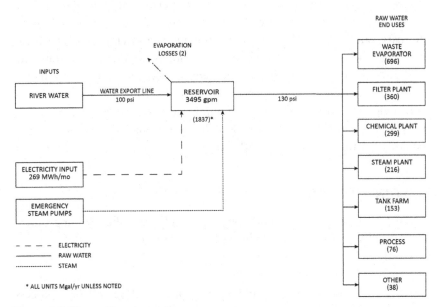

Figure 6.6 Site audit: water systems.

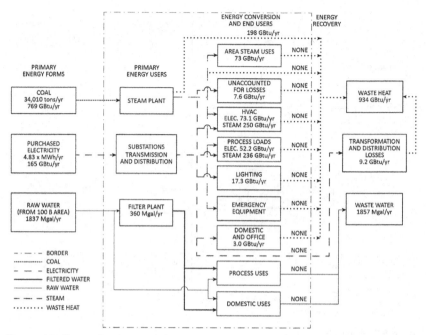

Figure 6.7 Site audit: summary sheet.

provides a composite view of energy losses, permitting evaluation of the overall site efficiency. Figure 6.7 shows an example of a site audit summary sheet. It is also useful to present the data in pie charts to show the relative shares of fuels used on site as well as to illustrate the breakdown by end-use system for each energy source (see Figure 6.8).

Energy management opportunities resulting from site audits cover a wide range of possibilities. As a preliminary guide, Table 6.2 provides a checklist of items to consider. This list is not all-inclusive but should provide a starting point for consideration.

BUILDING AUDITS

Figure 6.9 illustrates a methodology for conducting the building audit. The overall process may be subdivided into five steps:
- Compile historical energy use.
- Conduct building walk-through.
- Perform building audit—building envelope, lighting, HVAC, process equipment, or other special system assessments.

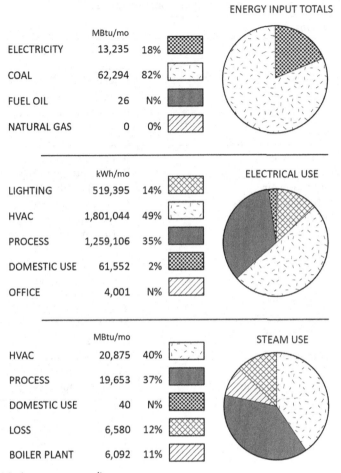

ENERGY INPUT TOTALS

	MBtu/mo		
ELECTRICITY	13,235	18%	
COAL	62,294	82%	
FUEL OIL	26	N%	
NATURAL GAS	0	0%	

ELECTRICAL USE

	kWh/mo		
LIGHTING	519,395	14%	
HVAC	1,801,044	49%	
PROCESS	1,259,106	35%	
DOMESTIC USE	61,552	2%	
OFFICE	4,001	N%	

STEAM USE

	MBtu/mo		
HVAC	20,875	40%	
PROCESS	19,653	37%	
DOMESTIC USE	40	N%	
LOSS	6,580	12%	
BOILER PLANT	6,092	11%	

Figure 6.8 Area energy audit summary.

- Compile building energy audit data and check.
- Summarize energy management opportunities.

Chapter 4 already described the steps necessary to compile historical energy use. If the building has meters, collect the meter data for later use as a check on the building audit results.

A building walk-through is useful as a prelude to conducting the energy audit. The walk-through permits the audit team to become acquainted with the facility and is particularly important if members of the team are not facility employees. It can also be used to divide up the areas to be audited and is an opportunity to determine what instrumentation or special equipment will be needed to conduct the audit. If the building

Table 6.2 Site audit energy management opportunity checklist

Electrical Systems

Line loss correction	Transformer heat recovery
Load management	Demand control
On-site generation	Cogeneration
Power recovery	Increase secondary voltage
Put major equipment on primary voltage	Power factor correction

Steam Systems

Improve boiler efficiency	Insulate lines
Preheat feedwater	Reduce pressure drops
Preheat combustion air	Cogenerate electricity
Topping or bottoming cycles	Stack heat recovery
Return condensate	Other heat recovery
Consider alternate fuels	Burn combustible waste

Water Systems

Improve pump efficiency	Off-peak pumping
Reduce pressure	Reduce pressure losses
Recycle water	Reduce water losses

Compressed Air Systems

Improve compressor efficiency	Reduce air losses
Reduce pressure	Recover compressor heat
Reduce inlet air temperature	Off-peak compression

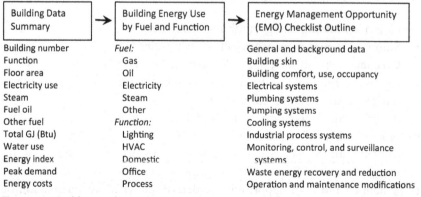

Building Data Summary	Building Energy Use by Fuel and Function	Energy Management Opportunity (EMO) Checklist Outline
Building number	*Fuel:*	General and background data
Function	Gas	Building skin
Floor area	Oil	Building comfort, use, occupancy
Electricity use	Electricity	Electrical systems
Steam	Steam	Plumbing systems
Fuel oil	Other	Pumping systems
Other fuel	*Function:*	Cooling systems
Total GJ (Btu)	Lighting	Industrial process systems
Water use	HVAC	Monitoring, control, and surveillance
Energy index	Domestic	systems
Peak demand	Office	Waste energy recovery and reduction
Energy costs	Process	Operation and maintenance modifications

Figure 6.9 Building audit methodology.

houses special processes or process equipment, arrange for the plant manager or process engineer to provide a briefing for the audit team.

The audit team should include the expertise necessary to perform the work. A team and its responsibilities might be constituted as follows: an architect, with responsibility for the building envelope; a mechanical engineer, for HVAC and processes; and an electrical engineer, for lighting and electrical loads. Other team members such as chemical engineers or process specialists may be added as needed. However, in many cases, a team comprising two or three engineers with energy efficiency training is sufficient.

Building audits should attempt to measure when and where all energy forms are used in the facilities and should also attempt to correlate the data with building use and occupancy patterns, weather conditions, site functions, number of shifts, and so forth. Building audits can help identify specific areas where there are opportunities for energy efficiency improvements and cost savings.

Examples of measurement to take during an energy audit include the following:

- Measure lighting intensity and room temperature at various points throughout the facility and over several daily cycles.
- Measure static pressure profiles in ducts and across heat exchange equipment.
- Monitor large electrical equipment to determine variations in the power factor and variations in load and use factors.

The composite of these measurements gives a rough picture of the facility's energy duty cycle and reveals inefficient equipment or applications. Investigating the transfer of various forms of energy and products between buildings, on a site level, and correlating this information with historical energy use, often reveals new possibilities for energy savings (by combining or eliminating multibuilding energy use).

Knowledge of the energy duty cycle is especially useful in those large industrial and research laboratory structures where the internal climate and circulation patterns must be maintained to meet safety or quality control requirements. Building simulation programs can be used to experiment with alternative energy duty cycles for the facility systems and to make optimal choices for the building HVAC system and processes before subjecting the operation and facility to specific changes (see Chapter 8).

Instrumentation for energy audits ranges from very simple to complex. Figure 6.10 shows a basic group of instruments that can be purchased for a few hundred dollars. Pictured are a volt—amp meter, a digital clamp-on ammeter, a sound/light meter with a digital temperature probe, a

Figure 6.10 Basic energy audit instrumentation.

thermometer, air velocity gage, digital camera, and binoculars. These instruments and tools facilitate reading of equipment labels and simple confirming measurements (illumination levels, temperatures, air velocities, etc.). They permit gross measurements which can be used to evaluate the need for (and benefit of) more detailed measurements.

Measurements and field audits (of both buildings and sites) are essential to the energy management process. The energy audit team *should not accept any individual's word concerning energy use anywhere.* Instead, they should make an independent appraisal, bringing to bear their unique expertise. In addition, the energy auditors should not necessarily believe what is shown in drawings and blueprints. Even "as-built" drawings are quickly out of date. The modifications made by maintenance personnel "to keep things running" somehow never get back into the drawings, and often thwart the actual purpose of the new "advanced energy management control system" that someone installed last year.

Of course, measurements can be overdone. Engineers are well known for collecting more data than can ever be analyzed economically. Similarly, measurements should be limited to those parameters that will lead to useful information.

For example, the U.S. Government at one time issued several "requests for proposals" to conduct energy audits on military facilities. One of the required tasks involved measurement (under load) of all electric motors with ratings more than a few horsepower. The objective of this task was to discover motor inefficiencies. Yet to an experienced observer, the results were obvious without expensive monitoring, although confirming measurements on a selected sample of motors might have been justified. These motors (that were part of many small air handlers and pumps) were all lightly loaded, and were operating with reduced efficiency and low power factor. A simple calculation also showed that the small savings that would result would not alone justify replacing these motors with properly sized motors. Why make measurements that would lead to obvious conclusions and no implementation? This was a case where it would have been more efficient to correct some of the fundamental deficiencies in these facilities, obtaining 80% of the energy savings at 20—30% of the cost of trying to do everything. Then, the energy management plan could include a stipulation to replace the inefficient motors upon failure, which would be cost-effective.

If more detailed measurements are justified, instrumentation costs increase rapidly, ranging from a few thousand dollars for a set of specific instruments to hundreds of thousands of dollars for a sophisticated field data acquisition systems. More detailed measurements might include those listed in Table 6.3.

Table 6.3 Examples of equipment for detailed audit measurements

Electric Power
Recording ammeters and volt meters
Demand meters
Power factor meters
Watthour meters
Data logger system

Thermal Measurements
Heat flux meters
IR scanners or thermometers
Recording thermographs
Digital thermometers
Combustion or flue gas analyzers
Psychrometers
Condensate meters

Mechanical Measurements
Load cells
LVDTs, velocity transducers, accelerometers
Tachometers

Flow Measurements
Airflow and velocity gauges
Manometers and pressure gages
Water flow meters

			Conversion Factors		
Plant Name:	XYZ Industries	By: CBS	Date:	1-Apr-13			Sheet: 1 of 1		kwh by	3.6	= MJ
Location:	Detroit, Mi	Period of s	1day	1 wk	1 mo X	1 yr			Btu/hr by	0.000293	= kW
Dept.	Office Bldg	Notes:	Test chiller performance						hp by	0.746	= kW
Symbols:	k= 10³; M = 10⁶										

			Power										
Fuel Type	Item No.	Equipment Description	Load (Btu/hr, kW, hp, etc.)	Conv. Factor to kW	kW	Efficiency	Est.% Load (100%, 50%, etc.)	Est. Hrs Use Per Period	kWh	Conv. Factor	Total Energy Use Per Period (MJ)	Notes
E	1.001	Chilled water pumps 2x15 hp	33.3	0.746	24.84	0.90	50	258	3,561	3.6	12,818	90 % eff motors
E	1.002	Air compressor 10 hp, 85%	11.76	0.746	8.77	0.85	30	730	2,260	3.6	8,137	85% effic
E	1.003	Chiller 160 ton	160	1	160.00	0.90	50	174	15,467	3.6	55,680	
E	1.004	Fluorescent lamps 50 x100W	5.00	1	5.00	1.00	100	258	1,290	3.6	4,644	

Figure 6.11 Simple building audit form.

In carrying out building audits, the most convenient approach is to use data sheets that have been prepared in advance and loaded onto laptop computers or tablets. The data can then be entered and processed digitally, eliminating the manual entry step required with paper audit forms. In this format, the data can also be readily uploaded to a central location for further analysis and review by other team members as well as for preparation of special reports.

The literature contains as many different forms for data entry as there are authors. Figure 6.11 shows an example of a simple form for entering data. It is sometimes desirable to tailor data entry forms to the needs of a specific audit scope or for each type of end-use. For example, in cases where HVAC accounts for a considerable share of facility energy use, it is useful to compile information about the building envelope as well as details about the HVAC equipment and control strategies. In contrast, for an industrial facility, it is more important to collect detailed information on furnaces, compressed air, steam, refrigeration, or other types of equipment used in production.

Individual building data collected, or calculated, and utilized would normally include:
- Building identification.
- Building function/use.
- Physical characteristics (floor area, window area, wall area, construction material).
- Building age and expected remaining life.
- Building energy consumption and costs.
- Process flow diagrams.
- Historical process data.

- Energy system data (type and capacity of air distribution, heating and cooling equipment, cold and hot air supply temperatures, lighting types and quantity, domestic hot water, process heat and cooling systems, mechanical equipment and motors, other types of production equipment, office equipment).
- Control equipment characteristics (reset temperatures for hot and cold air, humidification, economizer cycle, daylighting controls, variable speed drives, compressed air sequencing, and process equipment settings).
- Heat recovery and rejection approaches.
- Building occupancy data (number of people by shift for typical workdays, weekends, and holiday schedules).
- Seasonal variations in production.
- Recent and planned equipment and process upgrades.

All audit input data should represent the best information economically obtainable within the scope of the audit. Sources include fuel and utility records, meter and sub-meter readings, building energy management systems, supervisory control and data acquisition (SCADA) systems, generation and consumption records, and/or best engineering estimates based on visual observation, temporary metering, calculation, extrapolation, and discussions with the building operation and maintenance staff.

If possible, the data collected and utilized should cover a minimum of twelve recent consecutive months. It may be necessary to make adjustments if the available data period does not represent a typical year, from either climatic or energy usage considerations. Attempt to include total, minimum, and peak monthly qualities as well as typical demand profiles for average and extreme daily variations in the data collected for each energy and utility system.

When it comes to gathering data for specific end-uses, the basic approach is first to determine the rated capacity (i.e., useful energy delivered in units of Btu/h, kW, hp, tons of cooling, etc.) and efficiency of the device or system to be able to estimate the required input energy per unit time at rated full load conditions. The second step is to determine the load on the equipment and the factors that affect loading. For some types of equipment (such as nondimmable lamps), the load is 100% when the device is on. For other types of equipment (including motor driven systems), the load is often less than 100% and may vary with weather, production, or other drivers. The third step is to determine the operating hours of the equipment for a representative period of time (day, week, month, or year). Select the time period to agree with the period for which historical or meter data are available. The hours of operation can be established by analysis, observation, or measurement.

With the information gathered during the building audit, the energy use for each item of equipment item can then be computed using methods described in Chapter 7. The calculations are relatively simple for equipment with constant loads, but are often rather complex for systems with loads that vary or for systems that interact with one another.

Often it is of interest to examine certain energy uses in greater detail. Heating, ventilation, and air conditioning (HVAC) of buildings frequently fall into this category. In this case additional data on the building envelope is useful, since the building envelope influences heat losses and solar heat gains.

Building heat losses and gains can be evaluated by various computer simulation programs. Figure 6.12 is a "Building Profile Data Sheet" that summarizes much of the information needed for this purpose.

The final steps in the building audit procedure are to summarize building energy use and compare it with historical data, and to compile

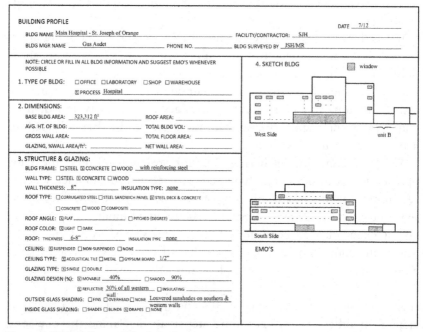

Figure 6.12 Building profile data sheet.

Fabrication Shop: 29,170 square feet
Contractor: RHO
Area: 200 W

The 277W fabrication shop is a tall, steel and wood-framed structure that is used for the fabrication, installation, and/or repair of mechanical equipment. As is the case with the neighboring 272W building, the 277W shop is uninsulated.

Over one-half of the total electrical energy is used for the HVAC systems, and one-third is used for process equipment. The largest loads are from the heat treatment ovens (86 and 36 kW) and the air compressors (75 and 40 hp.)

Building heating consumes all of the steam supplied to 277W. Audited steam use is 990 MBtu/mo. Audited electricity use is 39,311 kWh/mo.

ELECTRICAL ENERGY USE

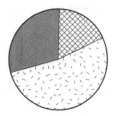

▨	LIGHTING	18%
░	HVAC	52%
▰	PROCESS	30%
▨	DOMESTIC USE	0%
▨	OFFICE	N%

Major Building EMOs:
- HVAC—B024/06.2: Install fans and flexible plastic ducts in each corner to blow warm air to the lower portions of each high bay during the winter. This should be done after adding insulation and decreasing infiltration. For an initial cost of roughly US$2,000, the anticipated savings are approximately 1,100 MBtu/yr.
- HVAC—B044/06.2: Install night setback.
- Building system—B111/06.2: Add roof and wall insulation.
- Lighting—B179/01.1: Improve lighting efficiency.
- Process—B015/01.3: Utilize the 36 and 86 kW ovens only during off-peak hours to defer roughly 122kW of demand. There is no capital expense for this modification.

Note: The "B" numbers are codes assigned to specific measures. For example, B179/01.1 is "Replace incandescent lamps with high pressure sodium."

Figure 6.13 Building summary sheet example.

the major energy management opportunities noted during the audit. One way to present the results is to display them in graphic form on a building summary sheet such as Figure 6.13.

Table 6.4 provides an illustrative checklist of items to consider when developing energy management opportunities. The table divides the opportunities by major building systems. Chapters 8 through 12 discuss these items in greater detail.

VARYING LEVELS OF AUDITS

The appropriate scale of the site or building audit depends on the energy management committee's goals. If the objective is to make major energy

Table 6.4 Building audit energy management opportunities

Building Envelope

Add insulation	Shade with trees or berms
Provide weather stripping	Add reflective coatings to windows
Upgrade windows (double glaze or better)	Change roof color
Reduce heated or cooled volume	Add vestibules

HVAC

Install economizer systems	Use night setback
Install heat recovery systems	Use night cool down
Use more efficient equipment	Advanced controls
Eliminate excess capacity	Change to variable volume
Improve chiller performance	Improve cooling towers

Lighting

Relamp (more efficient types)	Delamp
Improved controls (occupancy sensors)	Task oriented lighting
Improve lamp maintenance	Change room colors

Processes

Install more efficient equipment	Improved controls
Use heat recovery to heat water	Reduce steam, air, water leaks
Use more efficient processes	Insulate ovens and furnaces
Recover power from fluids	Reduce scrap and waste

saving capital improvements, the audit and analysis will need to be more rigorous to provide the high degree of confidence needed to justify high capital expenditures. However, if the goal is to do a relative comparison of one building to another (benchmark comparison), the audit and analysis process could be scaled down considerably.

The American Society for Heating, Refrigerating, and Air Conditioning Engineers (ASHRAE) has established three audit levels for commercial buildings:

- **ASHRAE Level 1 Audit:** Level 1 is a "walk-through" audit that consists of a more cursory review of historical data, a shorter onsite review and assessment of building and energy systems, and identification and preliminary analysis of energy efficiency opportunities. It is a good way to find major problems, to identify low or no-cost measures for immediate implementation, and to point out areas for more targeted assessments.

- **ASHRAE Level 2 Audit:** Level 2 is a more detailed audit that includes a greater degree of historical data collection and analysis, a longer and more comprehensive site visit with some measurements and data logging, and more thorough energy and financial analysis of identified opportunities to justify their implementation. It is typically the appropriate scale for low-to-medium cost measures.
- **ASHRAE Level 3 Audit:** Level 3 is a rigorous audit comprising detailed historical review and analysis, more onsite data collection, comprehensive analysis of baselines and energy savings, and detailed financial analysis to justify capital-intensive measures. It often includes modeling of buildings and specific end-use systems.

Oftentimes, the best method is to begin with a Level 1 audit and follow it with a Level 2 or 3 audit. This step-wise process helps convince management of the benefits of the audit process with minimal expenditure for the Level 1 phase. Then, the Level 2 or 3 phases can be used to substantiate analysis for the subset of high priority (i.e., management-approved) capital projects, rather than for the universe of potential measures.

We find an analogous approach works well for industrial facilities. After a preliminary facility walk-through, the next level is to do a "mid-level" audit of the facility, focusing on areas identified during the walkthrough. The mid-level audit includes collecting equipment inventories and nameplate information, taking spot measurements of the energy consuming systems with most promise for improvement, and discussing usage patterns and other energy information with facility and maintenance personnel. Then, the audit team can perform energy savings calculations and cost estimates based on the gathered information and the team's familiarity with energy efficiency measures and process improvements. This mid-level audit usually does not include detailed assessments of equipment, nor does it include data logging. But, it can be a cost-effective approach for identifying measures and developing reasonable estimates of energy savings, particularly for operation and maintenance improvements and medium cost retrofit measures such as energy efficient lighting and motors and drives.

Capital intensive measures at industrial facilities, such as process improvements and retrofits to support equipment, warrant comprehensive system assessments to validate their cost-effectiveness. Most of the time, the best place to begin is with the systems that support the industrial processes, including compressed air systems, chillers and refrigeration systems, boilers and steam or hot water distribution, and water treatment and disposal systems. This is because the process systems themselves are usually carefully tuned to meet production requirements and optimized for product quality, quantity, reduced down-time, etc., while energy optimization is not likely to be the primary priority. There may be opportunities for energy cost savings through process changes, but these opportunities need careful scrutiny to determine if they are suitable and justified, particularly because the production team may be guarded about changes that affect process equipment. Support systems, on the other hand, often account for the bulk of the energy consumption in an industrial facility and can be improved without changing the core processes. Many times support systems offer significant savings because facility staff have paid less attention to their energy consumption. Comprehensive system assessments typically involve more onsite measurements and data collection, including data logging of energy use and other parameters. They also involve more detailed engineering analysis and modeling to predict energy and financial impacts.

Example. A building and site energy audit of Punahou School in Hawaii.[1] The authors directed this detailed audit of an extensive campus comprising grades kindergarten through 12th. As of January 1995, the 76 acre Punahou campus consisted of 53 buildings with a total gross square footage of 628,650. The buildings ranged in age from 2 to 144 years old. Most of the buildings from the late eighteen hundreds and early nineteen hundreds had been renovated in the last forty years.

The scope of the work consisted of the following:
- Preliminary information gathering and planning, including analysis of historical utility data to establish baseline energy use.
- Field data acquisition to review existing conditions, including inventorying equipment, collecting technical information, and identifying end uses.
- Identification, evaluation, and prioritization of specific energy management opportunities.
- Analysis of energy management opportunities with projected savings.
- Economic/financial analysis including utility rebates.
- Development of operations and maintenance procedures and guidelines for renovation and new construction.

[1] Parmenter, K. and Smith, Craig B. (1997). *Energy Management Program: Punahou School, Honolulu, HI.* Final Report. Los Angles: DMJM Inc.

- Conceptual design of an energy management and control system to monitor energy use.
- Preparation of a campus-wide implementation plan.
- Additional measures to be implemented as part of a future campus expansion.

At the time of the audit in 1997, Punahou used 7.5 million kilowatt hours per year, with a peak demand of 1800 kW. The electric bill for fiscal year 1996—97 was $790,000, which was over 85% of the total utility costs of $900,000 per year (for water, gas, sewer, and electricity). Electricity usage was primarily for HVAC (52%) and lighting (28%), with other uses accounting for the remaining 20%. The school's electricity cost was $0.10/kWh.

We first collected data by inventorying equipment in the 50 buildings and facilities, studying building drawings and equipment plans, and interviewing maintenance personnel. The availability of HVAC and lighting inventories greatly simplified the initial data collection procedure. In addition, insight from maintenance personnel allowed us to focus our audit efforts to high priority areas.

Next, we inspected the buildings and equipment, determined operating hours, and took measurements of temperature, illumination levels at typical locations, and other parameters as necessary. While we placed emphasis on determining the capacity and condition of the heating, ventilating, and air conditioning (HVAC) systems and lighting, we also observed the physical condition of the buildings. During each building visit, the survey team noted potential energy efficiency improvements for future evaluation.

We compiled data from the individual building audits in a large database designed for Punahou School using Microsoft Access. We formulated the database to compute baseline annual energy use and demand for each type of equipment in every building on campus, for comparison with the historical energy use records. We also used it to estimate energy and demand savings for the potential energy efficiency improvements while we were still onsite. The calculations were based on engineering algorithms using equipment rating, loading, and operating schedules. As a result, energy use and demand subtotals for each building were readily available during the site visit. This allowed us to see the distribution of energy among the buildings, and, along with valuable input from experienced Punahou personnel, helped to target buildings with the most savings potential. Having the audit results completed while still onsite also enabled us to compare findings with utility bills, and to verify that no major energy using systems were overlooked before we left the site. In the process, we recognized that the campus had many associated "miscellaneous" uses of energy, for example, overhead projectors, computers, refrigerators in some classrooms, and so on. In most cases these items were small and their use infrequent, so we did not include them in the audit. Exceptions were made for major power users, such as the swimming pool water circulation pumps.

We studied buildings in greater detail if the possibility of a significant project or problems were evident. Five buildings were singled out for a detailed feasibility study for improving building air conditioning and lighting systems. For each of these buildings, we submitted an application to Hawaiian Electric Company (HECO) to secure co-funding for the feasibility study.

HECO also offered rebates for high efficiency lighting projects, improved controls, electronic ballasts, high efficiency chillers, and other qualifying energy efficient products. A substantial number of the Punahou projects were eligible for rebates from HECO's programs, both for renovation of existing buildings and for new construction.

After estimating the implementation costs for the identified energy management opportunities, we used projected energy cost savings calculated in the analysis database to determine the financial feasibility of each potential project. Simple payback was used as a screening criterion for acceptable projects.

Based on our findings, we reported that Punahou could reduce its electricity consumption by 23 percent, from 7.5 million kWh/yr to 5.8 million kWh/yr, and could cut demand from 1,800 kW to 1,550 kW. The overall simple payback for all recommended measures was 4.6 years. Projects required a future capital investment of $821,000; however, at least $69,000 of this could be obtained from Hawaiian Electric Power Company in the form of rebates. We also reported that without taking steps to improve energy use efficiency, Punahou School could expect to see a doubling of electricity costs in the next decade.

Even in 1997, Punahou recognized that to maintain its world class status in the face of rising operating costs, while still being able to offer a high quality education at competitive prices, it had to be world class in the efficiency and quality of its physical plant. The school established a specific goal of reducing energy use by 20% over a 1996 baseline. In addition, implementation of the measures described in this audit were viewed as a first step to enable Punahou to enter the 21st century with a campus that was among the most efficient in the U.S. in terms of energy use.

This is not the end of the Punahou story. Over the next two decades, the cost of electricity in Honolulu did not double, *it tripled*. Meanwhile, the school undertook a series of measures that built upon and went well beyond the recommendations of our report. In Chapters 8 and 9 we describe some of the HVAC and lighting measures that were recommended. In Chapter 12 we describe Punahou School today, a leading green school and a model for sustainability.

CONCLUSIONS

Building and site energy audits are a useful tool for the energy manager. They do not in themselves cause energy savings to occur—except in cases where inefficiencies are addressed immediately during the actual audit—but they do provide the basic data on which to establish an effective program. They also assist in setting priorities and provide a mechanism for evaluating the effectiveness of an energy management program. The energy literature has a number of handbooks that provide additional discussions of methods, analytical techniques, and case studies for building and site audits, including specific approaches for residential, commercial, and industrial facilities.[2,3,4,5]

[2] Thumann, Albert, Niehus, Terry, and Younger, William J. (2012): *Handbook of Energy Audits*, 9th Ed., Lilburn, GA: Fairmont Press.

[3] Kreith, Frank and Goswami, D. Yogi, Eds. (in press; publication date 2015): *Energy Management and Conservation Handbook*, 2nd Ed. Boca Raton, FL: CRC Press.

[4] Deru, Michael, Kelsey, Jim, Pearson, Dick, et al. (2011). *Procedures for Commercial Building Energy Audits*, 2nd Ed. Atlanta, GA: ASHRAE.

[5] Krarti, Moncef. (2011). *Energy Audit of Building Systems: An Engineering Approach*, Second Edition. Boca Raton, FL: CRC Press, Taylor & Francis Group.

CHAPTER 7

Energy Analysis

INTRODUCTION

There are several approaches for analyzing energy use and the energy savings and load impacts resulting from energy management activities. Techniques include engineering algorithms, building simulation and system modeling, and load analysis with metered data. The preferred analysis approach for a given activity depends on the type and complexity of the energy management measure, the energy use data available for analysis, the desired accuracy of the savings estimates, and cost considerations. Some techniques are most applicable for predicting energy and demand savings for proposed measures, while others are well suited for verifying savings after implementation of energy management projects. This chapter focuses on predicting energy and demand savings to aid in measure prioritization and Chapter 14 addresses verifying energy savings post implementation.

Regardless of the approach ultimately selected for analyzing energy use, two concepts that every energy manager must firmly grasp are *thermodynamic efficiencies* and *appropriate baseline conditions*. These factors come into play in every energy analysis, either directly or indirectly. Understanding how they affect energy and demand savings is critical to optimizing energy management decisions.

MEASURES OF EFFICIENCY

Ideally, energy management is the most cost-effective *efficient* use of energy. Thus, efficiency is an important concept for the energy manager. Efficiency may be considered from the point of view of the first or second laws of thermodynamics. *First law* efficiencies relate to the conversion of energy from one form to another and conservation of the overall *quantity* of energy, without direct consideration for the *quality* of energy. When evaluating the overall utilization of fuels or energy forms, *second law* considerations apply since they take energy quality into account and help define upper bounds of efficiency. Second law efficiencies are expressed in terms of a quantity known as *available work*.

Energy Management Principles.
DOI: http://dx.doi.org/10.1016/B978-0-12-802506-2.00007-0

In its most basic form, first law efficiency of a particular task may be thought of as the ratio of useful energy delivered to the task to the required energy input. However, there are many tasks for which this basic definition is inadequate or inappropriate. Therefore at least two other broad categories of energy efficiency have evolved. Herein they are referred to as "coefficients of performance" (COPs) or Energy Use Performance Factors (EUPFs).

Example: Electric motor efficiency. What is the first law efficiency of a single phase 1 hp electric motor rated at 240 V, 4.88 A full load current, and having a power factor of 80%? See Equation 7.1 for the answer. Note that we use energy per unit of time in this equation and in Equations 7.2 through 7.4 below.

$$\eta = \frac{Useful\ energy\ delivered(in\ this\ case,\ ``work")}{Energy\ input} = \frac{hp \times LF}{(\sqrt{\#phases})(V)(A)(pf)}$$

$$\eta = \frac{(1\ hp)(0.746\ kW/hp)(1.0)}{(\sqrt{1})(240\ V)(4.88\ A)(0.8)(10^{-3}kW/W)}$$

$$\eta = 0.796 = 80\%$$

[7.1]

where

η	=	efficiency, dimensionless (or %)
hp	=	motor horsepower, hp
LF	=	load factor, dimensionless
$\#phases$	=	number of motor phases, dimensionless
V	=	rated voltage, V
A	=	full load amperage, A
pf	=	power factor, dimensionless

This calculation indicates an operating efficiency of 80%. This is true for the stated conditions (i.e., for full load and when the power factor is 80%). For operation at other power factors, or for less than full load, efficiency is less. Power factor remains fairly constant at the full load rated value until the load falls below about 50−60%. When the load drops to about one-third of full load, the power factor can drop as low as 20−30%.

Tables 7.1a and 7.1b show typical full load electric motor efficiencies for premium motors. Note that the NEMA standards and the EU standards are in many cases identical. When these data are compared to motor efficiencies in the first edition of this book, efficiencies have improved from 76−85.5% (1.0 hp) and from 91−95.4% (100 hp).

Table 7.1A Nominal full load efficiencies: NEMA premium high efficiency electric motors

HP	OPEN FRAME			ENCLOSED FRAME		
	2 POLE	4 POLE	6 POLE	2 POLE	4 POLE	6 POLE
1	77.0	85.5	82.5	77.0	85.5	82.5
3	85.5	89.5	88.5	86.5	89.5	89.5
5	86.5	89.5	89.5	88.5	89.5	89.5
10	89.5	91.7	91.7	90.2	91.7	91.0
30	91.7	94.1	93.6	91.7	93.6	93.0
50	93.0	94.5	94.1	93.0	94.5	94.1
100	93.6	95.4	95.0	94.1	95.4	95.0
300	95.4	95.8	95.4	95.8	96.2	95.8

Source: NEMA MG-1 (2006) Table 12–12.

Table 7.1B European union IE3 premium motor efficiencies (3-phase cage induction motors)

kW	2 pole 50 HZ / 60 HZ	4 pole 50 HZ/60 HZ	6 pole 50 HZ/60 HZ
0.75	80.7/77.0	82.5/85.5	78.9/82.5
2.2	85.9/86.5	86.7/89.5	84.3/89.5
7.5	90.1/90.2	90.4/91.7	89.1/91.0
22	92.7/91.7	93.0/93.6	92.2/93.0
37	93.7/93.0	93.9/94.5	93.3/94.1
75	94.7/94.1	95.0/95.4	94.6/95.0
220	95.8/95.8	96.0/96.2	95.8/95.8

Source: IEC 60034-30 (2009).

As mentioned above, efficiency is best at or near full load. As the load on the motor drops from full load to less than 50% load, the motor efficiency begins to fall, dropping to 40–80% when the load is only 10–15%. The drop-off is greater for small motors.

This is the first point we wish emphasize in this chapter: *Efficiency is usually load dependent.* "Load," as used here, can mean a variety of things: temperature, pressure, force, work, etc.

Example: An electric resistance heater. The heater is rated at 240 V and 4.167 A and delivers 3412 Btus per hour of heat. What is its efficiency? Assume a load factor of 100% and a power factor of 100%.

$$\eta = \frac{\text{useful energy delivered (in this case, "heat")}}{\text{energy input}} = \frac{\dot{Q} \times LF}{(V)(A)(pf)}$$

$$\eta = \frac{(3412 \ Btu/h)(0.29307 \ wh/Btu)(1.0)}{(240 \ V)(4.167 \ A)(1.0)} \qquad [7.2]$$

$$\eta = 1.0 = 100\%$$

where

$\dot{Q}$ = Heat delivered per unit time, Btu/h (or W)

This calculation implies that all of the energy input—that is, electricity—is delivered to the load in the form of heat. Obviously, we have neglected the losses that arise in the process of converting fuel into electricity and any losses related to delivering heat to the load (such as radiation losses, vent or stack losses, etc.).

This leads to a second point: *Efficiency is only defined within certain specified system boundaries.*

Example: An electric incandescent lamp. The lamp is rated at 100 W and 120 V. This means that the input power is 100 W when fully loaded, that is, not dimmed. The light output is 1500 lm. The conversion factor from lumens to Watts is 1.496×10^{-3} Watts/lumen. This gives an efficiency for the incandescent lamp of the following:

$$\eta = \frac{\text{Useful energy delivered (in this case, "light")}}{\text{Energy input}} = \frac{lm \times LF}{\dot{E}_{in}}$$

$$\eta = \frac{(1500 \ lm)(1.496 \times 10^{-3} \ W/lm)(1.0)}{(100 \ W)} \qquad [7.3]$$

$$\eta = 0.0224 = 2.24\%$$

where

lm = lumens, lm

$\dot{E}_{in}$ = energy in, in this case lamp wattage, W

This is not too useful as a measure of efficiency, since the relationship of the input energy to the light delivered is not clear. A measure in common use is the ratio of the light output in lumens to the input power in Watts, called the *efficacy*:

$$Efficacy = \frac{1,500 \ lm}{100 \ W} = 15 \ lm/W \qquad [7.4]$$

Efficacy is example of an energy use performance factor; that is, a factor that measures how energy is used to meet a particular performance goal.

Example: A window air conditioner. This appliance uses 1000 W of input power to provide 10,200 Btu/h of cooling. Air conditioners use input energy (work) to transfer heat out of a lower temperature region (the interior space) to a higher temperature region (the outdoors) thereby cooling the interior space. Heat pumps in cooling mode operate the same way as air conditioners, but in heating mode they operate in reverse. When heating, heat pumps use work to transfer heat from the lower temperature outdoors to the higher temperature interior space.

Instead of using the symbol η, which is typically reserved for efficiencies ranging from 0 to 1.0, one approach taken to represent efficiency of air conditioners and heat pumps is to define a *coefficient of performance* (COP) given by:

$$COP = \frac{Performance\ achieved(i.e., quantity\ of\ heating\ or\ cooling\ delivered)}{Energy\ input(electricity\ in)} = \frac{\dot{Q}}{\dot{E}_{in}}$$

$$COP = \frac{(10,200\ Btu/h)(0.29307\ Wh/Btu)}{(1000\ W)} \tag{7.5}$$

$$COP = 2.99$$

Coefficients of performance are always greater than unity for heat pumps and can be either greater than or less than unity for air conditioners.

Another approach for measuring air conditioning performance (or performance of heat pumps in the cooling mode) is the energy efficiency ratio (EER), which is similar to the COP, but is not dimensionless:

$$EER = \frac{Quantity\ of\ cooling\ delivered}{Energy\ input(electricity\ in)}$$

$$EER = \frac{(10,200\ Btu/h)}{(1000\ W)} \tag{7.6}$$

$$EER = 10.2\ Btu/Wh = (COP)(3.412\ Btu/Wh)$$

Still another metric for cooling performance of air conditioners or heat pumps is the seasonal energy efficiency ratio (SEER), which is the ratio of total heat removed during the cooling season (Btu) to total electrical energy used during the cooling season (Wh).

In addition, the heating seasonal performance factor (HSPF) is a measure of heat pump performance in heating mode. It is the ratio of total space heating required during the heating season (Btu) to total electrical energy used during the heating season (Wh).

So much for first law efficiency. As can be seen from the above examples, efficiency (as commonly used) refers only to the ratio of work or heat output compared to energy input. This measure reflects the *quantities* of energy involved, but says nothing about the *quality*.

The quality of an energy form is a measure of its ability to perform useful work. For example, a gallon of oil has approximately 148 MJ (140,000 Btu) of heating value. This is roughly the same energy content as 1000 gallons of lukewarm water heated to 9°C (17°F) above ambient temperature. Although the *quantity* of energy is the same in both cases, the ability of the oil to perform useful work is much greater than the ability of the lukewarm water. The quality of the oil is much greater.

Availability (also called *available work* or *exergy*) is the metric used for quantifying energy quality. It represents the maximum amount of available work of a system relative to a reference state. It is also defined as the minimum work needed to bring a system at a reference state to an elevated state. For a control mass system (e.g., a piston and cylinder), availability is referred to as *non-flow availability* and can be expressed as follows:

$$B_{cm} = (U - U_0) + P_0(V - V_0) - T_0(S - S_0) + m\frac{v^2}{2} + mgz \qquad [7.7]$$

where:

B_{cm} = non-flow availability, J
U = internal energy, J
P = pressure, Pa
V = volume, m^3
T = temperature, K
S = entropy, J/K
$m\frac{v^2}{2}$ = kinetic energy, where m is mass (kg) and v is velocity (m/s), J
mgz = potential energy, where g is acceleration (m/s^2) of gravity and z is elevation (m), J

and the subscript *0* refers to the reference state

For a control volume system (e.g., a turbine), availability is called *flow availability* and can be expressed as follows:

$$B_{cv} = (H - H_0) - T_0(S - S_0) + m\frac{v^2}{2} + mgz \qquad [7.8]$$

where:

B_{cv} = flow availability, J
H = enthalpy, J

Flow availability is relevant to many thermodynamic cycles.

For a given energy, volume, and system composition, B decreases as the system entropy increases; B also decreases as the internal energy or enthalpy of the system approaches that of the reference state. (Note that the kinetic and potential energy terms can be neglected in many energy systems.)

If applied to a hydrocarbon fuel, B is the minimum useful work required to form the fuel in a given state from the water and carbon dioxide in the atmosphere. Since the minimum is also useful work of a reversible process, B also represents the maximum useful work that can be obtained by oxidation of the fuel and return of the products to the atmosphere.

In a relative sense, the quality (availability) of electricity and fuels such as oil, coal, and gas is quite high. Likewise, high pressure, high temperature steam has high availability. Conversely, hot water, low-temperature process heat, or low-pressure steam have relatively low availability.

A measure of thermodynamic *effectiveness* (or second law efficiency) of energy use for a process can be defined as the ratio of the increase in available work attained by the products in the process to the maximum available useful work of the fuel consumed. Another way of defining it is as the ratio of the theoretical minimum available work to accomplish a task to the actual useful work required for the task. We can think of it as the availability *recovered* divided by the availability *supplied*. The difference between what is supplied and what is recovered is lost or *destroyed* availability. Thus, the concept of availability provides a useful measure of efficiency that extends beyond first law efficiency constraints. In addition, availability analysis helps pinpoint process steps or areas where improvements in efficiency are possible.

Example: A steam boiler. The difference between the popular notions of efficiency and the concept of effectiveness is illustrated by steam boiler operation. An acceptable boiler is one that achieves efficiency of about 90%. That is, only 10% of the input energy dissipates in the flue gas or by heat transfer losses. From a first law point of view, we may be satisfied with 90% efficiency and consider we are doing the best by present technological standards. Yet, this overlooks consideration of whether we have utilized the fuel to its maximum potential. On the basis of thermodynamic availability, this "efficient" operating boiler has an effectiveness of only 40–45%, indicating some work is lost unnecessarily in making steam. For a more complete discussion of boiler losses, see Chapter 11, "Management of Process Energy."

Now we shall repeat the analyses performed in Equations 7.1, 7.2, and 7.5, this time computing the effectiveness rather than the efficiency. Table 7.2 summarizes first- and second-law efficiencies for common energy using processes. The reader should consult the literature for more detailed discussions of available work.

Table 7.2 First-law and second-law efficiencies for single-source-single output devices

End use	Source		
	Work E_{in}	Fuel: Heat of combustion $|\Delta H|$ available work B	Heat E_1 from hot reservoir at T_1
Work E_{out}	1. $\eta = E_{out}/E_{in}$ $\epsilon = \eta$ (e.g., electric motor)	2. $\eta = E_{out}/|\Delta H|$ $\epsilon = \dfrac{E_{out}}{B}\,(\simeq \eta)$ (e.g., power plant)	3. $\eta = E_{out}/E_1$ $\epsilon = \dfrac{\eta}{1-(t_0/T_1)}$ (e.g., geothermal plant)
Heat E_2 added to warm reservoir at T_2	4. $\eta(COP) = E_2/E_{in}$ $\epsilon = \eta\left(1 - \dfrac{T_0}{T_2}\right)$ (e.g., electrically driven heat pump)	5. $\eta(COP) = E_2/|\Delta H|$ $\epsilon = \dfrac{E_2}{B}\left(1 - \dfrac{T_0}{T_2}\right)$ (e.g., engine driven heat pump)	6. $\eta(COP) = E_2/E_1$ $\epsilon = \eta\dfrac{1-(T_0/T_2)}{1-(T_0/T_1)}$ (e.g., furnace)
Heat E_3 extracted from cool reservoir at T_3	7. $\eta(COP) = E_3/E_{in}$ $\epsilon = \eta\left(\dfrac{T_0}{T_3} - 1\right)$ (e.g., electric refrigerator)	8. $\eta(COP) = E_3/|\Delta H|$ $\epsilon = \dfrac{E_3}{B}\left(\dfrac{T_0}{T_3} - 1\right)$ (e.g., gas-powered air conditioner)	9. $\eta(COP)E_3/E_1$ $\epsilon = \eta\dfrac{(T_0/T_3) - 1}{1-(T_0/T_1)}$ (e.g., absorption refrigerator)

For the electric motor. We make the assumption that effectiveness is defined at the source of electricity and does not include the generation and delivery of electricity. In this case

$$\epsilon = Effectiveness = \frac{Work\ delivered}{Energy\ in,\ E_{in}} = \eta = 80\% \qquad [7.9]$$

This is the same result as before. If we instead defined the denominator as the maximum available energy in the primary fuel used to generate the electricity, the effectiveness would be lower due to generation, transmission, and distribution losses.

For the resistance heater. Assume the heater delivers warm air at 43°C (316 K) to a house where the outside air is 0°C (273 K). The available work usefully transferred is given as:

$$\dot{W}_{rev} = \dot{Q}\left(1 - \frac{T_0}{T_2}\right) \qquad [7.10]$$

where:

$\dot{W}_{rev}$ = the theoretical maximum available work of a heat engine operating between two heat transfer reservoirs in a reversible cycle, W

$\dot{Q}$ = heat output of heater, 3412 Btu/h or 1000 W

T_0 = heat sink temperature, 0°C or 273 K

T_2 = warm reservoir temperature, 43°C or 316 K

The maximum possible work usefully transferrable for the same function using the same energy input is $240\ V \times 4.167\ A = 1000\ W$, again assuming that effectiveness is defined at the source of electricity and does not include the generation of electricity. Effectiveness is therefore:

$$\epsilon = \frac{\dot{W}_{rev}}{\dot{E}_{in}} = \frac{\dot{Q}\left(1 - \frac{T_0}{T_2}\right)}{\dot{E}_{in}} = \eta\left(1 - \frac{T_0}{T_2}\right) \qquad [7.11]$$

$$\epsilon = 100\%\left(1 - \frac{273\ K}{316\ K}\right) = 13.6\%$$

This shows that a loss of available work results from using a premium high temperature energy form (electricity) to produce low-temperature heat. Effectiveness would be much greater (about $\epsilon = 68\%$) if T_2 were closer to 600°C. Effectiveness would also be greater if this were heat pump rather than resistance heater. In this case, ϵ would be equal to $(1 - T_0/T_2)$ multiplied by the COP, which is typically on the order of 3.0.

For the air conditioner. The air conditioner has an actual COP of 2.99. On typical hot summer day weather conditions, its effectiveness is given by:

$$\epsilon = \frac{\dot{W}_{rev}}{\dot{E}_{in}} = \frac{\dot{Q}\left|1 - \dfrac{T_0}{T_3}\right|}{\dot{E}_{in}} = COP\left(\frac{T_0}{T_3} - 1\right) \qquad [7.12]$$

$$\epsilon = 2.99\left(\frac{313\ K}{293\ K} - 1\right) = 20.4\%$$

where:

T_0 = heat sink temperature, 40°C or 313 K

T_3 = cool reservoir temperature, 20°C or 293 K

This indicates that the second law efficiency, or effectiveness, of an air conditioner is low when the outside air temperature is close to the temperature of the conditioned space. Once again, since electricity with high availability is being used to cool a space with small temperature differential relative to the environment (and thus a low availability for work) there is a significant loss in available work (Text Box 7.1).

FACTORS CONTRIBUTING TO INEFFICIENCY

Factors that contribute to inefficiency are losses that result in loss of energy or loss of availability. Table 7.3 summarizes typical examples.

These types of losses are present in any energy-using process. Measurements or data provided by manufacturers will provide an estimate of overall efficiency. The total energy lost in a particular process is generally the sum of the energy dissipated by several mechanisms. Sometimes it

Table 7.3 Causes of inefficiency

Losses of Energy

Mechanical energy	Friction
	Impacts
	Damping
Fluid energy	Line losses
	Fluid leakage
	Flow restrictions
	Fouling, corrosion
Heat energy	Convection
	Conduction
	Radiation
	Infiltration and exfiltration
Electrical energy	Resistance (i^2r) losses
	Eddy currents
	Hysteresis
	Discharges
	Line losses

Losses of Availability

	Combustion
	Pressure drops
	Temperature drops

is useful to perform a detailed analysis to determine the relative importance of each loss term. Then the energy management committee can establish the priority of potential corrective actions.

Figure 7.1 shows a process that uses electricity to drive a pump supplying two loads, one high-pressure and the other low-pressure. Figure 7.1a illustrates the losses present in this hypothetical system.

The losses depend on where we establish the system boundaries. For present purposes we take it on the secondary distribution side of the transformer. Please note, however, that for large industrial users this may not tell the whole story, since in-plant electrical distribution losses can be significant.

For the system boundaries shown, losses occur in the drive motor, the mechanical transmission that connects the motor to the pump, and in the pump itself. Other losses occur in the piping system, since it is a long distance to the high-pressure load, and in the loads themselves. Since one of the loads uses low pressure, the process has a pressure-reducing valve. This valve results in a loss of availability because high pressure fluids contain higher quality energy than low pressure fluids.

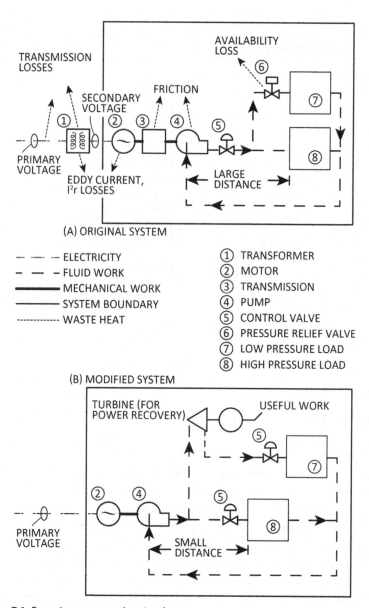

Figure 7.1 Pumping process showing losses.

Figure 7.1b shows a modified system. By selecting a different motor and driving it with primary voltage, the losses that would occur in the transformer can be eliminated. Since this is outside the system boundary, we neglect this consideration for the moment. Proper matching of the motor speed to the pump (or provision of a variable speed drive) could eliminate the need for mechanical transmission, thus eliminating another source of losses. Moving the pump closer to the high-pressure load reduces the fluid line losses (but may increase electrical distribution losses). Finally, there are several possible alternatives to service the low-pressure loads. One (not shown) is to provide two pumps, one for each load. Another (shown in Figure 7.1b) is to install a hydraulic turbine capable of extracting useful work, and then using the turbine exhaust stream to supply the low-pressure load. This second alternative permits recovery of some of the pumping power for other uses.

Before actually considering any of these remedies, the energy manager must conduct engineering and economic calculations to verify the feasibility and benefit for the specific case. In the case of an existing installation, some capital investment would be required, and therefore it would be necessary to carry out cost-effectiveness analysis to determine if the investment is justified. In a new design, it is possible that the more efficient system might even reduce capital costs, although it would generally be assumed that the more efficient system would have a higher first cost.

Figure 7.2 shows still another cut at system efficiency. In this example, we assume the plant contains a large integrated process industry and the system boundary definition includes primary fuel utilization.

The system in Figure 7.2A uses purchased electricity to supply pumping power to the types of high and low pressure loads discussed above. The same plant burns fuel to generate steam for use in low-pressure process heat applications.

Figure 7.2B shows a different possible configuration with a back pressure turbine to recover the available work that is otherwise lost in supplying low-pressure steam. With this approach, the steam turbine drives the pumps and the steam turbine exhaust supplies the process heat loads. This alternative reduces the need for electricity.

For an actual operation, the energy manager must get the answers to several questions before a system of this type would be implemented. These include the following:

• Does this system configuration really save energy?
• What are the costs?
• Do the timing and pump loads permit this approach?
• Are both loads physically located where combined service is possible?

The importance of these practical considerations cannot be overstressed. Often the energy manager will find large *potential* energy savings

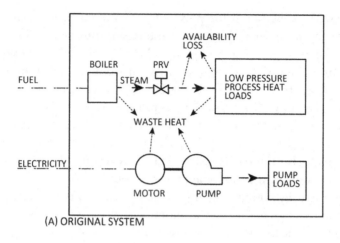

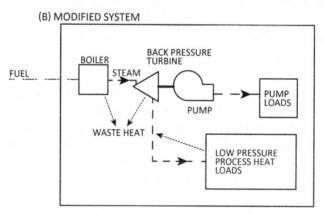

Figure 7.2 A pumping example in an integrated process plant.

that are difficult, if not impossible to implement. For this example, a case in point would be if the pumping loads occurred at night while the processing loads occurred during the day.

In summary, efficiency improvement is grist for the mill of the energy manager. Taken individually, many of the opportunities identified will seem inconsequential. It is only when converted to absolute terms and combined that they become worthwhile.

Consider the chain of events leading to the production of one horsepower (746 W) of mechanical work as shown in Table 7.4 The left-hand side of the table shows that there are at least seven steps between the raw fuel and the end-use. Most of these have fairly high efficiencies in relative

Table 7.4 The aggregate impact of inefficiency

Step description	RELATIVE IMPACT (per horsepower delivered per hour)			ABSOLUTE IMPACT (for a 1000 hp load operating 1000 hours per year)	
	Step efficiency (%)	Input energy (MJ)	(kWh)	Input energy (GJ)	Equivalent barrels of oil (Bbls)
End use	—	2.686	0.746	2,686	439
Conversion of mechanical to end use energy	70	3.838	1.066	3,838	627
Conversion of electrical to mechanical energy	90	4.262	1.184	4,262	696
Transformation of electrical energy	95	4.486	1.246	4,486	733
Distribution of electrical energy	90	4.986	1.385	4,986	815
Fuel input for generation of electrical energy	35	14.25	3.958	14,250	2,328
Refining, transportation of fuel	90	15.83	4.398	15,830	2,587

terms. Yet, in absolute terms (assuming a large industrial plant using electricity generated from oil) the impact of inefficiency is startling. Of the more than 2500 barrels of oil burned per year to supply this load, only the equivalent of slightly more than 400 barrels (17%) actually can be attributed to the final end use.

Some of the losses are thermodynamically unavoidable. Others are economically irretrievable. The challenge to the energy manager is to first identify those which he or she can do something about, find out how to do it, and then get management to agree to do it.

APPROPRIATE BASELINE CONDITIONS

When analyzing energy savings, a critical first step is to understand and define the baseline conditions—also known as the reference case—with which the proposed energy management project will be compared. Coming up with the most appropriate baseline is often a challenging task, particularly because the definition of *appropriate* depends on the audience scrutinizing savings estimates. What might be a perfectly acceptable baseline to use when selling an idea to corporate management may be completely unacceptable to use when estimating savings for a utility incentive program. This is particularly true when the proposed project involves replacing equipment at the end of its useful life or estimating savings for on-going maintenance measures.

For example, consider a project that involves replacing 30-year-old packaged air conditioning equipment in a commercial building. If the audience is corporate management, a likely goal of the energy savings calculations is to show management how much energy the proposed high efficiency air conditioning equipment will save relative to the old equipment currently installed. Therefore, using the currently installed equipment as the baseline may be the preferred way to justify expenditures for the new energy-efficient air conditioning systems.

Alternatively, if the audience for the energy savings calculations is a utility, it is important to understand that the utility only wants to give incentives for energy efficiency improvements that are actually *influenced* by the utility's energy efficiency program and enabled by their incentives. That is, the intention of the incentives is to improve the cost-effectiveness of higher efficiency alternatives so that customers make incrementally greater energy efficiency improvements than they would have made in absence of an incentive program. Therefore, if a customer has to replace AC equipment because it is reaching the end of its useful life, the utility only wants to pay the customer an incentive on the incremental savings of the new energy-efficient replacement relative to hypothetical replacement equipment that has minimum acceptable efficiency per current codes and standards. In that scenario, the appropriate baseline is new packaged AC equipment with the minimum acceptable EER.

Codes and standards exist for a multiplicity of end uses in various industries and they are periodically updated. The American National Standards Institute (ANSI) (www.nssn.org) and International Organization for Standardization (ISO) (www.iso.org) have searchable databases for standards. Refer to subsequent chapters in this book for discussions of codes and standards in lighting, HVAC systems, process equipment, and buildings.

For other types of end-use systems, such as large boilers and furnaces, the most appropriate baseline condition for projects involving equipment replacement may not be dictated by codes and standards. Instead, the most appropriate baseline may be refurbished equipment, if not existing equipment.

Baseline definition for maintenance measures can also be quite complex, especially if the audience for energy savings estimates is a utility. Below are two reason for this:

- **Maintenance practices are sometimes reactive rather than systematic:** For example, facility maintenance personnel may defer repairing leaks in steam lines until the leaks become disruptive to the process or present a safety concern. The appropriate baseline from the utility's perspective is not the catastrophic leak that is filling the facility with hazardous hot steam. Because, in this extreme case, the facility really has no other choice than to repair the problem, so the utility incentive does not really influence the decision to implement an energy management project. Instead, the more reasonable baseline from the utility perspective would be a small-grade leak that is easy to ignore in the absence of a financial incentive to repair it. Whether utility incentives are on the table or not, performing systematic instead of reactive maintenance actions is very often the most cost-effective solution. This type of routine maintenance should be incorporated in a facility's on-going energy management plan.

- **The performance of certain types of systems degrades over time:** Another issue that can muddy the baseline and energy savings calculations arises in systems with energy performance that degrades over time. Consider heat transfer surfaces prone to fouling. Over time, the efficiency of heat transfer decreases, gradually reducing the overall efficiency of the process up to the point when the degradation is no longer tolerable. Once cleaned, the efficiency of heat transfer increases again. Since the performance of the heat exchanger varies with the degree of fouling, what is the appropriate baseline? When the audience is corporate management, it may make sense for the current dirty condition to be the baseline and the ultra-clean condition to be the energy-efficient scenario, even though these savings will degrade over time. However, from the utility perspective, it may be more appropriate to use an average condition as the baseline to account for performance degradation over time.

Defining baselines for retrofit or early retirement projects is generally much simpler. In those cases, the appropriate baseline is the currently installed equipment, at least for the remaining life of that equipment, whether the audience is corporate management or a utility.

As described in Chapter 4, several other common factors influence energy use and should be considered when establishing the baseline:

- Weather.
- Operating hours and production schedule.
- Building occupancy.
- Production level.

ESTIMATING SAVINGS

There are various analytical techniques for estimating baselines and analyzing savings. Examples of common approaches include engineering algorithms, simulation software, analysis of historical data, and end-use monitoring or metering.

Engineering Algorithms

Engineering algorithms are well suited to estimating baseline energy use and savings for a wide range of projects. For no-cost or low-cost measures, they are often the best way to estimate savings since they can be readily applied with very little expense. They are also very useful for simple equipment replacement measures that have well-characterized savings. For more complex projects, engineering algorithms can provide a good way to approximate savings for high level comparisons with other potential measures and for preliminary prioritization of measures. However, complex and costly projects may warrant more detailed approaches that incorporate simulation modeling, analysis of historical data, and/or measurements of existing end-use systems.

The spectrum of potential engineering algorithms for estimating savings is far too extensive to cover here. Instead, we have provided examples of algorithms commonly applied to a few representative measures in Equations 7.13 through 7.18.

Lighting Replacement

$$kW_{savings} = \left[(N_{fixtures})(W_{fixture})\left(\frac{1\,kW}{1000\,W}\right)_{baseline} - (N_{fixtures})(W_{fixture})\left(\frac{1\,kW}{1000\,W}\right)_{proposed} \right] \times$$
$$CF \times IE_D$$

[7.13]

$$kWh_{savings} = \left[(N_{fixtures})(W_{fixture})\left(\frac{1\,kW}{1000\,W}\right)_{baseline} - (N_{fixtures})(W_{fixture})\left(\frac{1\,kW}{1000\,W}\right)_{proposed} \right] \times$$
$$h \times IE_F$$

[7.14]

where:

$kW_{savings}$ = peak demand savings, kW
$kWh_{savings}$ = annual energy savings, kWh

$N_{fixtures}$ = number of fixtures of the given type, baseline and proposed cases, dimensionless

$W_{fixture}$ = wattage of the fixture, baseline and proposed cases, W

CF = coincidence factor with peak demand, which is an estimate of the share of the lighting load operating during the peak demand hours, dimensionless

IE_D = demand interactive effects factor, which accounts for interactions between lighting and space conditioning systems, dimensionless

h = annual hours of operation of the lighting system

IE_E = energy interactive effects factor, which accounts for interactions between lighting and space conditioning systems, dimensionless

Motor Replacement

$$kW_{savings} = HP_{rated}\left(\frac{1}{\eta_{baseline}} - \frac{1}{\eta_{proposed}}\right)\left(\frac{0.746\ kW}{1\ hp}\right)LF \times CF \quad [7.15]$$

$$kWh_{savings} = HP_{rated}\left(\frac{1}{\eta_{baseline}} - \frac{1}{\eta_{proposed}}\right)\left(\frac{0.746\ kW}{1\ hp}\right)LF \times h \quad [7.16]$$

where:

$kW_{savings}$ = peak demand savings, kW

$kWh_{savings}$ = annual energy savings, kWh

HP_{rated} = nameplate horsepower of the motor, assuming motor size stays the same, hp

$\eta_{basline}$ = efficiency of the baseline motor under the given load, dimensionless

$\eta_{proposed}$ = efficiency of the proposed motor under the given load, dimensionless

LF = load factor, which is the ratio of the motor's actual load to the motor's rated full load, dimensionless

CF = coincidence factor with peak demand, dimensionless

h = annual hours of operation of the motor,

Unitary Air Conditioner Replacement

$$kW_{savings} = \left[Cap_{rated}\left(\frac{12}{EER_{baseline}} - \frac{12}{EER_{proposed}}\right)\right]CF \quad [7.17]$$

$$kWh_{savings} = \left[Cap_{rated} \left(\frac{12}{EER_{baseline}} - \frac{12}{EER_{proposed}} \right) \right] EFLH \qquad [7.18]$$

where:

$kW_{savings}$ = peak demand savings, kW

$kWh_{savings}$ = annual energy savings, kWh

Cap_{rated} = rated cooling capacity, assuming properly sized, ton

12 = efficiency conversion factor; note that $\frac{12}{EER} = \frac{kW}{ton}$

$EER_{baseline}$ = efficiency rating of the baseline cooling system, Btu/Wh

$EER_{proposed}$ = efficiency rating of the proposed cooling system, Btu/Wh

CF = coincidence factor with peak demand, dimensionless

$EFLH$ = annual equivalent full load hours of cooling, h

There are many resources available online to assist with energy and demand savings estimates. For example, the U.S. Department of Energy has some online tools on their website for common energy efficiency projects (http://energy.gov/eere/femp/energy-and-cost-savings-calculators-energy-efficient-products). The U.S. Department of Energy also sponsors the Industrial Assessment Center (IAC) (http://iac.rutgers.edu/database/). The IAC maintains a database of publicly available facility assessment data, including information on the size, type, and energy usage of industrial facilities along with listings of recommended energy projects and estimated energy and cost savings. Though it does not contain engineering algorithms, this database is a good resource for getting a rough idea of the potential savings from various industrial measures in specific industries. In addition, many state organizations and utilities have technical reference manuals with algorithms for estimating energy savings for the measures covered in their programs.

Building Simulation and System Modeling

For energy management measures that are sensitive to variations in weather, production, or other variables difficult to capture in simple algorithms, simulation software can be a valuable resource for analyzing energy use and savings. The decision to use a simulation model for calculating energy savings hinges on the complexity of the proposed project, the level of investment required, and the desired accuracy of the savings estimate. There are many tools available for modeling energy savings in buildings and specific end-use systems. The tools require users to input information for certain parameters such as building characteristics, system specifications, control strategies, production schedules, weather, and other relevant data. The models then simulate energy use for baseline scenarios and for proposed projects at a selected time interval—annual, seasonal, monthly, daily, hourly. Comparing the baseline and proposed cases yields energy and demand savings predicted for the projects. To improve the accuracy of the models, it frequently makes sense to calibrate them with actual energy or production data obtained from utility meters, sub-meters, facility data acquisition systems, or temporary data loggers.

Various types of software tools are available online for free or for purchase. For example, the U.S. Department of Energy maintains a directory of over 400 energy software tools, including those for building and system simulation: (http://apps1.eere.energy.gov/buildings/tools_directory/about.cfm). A few common tools for building simulation include EnergyPlus, eQUEST, and DOE-2. In addition, the U.S. Department of Energy provides a tool called AIRMaster + that is widely used for modeling compressed air systems (http://energy.gov/eere/amo/articles/airmaster). Also, ASHRAE, the American Society of Heating, Refrigerating and Air Conditioning Engineers has many resources available. Refer to Chapter 8.

ELECTRIC LOAD ANALYSIS

Electrical load analysis is a useful tool for the energy manager. If the facility has one or more interval meters, data is collected at intervals ranging from a few minutes to an hour. Utility-provided interval meters are generally set for 15 minute data collection intervals. These data may be accessible to the facility on the utility's website; if not it can be obtained from the utility service representative. Ready access to load data at high granularity greatly simplifies load analysis. Nevertheless, if interval data is not available, the energy manager can obtain considerable information by just using a standard utility-provided electric meter. When detailed measurements are required, additional instrumentation such as electric sub-meters or temporary data logging equipment may be required.

Standard meters range from single phase kWh meters common in residences to three-phase meters recording both demand in kW and energy use in kWh. In the absence of an interval meter, the following method can be used with a standard (disk type) kWh meter; if the meter records demand, demand readings could also be used.

The kWh meter reading is given by

$$E = (K_h)(P_t)(C_t)n \qquad [7.19]$$

where

E = electric energy use, kWh

K_h = meter constant, kWh per disk revolution

P_t = potential transformer ratio, dimensionless

C_t = current transformer ratio, dimensionless

n = number of revolutions of meter disk

Often the constants K_h, P_t, C_t are combined into a single multiplier (usually indicated on the meter) that, when multiplied times the meter reading, gives a value in kWh. Newer meters are digital and do not have a rotating disk. (See Figure 7.3). In this case, the meter reading is multiplied by K_h. In Figure 7.3, $K_h = 1.0$, so the meter reads directly in kWh.

To determine energy use during an interval of time of p hours, the meter would be read initially, the time recorded, and then the meter would be reread at the end of the interval. The difference in readings multiplied by the multiplier would give the kWh used during the period p.

Figure 7.3 Digital kWh meter.

To determine the average load over a period p, determine E in kWh as above and then use the relation:

$$L = \frac{E}{p} \qquad\qquad [7.20]$$

where

L = average load, kW

p = period, h

A daily load curve can be constructed by reading the meter every 15 minutes, every half hour, or every hour and then plotting the data against time. (If interval data is available, construction of the daily load curve would use the interval data directly.) Table 7.5 shows a set of representative data from a steel mill. Figure 7.4 is a graph of the data. It is instructive to prepare a daily load curve for a weekday, a weekend day, and both summer and winter days. The information contained in such graphs can then be analyzed to determine the following electrical load characteristics:

• Time of occurrence of peak demand.
• Energy use during lunch break.
• Ratio of maximum to least demand.
• Ratio of summer to winter demand.
• Percentage of total energy use occurring during off-shift hours.
• Percentage of total energy use occurring on weekends.

These data can be used to assess the following situations:

• Are major loads left on during the lunch break?
• Is lighting or other equipment left on during weekends or off-shift hours?
• Can the peak demand be reduced?

In an ideal situation, the load curve would be the shape of a rectangle. The load would be zero during off-shift hours, rise instantly to a maximum value when the working day starts, remain at a constant value until the end of the working day, and then decrease to zero again. Of course, this ideal is never achieved in practice. To quickly check the efficiency of the system or overall facility, calculate the percentage of energy consumed pre-shift, on-shift, and post-shift. If the off-shift energy use is 40% of the total or greater, conduct an energy audit to establish the components of this off-shift load and to evaluate whether this energy use is indeed necessary.

Table 7.5 Steel mill data for a summer day daily load curve

Time	Demand (kW)	Time	Demand (kW)
0:00	2385		
0:30	2400	12:30	5835
1:00	2805	13:00	6570
1:30	2925	13:30	7710
2:00	2925	14:00	7140
2:30	2910	14:30	5850
3:00	2910	15:00	4875
3:30	2910	15:30	5385
4:00	2880	16:00	4830
4:30	3000	16:30	5055
5:00	3345	17:00	4170
5:30	3300	17:30	4290
6:00	3270	18:00	4140
6:30	3300	18:30	4170
7:00	3300	19:00	4335
7:30	4515	19:30	4215
8:00	4950	20:00	4320
8:30	5340	20:30	6240
9:00	5640	21:00	4755
9:30	6225	21:30	4110
10:00	5595	22:00	4620
10:30	5820	22:30	5235
11:00	6170	23:00	5325
11:30	6325	23:30	3735
12:00	5610	24:00	3510

Table data:
- Total kWh: 110,160
- Pre-shift (00:00 hrs to 6:00 hrs): 17,290 kWh (15.7%)
- Shift 1 (6:00 hrs to14:00 hrs): 49,380 kWh (44.8%)
- Break (14:00 hrs to 15:00 hrs): 10,725 kWh (9.7%)
- Shift 2 (15:0 hrs to 23:0 hrs): 29,165 kWh (26.5%)
- Post-shift (23:00 hrs to 24:00 hrs) 3,600 kWh (3.4%)
- Maximum demand on-peak: 7,710 kW (13:30)
- Maximum demand off-peak: 5,325 kW (23:00)

Parameters for Electric Load Analysis

A typical industrial electrical load consists of lighting, motors, chillers, compressors, and other types of equipment. The sum of the capacities of this equipment, in kW, is the connected load. The actual load at any point in time is normally less than the connected load since every motor is not

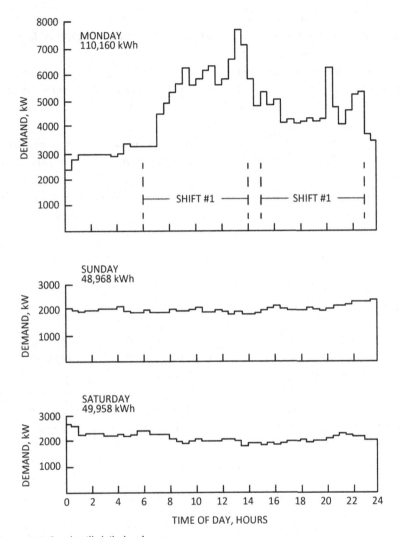

Figure 7.4 Steel mill daily load curves.

turned on at the same time, motors frequently are not loaded to rated horsepower, only part of the lights may be on at any time, etc. Thus, the load is said to be diversified, and a measure of this is the *diversity factor*:

$$DV = (D_{m1} + D_{m2} + D_{m3} + etc.)/D_{max} \qquad [7.21]$$

where

$D_{m1} + etc.$ = sum of maximum demand of individual loads, kW
D_{max} = maximum demand of plant, kW

If the individual loads do not occur simultaneously (usually they do not) the diversity factor will be greater than unity. Typical values for industrial plants are 1.3 to 2.5.

If each individual load operated to its maximum extent simultaneously, the maximum *demand* for power would be equal to the connected load and the diversity factor would be 1.0. However, as pointed out above, this does not happen except for special cases.

The demand varies over time as loads are added and removed from the system. It is usual practice for the supplying utility to specify a demand interval (usually 0.25, 0.5, or 1.0 h) over which it will calculate the demand and compute the demand charge using the relationship:

$$D = E/p \qquad\qquad [7.22]$$

where
D = demand, kW
E = kilowatt hours used during period p, kWh
p = demand interval, h

The demand calculated in this manner is an average value, being greater than the lowest instantaneous demand during the demand interval, but being less than the maximum demand during the interval.

Utilities are interested in *peak demand* since this determines the capacity of the equipment they must install to meet the customer's power requirements. Peak demand is measured by a demand factor, defined as

$$DF = D_{max}/CL \qquad\qquad [7.23]$$

where
DF = demand factor, dimensionless
D_{max} = maximum demand, kW
CL = connected load, kW

The demand factor is normally less than unity; typical values range from 0.25 to 0.90.

Since a customer normally pays a premium for the maximum load placed on the utility system, it is of interest to determine how effectively the maximum load is used. Most effective use of the equipment would be to have the peak load occur at the start of the use period and continue unchanged throughout it. Normally, this does not occur, and a measure of the extent to which the maximum demand is sustained throughout the period (a day, month, or year) is given by the *Hours Use Of Demand*:

$$HUOD = E/D_{max} \qquad [7.24]$$

where

$HUDO$	=	hours use of demand, h
E	=	energy use in period p, kWh
D_{max}	=	maximum demand during period p, kW
p	=	period over which HUOD is determined, for example,

one day, one month, or one year (p is always expressed in hours)

The *load factor* is another parameter that measures the plant's ability to use electricity efficiently. In effect, it measures a ratio of the average load for a given period of time to the maximum load that occurs during the same period. The most effective use results when the load factor is as high as possible once E or $HUOD$ have been maximized (it is always less than one). The load factor is defined as:

$$LF = E/(p)(D_{max}) \qquad [7.25]$$

where

LF	=	load factor, dimensionless
E	=	energy use in period p, kWh
D_{max}	=	maximum demand during period p, kW
p	=	period over which HUOD is determined, h

Other ways to determine load factor use the relations:

$$LF = HUOD/p = L/D_{max} \qquad [7.26]$$

where the symbols have the same definitions as above.

So far the discussion has dealt entirely with power and has neglected the reactive component of the load. In the most general case the apparent power in kVA that must be supplied to the load is the vector sum of the active power in kW and the reactive power in kVAR:

$$|S| = \sqrt{\dot{W}^2 + VAR^2} \qquad [7.27]$$

where

S	=	apparent power, kVA
$\dot{W}$	=	active power, kW
VAR	=	reactive power, kVAR

In this notation the apparent power can be considered a vector of magnitude S and angle θ, where θ is commonly referred to as the phase angle:

$$\theta = \tan^{-1}(VAR/\dot{W}) \qquad [7.28]$$

Another useful parameter is the power factor pf:

$$pf = \cos\theta = |\dot{W}/S| \qquad [7.29]$$

The power factor is always less than or equal to unity. A high-value is desirable since it implies a small reactive component to the load. A low value means the reactive component is large.

The importance of the power factor is related to the reactive component of the load. Even though the reactive component does not dissipate power (it is stored in magnetic or electric fields), the switchgear and distribution system must be sized to handle the current required by the *apparent power*, or vector sum of the active and the reactive components. This results in greater capital and operating expense. The operating expense is increased due to the standby losses that occur in supplying the reactive component of the load. In addition, low power factor can result in overloaded transformers, as well as unnecessary losses in motors and in wiring. Low power factor also causes poor voltage regulation and excessive voltage drops.

Power factor can be improved by adding capacitors to the load to compensate for part of the inductive reactance. The benefit of this approach depends on the economics of each specific case and generally requires a careful review and analysis. In principle, the negative capacitive reactance in effect "cancels" the positive inductive reactance, reducing the reactive component of the load (the kVARs). However due to variations in load parameters it is generally impossible to exactly balance out the power factor (in any event it may not be economically desirable). Note also that when capacitors are installed, there is a slight increase in real power consumption, since capacitors have some leakage resistance and therefore i^2R losses.

The following parameters need to be considered when installing capacitors:

- Location (either at the load or at the supply).
- Safety (provisions must be made for safely discharging stored energy).
- Overvoltage.
- Economics.

There are two approaches for connecting capacitors: at the load, or at the switchboard or distribution panel. Connecting at the load is convenient because the capacitors then switch on and off with a load. Also, reactive current losses are lower in this manner. The second method has the advantage that larger (and therefore cheaper unit cost) capacitors can be used, but the savings in the branch circuits are not obtained. This approach primarily leads to savings in the primary feeders and main transformer, but not on the load side of distribution.

CONCLUSIONS

First and second law thermodynamic efficiencies are hugely important concepts to grasp. It is not enough to just consider minimizing loss in the *quantity* of energy delivered to a task, but the loss in energy *quality* must be considered as well. Cost-effectively maximizing overall resource efficiency to optimize use of available energy is the essential ingredient of all energy management activities. With a firm understanding of efficiency, in addition to an appreciation for considerations affecting appropriate baseline conditions, the energy manager has numerous potential approaches and tools at his or her disposal to estimate energy use and savings from energy management projects. One approach is to use engineering algorithms; another is to use simulation software to model key building systems or processes; still another is to measure loads and baseline energy use directly with load monitoring equipment or whole-building meter data and use it along with algorithms or software to estimate energy and demand savings from proposed measures. Generally, it costs more to get more accurate savings estimates. So, when possible, it makes sense to conduct rough estimates first to see if the measure is worth pursuing before committing to a more rigorous analysis approach.

CHAPTER 8

Management of Heating and Cooling

INTRODUCTION

Depending upon the purpose of the building, heating, ventilating, and air conditioning (HVAC) systems may be among the most important energy users. In commercial buildings, HVAC and lighting will normally dominate the energy use. In industrial facilities, process energy will usually (but all not always) be more important.

In a typical central HVAC system, the largest electricity users are chillers and air distribution fans, followed by cooling tower fans and water pumps, unless an electric boiler or some form of electric heating is used. Central heating systems in large facilities generally use fuel-fired boilers (gas, oil, or coal). In a decentralized or rooftop system, the breakdown is similar except for the lack of cooling tower fans and pumps and the substitution of direct heating for boilers.

Very large reductions in energy use by HVAC equipment can and have been realized by energy management initiatives. Often these improvements are accomplished with existing equipment. There are examples of energy use being cut in half with no noticeable effect on the comfort of building occupants. This is the essence of energy management—not doing without, but doing more with what is available.

This chapter begins by presenting the general principles for energy management in space heating and cooling systems. It then covers the requirements for human comfort and health as they pertain to HVAC system design. The discussion continues with a summary of the basic principles governing HVAC equipment operation, energy analysis techniques, codes and standards, and an overview of typical air distribution systems. It concludes by focusing on specific opportunities for optimizing energy use in individual system components as well as at the overall system level.

Energy Management Principles.
DOI: http://dx.doi.org/10.1016/B978-0-12-802506-2.00008-2

GENERAL PRINCIPLES OF ENERGY MANAGEMENT IN HVAC SYSTEMS

We have identified eleven general principles for energy management in HVAC systems (see Table 8.1). They range from improving controls, to installing high efficiency equipment, to taking advantage of passive heating and cooling. The discussion below explains each principle in greater detail.

Optimize Controls

The first step in improving the efficiency of an HVAC system is to look at what can be done with straightforward changes. Many times making relatively simple changes in the way the systems are controlled can have a significant impact with very low cost. Basic operational changes include resetting thermostats or installing time-based controls to ensure equipment is only heating or cooling where and when needed. We assume most facilities have already implemented these types of measures, but sometimes even the most basic items are overlooked, or the control procedures or settings slip over time.

Other control measures include sequencing equipment to minimize part-load operation and altering temperatures or other settings at the system component level to maximize efficiency; though these require more insight into the fundamentals of HVAC operation, they are certainly within reach for the energy manager and can be accomplished with either manual or automatic controls.

Generally speaking, the more manual the process is, the less likely it will persist consistently over time. Programmable thermostats and other forms of automated advanced control systems are readily available today to help optimize control settings for energy efficiency.

Manage Peak Demand

Another application of HVAC controls is to manage peak demand during periods of utility system constraints. Effective peak demand control helps facilities reduce energy costs and helps utilities supply energy more reliably and economically. Demand response programs are now widely offered by utilities to encourage customers to reduce loads in response to peak demand events. For many utilities, system peaks occur during summer weekday afternoons when air conditioning loads are high, so actions aimed at reducing on-peak space cooling demand during event periods

Table 8.1 Principles for energy management in HVAC systems

Principle	Description
1. Optimize controls	Use controls to provide heating and cooling only *when* it is required. Also use controls to optimize equipment operation.
2. Manage peak demand	Employ techniques to manage HVAC equipment demand during utility on-peak periods and/or in response to peak demand events called by the utility (i.e., *demand response*).
3. Optimize capacity	Review the system capacity and eliminate excess capacity. Note that other energy management activities (reducing heat loads from lighting, for example) could impact HVAC energy use.
4. Reduce the load	Minimize heating and cooling loads by reducing infiltration, solar heat loads, etc.
5. Heat and cool people, not buildings	Provide heating and cooling where people work, rather than in aisles, etc. Do not condition unoccupied spaces unless necessary for a given application. Also provide space conditioning and adequate ventilation based on occupancy (e.g., *demand-controlled ventilation*).
6. Use efficient processes	Select the most efficient heating and cooling processes for the building, its occupants, and for the climate. Customize systems for intended uses (e. g., once-through systems may have been required for previous uses—chemicals contaminants, etc.—but new "office" or other use would allow recirculation).
7. Use efficient equipment	Select the most efficient equipment for the heating and cooling process selected.
8. Operate equipment efficiently	Make certain existing equipment is operating as efficiently as possible; correct deficiencies.
9. Use passive concepts	Make the building and the climate do as much of the heating/cooling work as possible.
10. Employ heat recovery	Heat can be recovered from equipment, building exhausts, and other sources.
11. Provide energy storage capability	Energy storage can permit off-peak use of equipment, load leveling, and more efficient utilization of equipment.

are particularly effective. Control strategies include pre-cooling work spaces prior to the event period and then increasing temperature set-points during peak hours. Other methods are to cycle constant air volume equipment, or to reset pressure in variable air volume systems to lower airflow, thereby reducing demand. Facilities can employ analogous control strategies to reduce space heating demand to help winter-peaking utilities, such as pre-heating work spaces and decreasing temperature setpoints in response to events. We discuss some of these control strategies in greater detail later in the chapter.

Optimize Capacity

The third general principle is to reduce system capacity to the lowest acceptable level. (This will vary, depending on local codes and standards and the design and function of the building.) For example, ventilation capacity should be designed to maintain health and comfort on a day where maximum space conditioning is required. If there is excess capacity, as is typically designed into buildings, the additional fan horsepower is wasted, and more heating and cooling than is necessary may be occurring. Reducing ventilation by 20% can theoretically cut fan energy use in half, as we explain later.

Reduce the Load

Reducing the load seems obvious, but is frequently overlooked. Infiltration, solar heat gains, equipment heat, conduction or radiation losses or gains, all can add to or detract from the building heat load. Insulation, weather stripping, reflective window covers, shading, different colored paints, venting equipment waste heat, and more efficient lighting, are some measures that can be employed to reduce heat gain or loss in the building.

Heat and Cool People, Not Buildings

In most cases, the purpose of an HVAC system is to heat or cool *people, not buildings*. Recognition of this simple fact can sometimes lead to changes in operational patterns that make large savings possible. Examine the basic heating and cooling requirements for the facility and avoid space conditioning in unoccupied areas. Also, consider applying demand-controlled ventilation strategies whereby ventilation rates are adjusted based on CO_2 levels within the conditioned space.

Of course, there are a few exceptions where HVAC equipment must serve non-human loads. Examples include data centers, cold food storage facilities, areas where chemicals or pharmaceuticals require temperature control, or for frost protection during the winter.

Use Efficient Processes

The sixth principle is to select the most efficient process for heating or cooling. In new construction, well insulated buildings combined with passive design features can greatly reduce heating energy. In certain climates, evaporative cooling (either "wet" or "dry") will use less energy than refrigeration. In addition, heat pumps are an efficient alternative for supplying heating and cooling in some types of applications. There are several types of heat pumps and they are typically designated by their heat source. Examples include air-source, ground-source, water-source, and waste heat recovery heat pumps.

Use Efficient Equipment

Using the most efficient equipment for a given HVAC process is another way to maximize efficiency. First costs are higher, but the energy savings pay off over time. Selecting high efficiency equipment can be particularly cost-effective in new design or upon equipment failure. There is a fairly wide range of efficiencies for chillers, boilers, unitary AC systems, furnaces, and electric motors. Likewise, variable speed motor drives for chillers, pumps, and fans may permit energy savings.

Operate Equipment Efficiently

The way equipment is operated has a dramatic effect on overall system efficiency. This principle overlaps with optimizing controls. For example, it makes no sense to have two high-efficiency chillers and then operate them at part load if one could carry the full load. Also, energy-saving economizer systems with inoperative or improperly operating controls will not perform as intended.

Use Passive Concepts

Chapter 12 discusses the use of passive design concepts in greater detail. The approach is to plan the building design (or its retrofit) in such a way as to optimize the use of energy flows to and from the environment.

Employ Heat Recovery

For space heating applications, recovering heat is particularly important (and economically attractive) in colder climates. There are many sources of waste heat in industrial facilities. From an energy *quality* perspective, high temperature waste heat from direct-fired sources such as process furnaces and incinerators are best reserved for generating steam in waste heat boilers or for providing medium temperature process heat before being considered for lower temperature space conditioning applications.[1] But heat from medium and lower temperature sources like exhaust from steam boilers or gas turbines and cooling water from air compressors, engines, pumps, and other machinery can sometimes be economically extracted with heat pumps for space heating or can be used to preheat ventilation air. In commercial buildings, sources of waste heat include condenser water from refrigeration systems and chillers, and flue gas from combustion equipment, and exhausted ventilation air.

Provide Energy Storage Capability

Cooling system loads can be shifted to off-peak hours when coupled with thermal energy storage using chilled water storage tanks, ice banks, or eutectic salts. Depending on the application, this configuration can reduce energy costs and improve efficiency. Thermal energy storage may also allow for smaller capacity equipment since loads can be spread out over a long period of time ("load leveling"). Moreover, it benefits utility companies by permanently reducing on-peak loads.

We examine how to implement these eleven general principles for energy management in HVAC systems later. First, we present a brief overview of the requirements for human comfort and health to form the basis for the goals of building heating and cooling systems.

THE REQUIREMENTS FOR HUMAN COMFORT AND HEALTH

Human comfort and health in buildings depend on indoor air temperature, air velocity, the mean radiant temperature (MRT) of the surrounding surfaces, relative humidity, and dilution or control of airborne pollutants. The definition of an optimal indoor environment varies with the individual, his or her age, type and amount of clothing, and level of activity. Thus, it is impossible to firmly specify all of these variables so as to please every occupant, especially considering that the individual occupants and overall building occupancy fluctuate over time. Therefore,

---

[1] Parmenter, K., E. Fouche, R. Ehrhard. (2007). *Tech Review: Industrial Heat Pumps for Waste Heat Recovery*. Lafayette, CA: Industrial Technology Application Service, Global Energy Partners.

HVAC systems are designed to provide a range of conditions intended to be adequate for comfort and health for most occupants most of the time.

Heating and cooling needs can be understood by considering how homeotherms (warm-blooded animals) maintain a relatively constant and high body temperature that is largely independent of the surrounding environment, except in very extreme weather conditions. Homeotherms regulate temperature by employing metabolic processes and exchanging heat with the surrounding environment. When air temperatures are low, metabolic processes speed up to counteract the heat loss to the environment; when temperatures are high, heat rejection increases to stabilize body temperature. The relatively high body temperature of homeotherms facilitates rejection of heat to the surroundings.

There are four mechanisms to transfer heat with the surroundings: conduction, convection, radiation, and evaporation. For humans (body temperature of 37°C [98.6°F]), convection and radiation dominate when the surroundings are at temperatures less than about 27°C (80°F). Above 27°C, evaporation of the water in sweat begins to be of increasing importance for heat rejection; at ambient temperatures of 37°C, evaporation dominates. Heat exchange by conduction is usually small unless portions of the body are in contact with cold or hot surfaces.

Many commercial and industrial facilities have higher cooling loads than heating loads to counteract heat produced by humans, machinery, and solar radiation. Cooling is often required even in the winter, depending on climate. To cool people in indoor environments, convection is usually the most important mechanism, and benefits from slight air currents resulting either from natural convection, wind, or mechanical cooling. For radiation to be important for cooling, the body (influenced by the clothing worn) must radiate energy to nearby surfaces that are at lower temperatures. The MRT is a measure of these temperatures, and is typically in the range of 20–27°C. To warm people in indoor spaces, convection is again the predominant mechanism, but radiation also plays a role, particularly in buildings with radiant heaters. Humidity (over a broad range) has a second-order effect on heat transfer and comfort.

Table 8.2 shows examples of the rate at which heat is generated by humans for five different activities, ranging from sleeping to heavy labor. This heat must be accounted for in the design of heating and cooling systems.

Figure 8.1 shows the approximate comfort zone as well as an extended zone that might be tolerable (slightly cool or slightly warm) depending on the level of activity and type of clothing worn.

Table 8.2 Examples of rate of heat produced by humans (Varies by person)

Activity	Metabolic rate, met	Approximate heat rate	
		Watts	Btu/hr
Sleeping	0.7–0.9	75–95	250–320
Sitting quietly	1–1.1	105–115	355–390
Desk work	1.2–1.8	125–190	425–640
Light to medium activity	2–4	210–420	710–1425
Medium to heavy activity	5–10	525–1050	1780–3565

Note: 1 Met = 58.2 W/m^2 = 18.4 Btu/h-ft^2. The surface area of a typical human body is about 1.8 m^2 (19.38 ft^2).

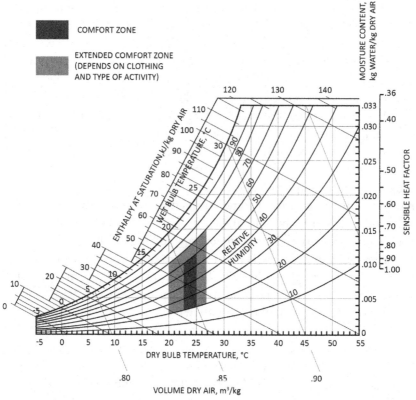

Figure 8.1 Psychometric chart.

In addition to ensuring the air temperature, air velocities, and humidity levels are in a comfortable range, it is essential to control contaminants in indoor air to prevent building-related illnesses, reduce absenteeism, protect against litigation, and maintain an environment that fosters worker productivity. There are many types of potential airborne contaminants, including radon, carbon monoxide, volatile organic compounds, bacteria, viruses, fungi, allergens, lead, asbestos, and of course carbon dioxide (CO_2). The three basic methods for controlling airborne pollutants are to (i) eliminate the source, if practical; (ii) use filtration and other air purification technologies to capture contaminants; and (iii) provide adequate ventilation to dilute.

HVAC system designs generally focus on the second and third approach. Specifically, building codes and standards require minimum rates of ventilation with outside air. CO_2 is often the air contaminant used to measure indoor air quality since it is produced when building occupants exhale. High concentrations of CO_2 (greater than about 1300 ppm) indicate ventilation rates are too low and air quality is poor. Poor air quality is unpleasant because of odor retention, but more importantly is can cause myriad health problems and is linked to low productivity. Very low concentrations of CO_2 (less than about 900 ppm) may be a sign that ventilation rates are excessive. Too much ventilation greatly increases energy use, since incoming air must be heated or cooled.

Based on the preceding discussion some concepts for heating and cooling emerge. For efficient heating:
- Temperatures should be comfortable in occupied areas.
- MRT should be high in occupied areas (i.e., don't put individuals in front of a large window).
- Eliminate cold airflows (drafts) if possible.
- Keep humidity up to acceptable levels.
- Isolate individuals from conduction losses.

For efficient cooling:
- Temperatures should be comfortable in occupied areas.
- MRT should be low in occupied areas.
- Maintain sufficient air movement.
- Don't let humidity get too high.

For efficient ventilation and healthy air quality:
- Ensure adequate ventilation, but don't over-ventilate.
- Attempt to eliminate sources of pollutant.
- Remove pollutants with air purification technologies.

BASIC PRINCIPLES GOVERNING HVAC SYSTEM OPERATION

The main energy-using components of a commercial or industrial HVAC system consist of a cooling source, heating source, fans, pumps, and, in some cases, a cooling tower. The types and configurations of the components employed vary by application and facility size. Large facilities often use a central plant with a chiller system for providing cooling and a boiler system for heating. Figure 8.2 shows elements of this type of central HVAC system.

The subsections below describe the basic principles governing operation of HVAC system components.

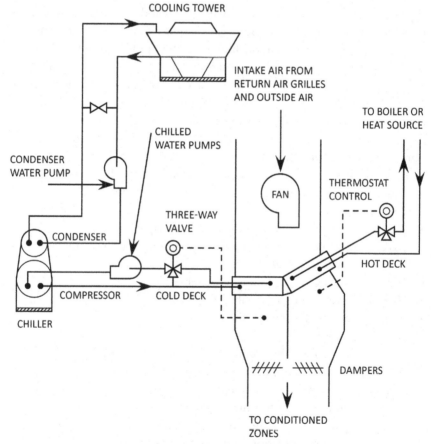

Figure 8.2 Components of a typical central HVAC plant.

Space Cooling Systems

In most commercial and industrial applications, cooling systems and associated auxiliary equipment account for the majority of the HVAC energy use. Cooling systems range from small individual air conditioning units, to unitary systems, to large central plants. Individual air conditioning units are self-contained systems that are mounted in windows or on an external wall and provide cooling to a space without the use of ducts. Unitary air conditioning (or heat pump) systems include packaged rooftop units and split systems, often with multiple units cooling separate zones of the building. Central systems generate cooling in a central chiller and then distribute the cooling with chilled water systems to air-handling or fan-coil units.

Vapor Compression Refrigeration Cycle. The majority of cooling systems are based on the *vapor compression refrigeration* cycle. Figure 8.3 shows how the vapor compression cycle compresses, condenses, expands, and boils refrigerant to provide cooling. The bullet points below describe each step in the cycle. Note that in the ensuing discussion we assume the reader has a basic familiarity with thermodynamics and fluid mechanics.

- States 1 and 2 on the figure represent the **compression** portion of the cycle, which requires input of electrical energy. In the compressor, refrigerant vapor leaving the evaporator is compressed from its low evaporating pressure to the required condensing pressure. The temperature of the vapor also increases. The energy used during this stroke is determined by subtracting the *enthalpy* of the State 2 from that of the State 1.

- **Condensing** and cooling of high-pressure refrigerant occurs between States 2 and 3 at a constant pressure. Heat absorbed in the evaporator and then added by the compressor is removed from the vapor by either an air-cooled or water-cooled condenser. Large central chiller systems are generally water-cooled and use cooling towers to reject the heat. The refrigerant leaving the condenser is now a liquid. The amount of heat *rejected* in this step is determined by subtracting the enthalpy of the State 2 from that of State 3.

- Between States 3 and 4 there is a constant enthalpy **expansion** through the expansion valve that reduces the liquid refrigerant pressure before it enters the evaporator.

- In the cooling part of the cycle between States 4 and 1, the liquid refrigerant **evaporates** (boils) and absorbs heat in the evaporator.

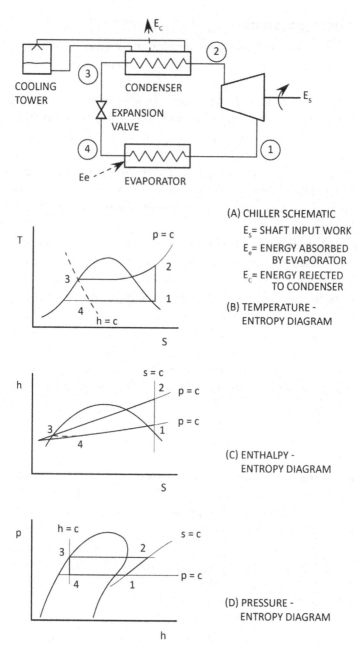

Figure 8.3 Vapor compression refrigeration cycle.

Within the two-phase region, this is a constant temperature/constant pressure process, but when all the refrigerant has evaporated, it begins to increase in temperature as it continues to absorb heat. This is called super heating the refrigerant and is required to assure that no liquid refrigerant gets back to the compressor, where it could cause damage. The amount of heat absorbed by the evaporating refrigerant is determined by subtracting the enthalpy of State 1 from that of the State 4.

In central chiller plants (*chilled water* systems), water is pumped through the evaporator where it cools and then is piped to cooling coils in the air distribution system. This is the approach sketched in Figure 8.2. In smaller or packaged systems the cooling tower shown in Figure 8.3 is not required.

An alternative approach, suitable when the cooling coils can be located close to the compressors, is the *direct expansion* (DX) unitary system. The cooling technology behind DX systems is similar to chillers except that the refrigerant expands in the evaporator coils and cools air directly as it moves across the coils, rather than using chilled water as a heat transfer medium. The most common DX air conditioning (or heat pump) systems for commercial and industrial applications are forced air packaged units (especially rooftop units) or split systems. These forced air systems use air handlers and duct systems to distribute conditioned air to zones. In split systems, the components of the refrigeration cycle are split into an outdoor unit and an indoor unit. The outdoor unit consists of the compressor, condenser coil, and cooling fan. The indoor unit consists of the evaporator and supply fan (blower). In packaged systems, all the equipment is contained within one package that is located outdoors. DX units eliminate the need for chilled water pumps and also eliminate the efficiency losses associated with transfer of heat to and from chilled water since the air is cooled directly with the evaporator. A typical DX system is designed to serve one air handling unit, so large buildings will have multiply DX systems to condition the entire space. In contrast, one chiller can serve multiple air handling units.

Still another approach is to use a *ductless* space conditioning system. Ductless systems are a type of split DX systems, but instead of distributing conditioned air by using air handlers and ducting, ductless systems condition the air directly in the zone. In cooling mode, indoor units located within or near the conditioned space act like evaporators and cool the air locally without ducts; in heating mode, the indoor units act like condensers and heat the air locally. There can be multiple indoor units for a single outdoor unit. They range in scale from *mini-split systems* with one outdoor and one indoor unit to *multi-split systems* with one outdoor unit and dozens of indoor units. An emerging technology is the *variable refrigerant flow* system.[2] It is a more advanced version of a ductless multi-split system that allows even more indoor units for a given outdoor unit and permits some zones to be cooled while others are simultaneously heated. In addition, it incorporates energy efficient features such as variable speed drives and heat recovery between units in cooling mode and others in heating mode.

The *capacity* of a cooling system is frequently expressed in tons, where one ton of refrigeration is defined as the transfer of heat at the rate of 3.52 kW (12,000 Btu/hr) per ton, which is approximately the rate of cooling obtained by melting ice at the rate of one ton per day.

Cooling systems—particularly chillers—are classified according to the type of compressor technology employed. There are four main types of compressors used in commercial and industrial HVAC systems: centrifugal, screw, scroll and reciprocating. Centrifugal compressors are often used in larger capacity chiller applications ($\sim$70–3500 tons or greater) with most centrifugal chiller systems being larger than 300 tons. They have the best full load efficiency of the compressor technologies. Screw compressors are also used in larger chiller systems ($\sim$40–1250 tons), but are most commonly found in systems with capacities of 300 tons or less. Screw compressors work well at partial loads and have recently taken over a large share of the HVAC compressor market from centrifugal systems. Scroll and reciprocating compressors are typically used in smaller capacity applications ($\sim$10–500 tons), where multiple small compressors might be used in a single packaged chiller system or rooftop unit. More and more scroll compressors are replacing reciprocating compressors because they have greater reliability. Both scroll and reciprocating compressors are less efficient than centrifugal and screw compressors.

The overall performance of a cooling system depends largely on the compressor efficiency. As noted above, compressor efficiency is a function of compressor type and size. It also depends on the original design and ongoing maintenance of the compressor components as well as inherent

---

[2] Amarnath, A., M. Blatt. (2008). "Variable Refrigerant Flow: An Emerging Air Conditioner and Heat Pump Technology." *2008 ACEEE Summer Study on Energy Efficiency in Buildings.* Aug. 17–22, 2008. Pacific Grove, CA. Washington, DC: American Council for an Energy-Efficient Economy.

losses in the compression process, which are expressed in terms of the compressor *volumetric efficiency*. Other factors affecting overall cooling system performance include cycle temperatures, refrigerant type, and maintenance of expansion devices, evaporators, and condensers. The efficiency of ancillary equipment such as cooling fans and chilled water pumps also plays a role.

The mechanical condition of compressors is determined as follows. If the actual compression process deviates considerably from isotropic compression, it could mean that significant throttling is occurring within the compressor. In small and poorly designed compressors, this can be due to inadequate valve sizing and poor design. In larger units, it is due to valve malfunction or blow-by past rings. As some of the refrigerant throttles back from the discharge to the suction port, it maintains much of its heat of compression, and must be recompressed, requiring excess energy use. In centrifugal compressors, the surfaces between the impeller and housing are machined to high tolerances to provide a very close fit. After many years of operation, the surfaces wear, resulting in substantial back leakage and inefficient operation.

The temperature difference across a heat exchanger is determined by comparing the refrigerant condensing or evaporating temperature to the leaving air or water temperature. Refrigerant to water heat exchangers and direct expansion evaporators are typically sized for a 6°C (10°F) temperature difference, while air-cooled condensers are typically designed for 12°C (20°F). Small low-cost packaged and window units are often designed with as much as 12°C (20°F) across the evaporator and 17°C (30°F) across the condenser. Temperature differences much higher than these indicate fouling of the heat exchange surface and a need for maintenance. Every degree centigrade of excess temperature difference requires close to 2% additional energy or 0.02 kW/ton.

The superheat setting is determined by subtracting the saturated liquid temperature corresponding to the evaporating temperature from the refrigerant temperature out of the evaporator. This difference is typically 5−7°C and is maintained by the expansion valve. Where this temperature difference is low, there is a chance that some of the refrigerant will return to the compressor as a liquid and cause damage. Where this difference is high (12−17°C or 20−30°F), most of the evaporator is being used to superheat the refrigerant, meaning that a much lower evaporating temperature is required to provide the same amount of cooling, which lowers efficiency. Thus, adjusting the superheat setting of the thermostatic expansion valve will result in an increase in evaporating temperature. Savings will be about 0.02 kW per ton per degree centigrade rise in the evaporating temperature.

Vapor Absorption Cycle. The vapor absorption cycle is another refrigeration cycle sometimes used for space cooling (and heating, in the case of absorption heat pumps). Instead of using an electric compressor, it uses a thermal compression process. This process has the following steps, several of which are similar to stages of the vapor compression cycle:

- A source of heat boils water out of a lithium bromide/water (or ammonia/water) solution and compresses the water vapor so that it now has a high temperature and pressure. This takes places in a *generator*.
- Next, the high pressure, high temperature water vapor rejects heat by condensing to a liquid in a condenser.
- The high pressure liquid then goes through an expansion valve where its pressure and boiling temperature reduce.
- The low pressure liquid next enters an evaporator and is boiled at low temperature and pressure. The evaporator is where cooling of the conditioned air takes place.
- The water vapor then enters an *absorber*, where it is absorbed back in the bromide/water (or ammonia/water) solution.
- Next, the solution is pumped back into the generator and the cycle continues.

Vapor absorption cycles require more energy than vapor compression cycles, but they can be cost-effective (and have high second-law efficiencies) when a suitable source of low-grade waste heat is available.

Evaporative Coolers. Direct evaporative coolers (also known as swamp coolers) are used in smaller air conditioning systems especially in dry, hot climates. These systems operate by evaporating water into air, thereby cooling the air. Air gives up heat (the latent heat of vaporization) at the rate of roughly 2.3 MJ/kg (approximately 1000 BTUs/lb) of water evaporated, depending on the temperature of evaporation. The process takes place along the line of constant wet bulb temperature and therefore the ambient wet bulb temperature is a lower limit on cooling that can be achieved. In actual practice, saturated conditions are avoided. Evaporative systems use less energy than refrigerated systems but require a significant amount of water as well as larger flows of outside air to provide adequate cooling. Another approach is indirect cooling of supply air by an evaporative cycle. (See Figure 8.4.)

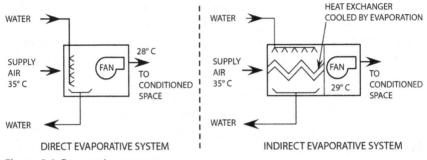

Figure 8.4 Evaporative systems.

Figure 8.5 Electric infrared radiant heater.

Space Heating Systems

Heating systems represent the next largest share of HVAC energy use in most commercial and industrial spaces. There are several types of space heating systems. Many large facilities use *boilers* to produce hot water or steam. In these instances, hot water or steam is piped to finned heat exchanger coils located in the ducts (Figure 8.2). The supply air warms up while passing over this heat exchanger and is then directed to occupied spaces. A thermostatically controlled valve maintains appropriate temperatures within the heated space. Smaller buildings may use electric or fuel-fired *forced air furnaces* or *heat pumps*. Other types of heaters occasionally encountered for local areas include gas or electric *unit heaters, radiant heaters*, and *electrical resistance heating elements* that can be placed in air conditioning ducts or in rooms (floors, ceilings, or along baseboards).

Radiant heaters have the advantage that most of the input energy is converted to radiated heat, so losses by conduction or convection can be low. Electric infrared radiant heaters have the further advantages that the heat output can be varied by rheostat controllers and no carbon monoxide is produced. Figure 8.5 is an example.

Of particular importance to HVAC applications are the following aspects of heat sources:

- Containment of heat (insulation of lines, prevention of leaks, condensate or hot water recovery).
- Efficient transfer of heat (heat exchanger maintenance).
- Heat recovery (from ventilation air and other sources).

We discuss these topics elsewhere in this chapter and in Chapter 11.

Fans

Fans are another major energy using component of HVAC systems. Most fan power is for delivering conditioned or ventilation air to occupied spaces. Fans for this type of air distribution are located at various points in the HVAC system and may be installed in central *air handling units*, small *terminal units* or within *duct work*. These fans circulate *supply, return, exhaust, and outside makeup* air. Supply air is generally a combination of return air (air that has returned from the conditioned space) and fresh outside makeup air. The makeup air replaces the portion of return air that has been exhausted. Fans are also used in air-cooled condensers for cooling the refrigerant and in cooling towers for cooling the hot condenser water. Fans commonly encountered are of two types: centrifugal or axial-flow.

Fans cause motion of air or gas streams by imparting energy to overcome the resistance of the flow path. The power required depends on the volume of gas moved, the pressure difference across the fan, the gas density, and the fan mechanical size, design, and efficiency.

The total pressure rise measured across a fan is a sum of two components: the static pressure and the velocity pressure. The power output of a fan is expressed in terms of its *air power* and is equivalent to the work done by the fan on the air. For HVAC applications, the increase in air density from inlet to outlet can generally be ignored; therefore, the gas can be treated as though it were incompressible. With this assumption, and by making use of Bernoulli's law, the following three *fan affinity laws* can be established for a fan with a speed change from N_1 to N_2.

- **Volumetric Flow Rate:** The volumetric flow rate of air flowing through a fan, $\dot{V}$, varies directly with impeller speed:

$$\dot{V}_2 = \left(\frac{N_2}{N_1}\right)\dot{V}_1 \quad \text{m}^3/\text{sec} \qquad [8.1]$$

Where:

N = fan speed, RPM

- **Pressure:** The pressure developed by the fan, P, (either static or total) varies as the square of the impeller speed:

$$P_2 = \left(\frac{N_2}{N_1}\right)^2 P_1 \quad \mathrm{N/m^2} \qquad [8.2]$$

- **Power:** The power needed to drive the fan, Q, varies as the impeller speed cubed:

$$Q_2 = \left(\frac{N_2}{N_1}\right)^3 Q_1 \quad \mathrm{W} \qquad [8.3]$$

These fan laws indicate that for a given air distribution system (specified ducts, dampers, etc.), doubling the airflow requires eight (2^3) times more power. Conversely, cutting the airflow in half requires one-eighth ($1/2^3$) the original power. This is an important fact for HVAC systems because even a small reduction in airflow (say 10%) can result in important energy savings (27%). Note that similar laws apply for pumping power, as discussed below and in Chapter 11.

How the airflow is reduced is critical in realizing the savings. In the case of constant flow applications, sizing the motor to provide exactly the needed airflow maximizes savings. In existing systems, simply changing pulleys to provide the desired speed will also result in energy reductions according to the *cubic law* in Equation 8.3. However, the efficiency of existing fan motors tends to drop off below the half-load range, so there is a practical limit to saving energy by simply changing pulleys.

In applications requiring variable volume air delivery, methods to control airflow rate include inlet vane control, outlet dampers, variable speed motors, controlled pitch fans, or cycling. Figure 8.6 shows the relative efficiency of these approaches.

Variable speed control is the most efficient practical method of controlling fans and saving energy because the motor and fan speed vary according to the airflow requirements and other mechanical means for reducing airflow are not required. Today highly efficient, compact, variable speed AC motor drives make this option attractive. On the other hand, outlet damper controls waste energy since the motor and fan operate at a constant speed while throttling by the dampers controls airflow.

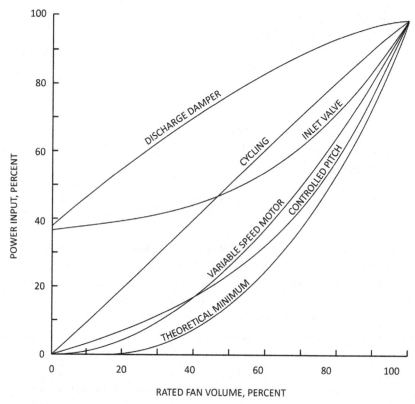

Figure 8.6 Fan power comparison for various types of part load controls.

The excess pressure drop caused by the dampers puts more load on the fan and lowers efficiency. Ultimately this also contributes additional heat that must be removed by an air conditioning system, in addition to wasting fan energy.

Manufacturers provide *fan curves* and data tables to summarize fan performance. Figure 8.7 is an example of a typical fan curve. These curves typically show the relationships between airflow and static pressure for various fan speeds. Sometimes they also include curves for brake horsepower, system resistance curves, and efficiency.

Pumps

Pumps represent the last major energy-using device in HVAC systems. They are used to move hot and cold water for space heating and cooling as well as for various auxiliary pumping needs. Pressure drops through the relevant system components dictate pumping power.

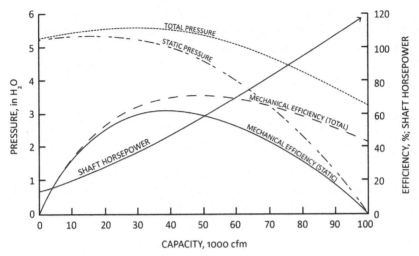

Figure 8.7 Typical fan performance curve.

Most pumps of interest in HVAC operations are centrifugal or axial-flow types. In such pumps, liquid enters the pump under atmospheric or higher pressure and is forced into a set of rotating blades or vanes called the *impeller*. The impeller then discharges the liquid at higher pressure and higher velocity.

In the general HVAC case, the work done on a fluid, Q, consists of three terms:

- Kinetic energy, $\dot{m}\frac{v_2^2 - v_1^2}{2}$, where $\dot{m}$ is mass flow rate and v is velocity.
- Flow energy, $\dot{m}\frac{P_2 - P_1}{\rho}$, where P is pressure and ρ is density.
- Potential energy, $\dot{m}(z_2 - z_1)g$, where z is the elevation and g is acceleration of gravity.

When divided by mg, each term has units of length. When combined, they are often called the total *head* developed by the pump and can be expressed in units of meters (feet). With this understanding, the fluid power of the pump is given by:

$$Q = \dot{m}gH = \dot{V}\,\rho gH = \dot{V}\,\Delta P \quad \text{W} \qquad [8.4]$$

Where:

$$
\begin{aligned}
Q &= \text{pumping power, W} \\
\dot{V} &= \text{volumetric flow rate, m}^3/\text{sec} \\
\dot{m} &= \text{mass flow rate, kg/second} \\
\Delta P &= \text{total pressure developed by the pump, N/m}^2 \\
p &= \text{fluid density at pump inlet, kg/m}^3 \\
g &= 9.81 \text{ m/sec}^2
\end{aligned}
$$

The velocities within the centrifugal pump depend on the geometry (fixed by the impeller diameter and blade angles) and on the impeller speed, and therefore vary with the pump speed. Thus, pumps can be shown to follow laws similar to the fan laws described previously (Equations 8.1–8.3). In other words:

- Flow varies directly as speed N.
- Pressure (head) varies as N^2.
- Power varies as N^3.

Interestingly, the effect of a reduction in impeller diameter is approximately the same as a reduction in speed. Thus Equations 8.1–8.3 can be rewritten as a function of impeller diameter, where D_1 and D_2 represent two impeller diameters:

$$\dot{V}_2 = \left(\frac{D_2}{D_1}\right)\dot{V}_1 \quad \text{m}^3/\text{sec} \qquad [8.5]$$

$$P_2 = \left(\frac{D_2}{D_1}\right)^2 P_1 \quad \text{N/m}^2 \qquad [8.6]$$

$$Q_2 = \left(\frac{D_2}{D_1}\right)^3 Q_1 \quad \text{W} \qquad [8.7]$$

Where:

D = impeller diameter, m

This is a practical result of some significance in retrofit situations, as shall be discussed in Chapter 11. For low speed pumps impeller trimming is feasible for speed reductions of up to 25–30%, and is less feasible for higher speed pumps. Avoid excessive trimming to prevent a large mismatch between the casing and impeller size, because that affects internal flow and can reduce efficiency.

Equally important is the fact that pumping power, as in the case of fans, varies as the cube of the volumetric flow rate, $\dot{V}$. Therefore small changes in flow rates can have a large effect on pumping power. (Refer to Equation 11.8 in Chapter 11.)

As with fans, manufacturers summarize pump performance in tabular or graphical form. The graphical presentation of pump data takes the form of pump curves (Figure 8.8) that resemble the fan curves illustrated in Figure 8.7.

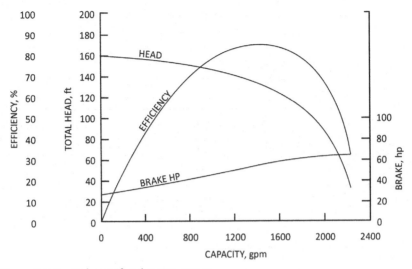

Figure 8.8 Typical centrifugal pump curves.

Cooling Towers

Cooling systems with water-cooled condensers like the one illustrated in Figure 8.2 use cooling towers to dispose of heat from condensers or other equipment. The condensers are usually shell and tube heat exchangers, with refrigerant circulating through the shell and cooling water flowing through the tubes. Condenser water pumps circulate the hot condenser water to the cooling tower where it is cooled by air. Typical tower types are induced draft, forced draft, cross-flow induced draft, and hyperbolic. Induced draft towers (Figure 8.9) are the most common in HVAC systems because they have a lower fan power requirement, but they are louder and less compact that forced-draft systems. Induced-draft towers use a propeller fan at the top of the cooling tower to pull air through the tower while water flows downward.

In the cooling tower, the condenser water sprays in fine droplets or dribbles down over fill media into a rising flow of air to exchange heat through direct water-to-air contact. There is sensible cooling when the air is at a lower temperature than the water. However, the greater effect is latent cooling brought about by vaporization of some of the condenser water. After the cooled water has made its way down to the basin at the bottom of the tower, it returns to the condenser. Some water is lost in the vaporization process and must be replaced by makeup water.

Cooling tower performance is defined by *cooling range* [the ΔT between water temperature at States 1 and 2 in Figure 8.9] and *approach*

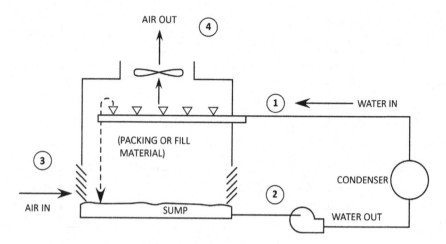

Figure 8.9 Induced draft cooling tower sketch.

[the ΔT between the water temperature at State 2 and the entering air wet bulb temperature at State 3]. For given tower conditions, having a high incoming water temperature maximizes heat transfer and reduces cooling tower fan power requirements. However, chillers operate more efficiently at lower condensing temperatures, so the cooling tower efficiency advantages of higher condensing temperatures must be weighed with the chiller efficiency advantages of lower condensing temperatures, as discussed later.

Potential problems with cooling towers are fouling (that affects airflow and heat transfer), excessive makeup water, improper temperature controls, and mechanical damage leading to leaks and inefficient operation.

Water-cooled systems are more efficient for heat transfer than air-cooled systems because air-cooled systems are limited by the dry bulb temperature, resulting in higher condensing temperatures and therefore lower chiller efficiency. However, since water-cooled systems require cooling towers and associated water pumps, piping, fans, and water treatment systems, they have higher installation and maintenance costs than air-cooled alternatives.

ANALYZING HEATING AND COOLING LOADS IN BUILDINGS

Due the importance of HVAC energy use and its potential for savings, analyzing heating and cooling energy use in buildings as a function of building characteristics, internal loads (people, lighting, and other equipment) and external loads (weather) is frequently an important part of an energy management program. There are three general approaches commonly used: (i) degree-day method; (ii) bin method; and (iii) building simulation software. These approaches vary in complexity, but all use algorithms or measured energy use data to model heat losses and gains of the building, performance characteristics of the HVAC equipment, the influence of external weather patterns, and effects of energy management measures. We briefly describe each of these three methods below. ASHRAE's *Fundamentals Handbook* has more detailed information on performing cooling and heating load calculations and on energy estimating and modeling methods.[3]

A fourth approach—analysis of historical energy use data using statistical techniques—may also shed some light on the influence of weather and energy management measures on heating and cooling loads, but if the energy data is for the building as a whole, there may not be enough granularity in the data to model and predict heating and cooling loads accurately unless detailed building characteristics are also captured in the models. However, historical energy use data (particularly sub-metered data for HVAC components) and other measured trend data (such as supply air temperatures) can very effectively supplement the other three approaches, and analysts often use this type of data to build more robust models, validate savings estimates, or to calibrate models.

Example. Submetering a rooftop air conditioning unit helped calibrate a building simulation model. To analyze the annual energy use of several high efficiency packaged rooftop cooling units with differential enthalpy controls installed at a middle school on the east coast of the U.S., our project team carried out a five-step analysis process. First, we collected data from the site, including building construction, operating hours, and specifications for each of the rooftop units. Second, we installed temporary data logging equipment on two of the units to sub-meter energy usage for a period of two weeks during the summer. (See Figure 8.10 for a graph of electricity demand versus time for one of the rooftop units on a September school day during the measurement period). Third, we used the site characteristics, weather data, and equipment specifications to create a DOE-2 building simulation model for the

---

[3] ASHRAE. (2013). *2013 ASHRAE Handbook—Fundamentals*. Atlanta, GA: ASHRAE. www.ashrae.org.

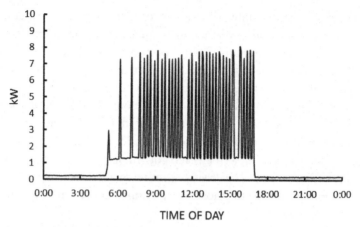

Figure 8.10 Demand profile for rooftop air conditioning.

site. Fourth, we did an initial calibration of the model with the sub-metered data collected for two of the rooftop units. Fifth, we did a final calibration of the model using 12 months of billing data for the site. This five-step process allowed us to build a robust model to estimate annual energy use as a function of weather. It also allowed us to predict what the energy use would have been for rooftop units with lower efficiency levels to show the savings achieved with the higher efficiency systems.

Degree-Day Method

For smaller or less complex buildings dominated by weather and not internal loads, the degree-day method is probably the simplest approach for energy analysis and is especially appropriate for preliminary calculations. Generally it is used for heating loads and is often not recommended for cooling loads because of complications such as latent cooling loads and greater complexity in cooling technologies and performance variations with weather. However, there are variations in the approach that make it more reliable for analyzing cooling systems. For example, the Chartered Institution of Building Services Engineers provides a thorough description of degree-days analysis, including application to cooling systems, and is a good resource for more detailed information on degree-day methods.[4]

---

[4] CIBSE. (2006). *Degree Days: Theory and Application.* TM41: 2006. London: Chartered Institution of Building Services Engineers. Available for download here: http://www.degreedaysforfree.co.uk/pdf/TM41.pdf.

For heating systems, the approach uses straight-forward algorithms and can be carried out by hand or using spreadsheets. The basis of analysis is that heating loads are proportional to the temperature difference between the indoors and outdoors, where the "constant" of proportionality is the overall heat loss coefficient (though, this coefficient is not necessarily constant). The first step is to determine coefficients of heat loss through the building envelope (walls, roof, floor, doors, windows, etc.). These coefficients relate to heat transfer through the building materials (*skin*) as well as to infiltration and exfiltration of air through cracks and openings in the building envelope. Losses associated with air exhausted and make-up ventilation air should also be account for. The individual coefficients combine to form an overall coefficient that is expressed in units of Watts per Kelvin (Btu per hour per °R). (See Chapter 12 for more details on calculating heat losses.) The second step is to determine the appropriate value of *degree-days* to include in the algorithms. Degree-days are a way to represent the indoor-outdoor temperature difference, averaged over a prescribed period of time.

Knowing the overall heat gain coefficient and the degree-days allows one to estimate the average heating load. Then, the average heating energy requirement can be calculated by dividing by the average overall efficiency of the heating system during the heating season. The building parameters and weather can be varied to study and compare different configurations.

Degree-Days. The number of heating degree-days (HDDs) is a measure of how much lower the outside air temperature is relative to a *base temperature*, and for how many days during the period. Similarly, the number of cooling degree-days (CDDs) is a measure of how much higher the outside air temperature is relative to a base temperature, and for how many days during the period (single day, month, year). For a given day, the degree-day value is calculated by determining the difference between the base temperature and the 24-h average temperature, where the average temperature is the average of the day's high and low temperatures. For example, an average temperature of 15°C or 59°F would correspond to three heating degree-days Centigrade or six heating degree-days Fahrenheit for a base temperature of 18°C (65°F). The number of heating degree-days per period is obtained by summing the degree-days for the entire period. Sometimes monthly values based on average monthly temperatures are used instead of daily values. Outside air temperature and degree-day data for various localities are readily available online.

Base Temperature. The base temperature represents the temperature at which the building requires no space conditioning—that is, the heat loss balances with the heat gain from internal loads and the sun. Common base temperatures to use are 18°C (65°F) or 15.5°C (60°F). However, it may be appropriate to establish a more accurate base temperature for the specific building in question. The base temperature may also be different for CDD and HDD calculations.

The degree-day approach has the advantage of being very simple to apply. But since it is based on average weather conditions and a constant base temperature, as well as on an average system efficiency for the entire period, it fails to capture changes in performance under different weather and operational conditions. However, there are ways to modify the basic method to increase its accuracy.

Bin Method

The bin method is more accurate than the degree-day method in that it accounts for the performance of the heating or cooling systems under different weather conditions across the seasons. The basic approach is similar to the degree-day approach, but instead of using an average value for degree-days for a given period, it uses the number of average hours of occurrence for each different temperature bin during the period. The choice of temperature range associated with each bin depends on the detail required for a given estimate. For example, bins may be divided up into 5 degree groupings (e.g., 0–5°C, 5–10°C, 5–15°C), or bins may have larger or smaller temperature ranges. For each bin, there will be an average number of hours that the weather was in that temperature range.

The calculations then can capture variations in performance of the cooling or heating systems for the different temperature bins, and the results for each bin are added to get the overall energy use for the analysis period. Examples of performance variation include the fact that the efficiency of some HVAC equipment varies with temperature, so using a temperature-specific efficiency improves accuracy. Also, part load performance for some of the temperature bins will have a different efficiency than full-load performance for other temperature bins. So, instead of using one set of average weather conditions and average performance to estimate energy use, the energy use for each bin can be calculated and summed to get a more accurate estimate of total energy use across the period. However, part load performance and efficiency at different temperatures may be difficult to determine, which adds complexity to this approach. Metering HVAC equipment and collecting trend data over a range of weather and operational conditions can help determine performance variations.

Bin data is available online for various locations. For example, the *Engineering Weather Data CD_ROM* has pre-made bin data for about 800 weather stations worldwide and can be obtained from National Climatic Data Center (http://www.ncdc.noaa.gov/).

Building Simulation Software

Modeling buildings with advanced simulation software is a more sophisticated way to estimate cooling and heating loads and energy use. Simulation software allows more accurate modeling of the interaction between building systems with each other, the building envelope, and the environment and can give a good indication of the building's overall load profile as well as energy use by specific end-uses, including HVAC equipment. Software is also capable of capturing effects of changes in equipment and controls to predict energy and demand savings.

Simulation software requires user input of building characteristics, system specifications, control strategies, building schedules, weather, and other relevant data. The models then simulate energy use for baseline scenarios and for proposed projects at a selected time interval—annual, seasonal, monthly, daily, hourly. Flexibility of modeling inputs varies with the simulation engine and the user interface to the engine. For example, one widely used simulation engine is DOE-2. It is a very comprehensive system and gives the modeler the flexibility to customize inputs even for very complex buildings. However, this flexibility comes at an expense since it is very time-consuming to develop a comprehensive, customized building model from scratch. In addition, it takes a skilled modeler. As a result, user interfaces such as eQUEST and VisualDOE have been developed to facilitate model development. Starting with prototypes of common buildings and libraries of system components that can be used directly or modified also helps speed up the modeling process.

There are numerous tools in use today. As noted in Chapter 7, the U.S. Department of Energy maintains a directory of over 400 energy software tools, including those for building and system simulation (http://apps1. eere.energy.gov/buildings/tools_directory/about.cfm).   Some   of   these programs can also be used to make comparisons of system capital and operating costs, and are useful for economic analyses. In addition to the tools already mentioned, a few other common tools for simulating building HVAC systems include EnergyPlus, HAP, and TRNSYS.

Other developments leverage the power of simulation models and the simplicity of bin analysis to create more sophisticated tools that are spreadsheet-based and easier and faster to use than simulation software, but more accurate than traditional bin analysis. For example, one such tool has recently been developed for heat pump systems.[5]

CODES AND STANDARDS

A number of codes and standards affect the design and operation of HVAC systems. Among these are local building codes, technical society standards, and codes and standards prepared by international bodies. The primary difference between codes and standards is that codes are lawful requirements and standards are recommendations not enforceable by law unless they have been adopted in codes by governing bodies. Examples of codes and standards related to HVAC energy use are as follows.

Codes

The International Energy Conservation Code (IECC) published by the International Code Council regulates "the design and construction of buildings for the effective use and conservation of energy over the useful life of each building." The latest version was published in 2012;[6] a new version is being developed for 2015. The code covers new construction and changes to systems and equipment in existing buildings, including HVAC equipment.

Other international codes pertaining to HVAC systems—but not specifically to energy efficiency—include the International Association of

[5] Dunn, J., Leichliter, K., Djunaedy, E., Van Den Wymelenberg, K., University of Idaho Integrated Design Lab. (2014). "Merging the Power of Simulation with the Simplicity of a Spreadsheet: Heat Pump Savings Calculator." *2014 ACEEE Summer Study on Energy Efficiency in Buildings*. Aug. 17-22, 2014. Pacific Grove, CA. Available for download here: https://www.aceee.org/files/proceedings/2014/data/papers/11-1078.pdf.
[6] ICC. (2012). *2012 International Energy Conservation Code*. Washington, D.C.: International Code Council. www.iccsafe.org.

Plumbing and Mechanical Officials' Uniform Mechanical Code[7] and the International Code Council's International Mechanical Code.[8]

Numerous other organizations at the national and local levels publish codes relevant to HVAC systems. For example, the Chartered Institution of Building Services Engineers (www.cibse.org) sets forth guidance and codes that encompass environmental design, HVAC, and energy efficiency in buildings for the United Kingdom, Europe, and elsewhere.

Standards

Standards organizations provide recommendations and guidance in several areas related to HVAC systems. Some of these organizations provide standards at the national level, while others such as the International Standards Organization (www.iso.org) publish international standards. There is also a search engine for locating standards (www.nssn.org).

For example, ANSI/ASHRAE/IES Standard 90.1-2013 sets forth minimum efficiency requirements for new construction, and new systems and equipment in existing construction.[9] In particular, for HVAC systems, the standard provides comprehensive lists of minimum efficiencies for various types of equipment, including room air conditioners, unitary air conditioners and heat pumps, chillers, condensing units, boilers, furnaces, unit heaters, fans, etc. It also address economizers, temperature set-points, elimination of simultaneously heating and cooling, other energy efficient control strategies, insulation and sealing of ducts and plenums, energy recovery, etc.

Some standards focus on acceptable thermal conditions for comfort and health:

- ISO 16813:2006 addresses aspects of thermal comfort as well as indoor air quality, energy efficiency and HVAC systems.[10]
- Canadian Centre for Occupational Health and Safety (CCOHS) recommends indoor air temperatures in offices be between 23°C and 28°C (73−82°F) during the summer and 20 to 25.5°C (68−78°F)

---

[7] IAPMO. (2012). *Uniform Mechanical Code.* Ontario, CA: International Association of Plumbing and Mechanical Officials. www.iapmo.org.

[8] ICC. (2015). *2015 International Mechanical Code.* Washington, D.C.: International Code Council. www.iccsafe.org.

[9] ANSI/ASHRAE/IES. (2013). Standard 90.1-2013. *Energy Standard for Buildings Except Low-Rise Residential Buildings.* Atlanta, GA: ASHRAE. www.ashrae.org.

[10] ISO. (2006). ISO 16813:2006. *Building Environment Design − Indoor Environment − General Principles.* Geneva: International Organization for Standardization. www.iso.org.

during the winter.[11] The reason for warmer temperatures during the summer is to minimize the temperature difference between the indoor and outdoor environment and to account for the fact that people generally wear lighter clothing during the summer. CCOHS also notes that relative humidity should be above 20% to prevent drying of skin and mucous membranes, and below 70% to prevent condensation and subsequent fungal growth. In addition they state that air velocities below 0.25 m/s (50 ft/min) do not create distraction for occupants.

- The Health and Safety Executive in the United Kingdom also has standards for workplace conditions that state workroom temperatures should be reasonable for the given application without the need for special clothing.[12] They suggest the temperature should be at least 16°C (61°F) for most applications, or at least 13°C (55°F) if the work involves high physical activity. Sedentary activities should have higher temperatures.
- In the U.S., the governing standard for indoor thermal environmental conditions is ANSI/ASHRAE Standard 55. The most recent version was released in 2013.[13]

Other standards address indoor air quality:

- ANSI/ASHRAE Standard 62.1-2013 is a commonly referenced standard for determining ventilation requirements in buildings.[14] The standard covers different ways to comply with requirements, including the ventilation rate procedure, the indoor air quality procedure, and the natural ventilation procedure. The standard has evolved over time to reflect advances in building envelope designs. Today's buildings are much tighter and energy efficient, but less air infiltration comes at the expense of needing greater mechanical ventilation to maintain air quality.
- ISO 16814:2008 is an international standard that specifies methods for meeting target levels of acceptable indoor air quality.[15]

[11] CCOHS. (2011). CSA-Z412-00 (R2011). *Guideline on Office Ergonomics*. Hamilton, Ontario: Canadian Centre for Occupational Health and Safety.

[12] HSE. (2013). *The Workplace (Health, Safety and Welfare) Regulations 1992*. Bootle, Merseyside: Health and Safety Executive. www.hse.gov.uk.

[13] ANSI/ASHRAE. (2013). Standard 55-2013. *Thermal Environmental Conditions for Human Occupancy*. Atlanta, GA: ASHRAE. www.ashrae.org.

[14] ANSI/ASHRAE. (2013). Standard 62.1-2013. *Ventilation for Acceptable Indoor Air Quality*. Atlanta, GA: ASHRAE. www.ashrae.org.

[15] ISO. (2008). ISO 16814:2008. *Building Environment Design — Indoor Air Quality — Methods of Expressing the Quality of Indoor Air for Human Occupancy*. Geneva: International Organization for Standardization. www.iso.org.

- The World Health Organization (WHO) Regional Office for Europe also publishes standards for indoor air quality.[16,17]

Energy managers should be aware of all applicable codes and standards for HVAC systems and take them into consideration in formulating the energy management program.

DESCRIPTION OF TYPICAL AIR DISTRIBUTION SYSTEMS

There are many types of HVAC systems with different configurations for handling and distributing conditioned air to spaces (or zones) within the building. A few of the main categories of air distribution systems include terminal reheat, dual duct, multi-zone, and variable air volume. Some systems are also based on modifications to these main types, including hybrid configurations that combine features.

Reheat Systems

Terminal or zone reheat systems provide chilled air to each zone. The air is then reheated depending on the temperature requirements of the zone using some type of heating system (e.g., electric resistance or hot water coils). (See Figure 8.11). Zones with large internal heat sources (lights, people, and equipment) will call for little or no additional heating. Other zones may require considerable additional heating. Since all air is first cooled, some duplicate energy use is bound to occur. Therefore, these

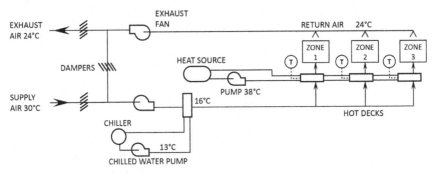

Figure 8.11 Terminal reheat system.

[16] WHO. (2010). *WHO Guidelines for Indoor Air Quality: Selected Pollutants.* Copenhagen: World Health Organization, Regional Office for Europe.
[17] WHO. (2009). *WHO Guidelines for Indoor Air Quality: Dampness and Mould.* Copenhagen: World Health Organization, Regional Office for Europe.

systems are very inefficient and building codes often restrict their use, unless they are accompanied with features to improve efficiency. One modification is to use controls that increase the supply air temperature when the cooling load in the warmest area of the building decreases. Known as *temperature reset* or *variable temperature*, this helps lower simultaneously heating and cooling in the zones with lowest cooling loads (colder areas) of the building.

Dual Duct System

Dual duct systems supply hot and cold air to each zone in separate ducts that are parallel to each other. Local mixing boxes at each zone have a damper for hot air and another for cold air. The hot air damper opens during heating and the cold air damper opens during cooling. A source of inefficiency of dual duct systems is that simultaneous heating and cooling can occur during moderate temperatures if hot and cold air are mixed to achieve the desired temperature and flow rate of air. Dampers are also prone to leakage, which can lead to excessive energy use. Figure 8.12 is a sketch of this system. Thermostats in each zone control the quantities of hot and cold air delivered to the zone to achieve the desired temperature.

Multi-zone System

The multi-zone system is similar to the dual duct system except that the mixing boxes are located at the fan instead of being distributed throughout the building. Dampers for each zone control the mixing of hot and cold air to achieve the desired temperature. Thus, the unit can provide

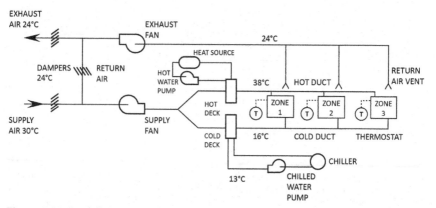

Figure 8.12 Dual duct system.

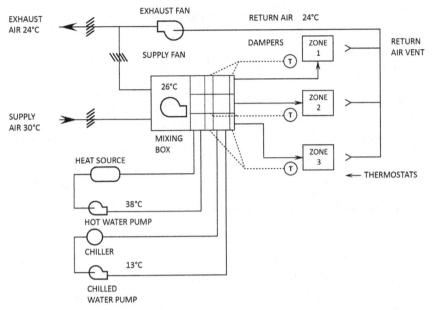

Figure 8.13 Multi-zone system.

warm air to one zone while providing cool air to another zone (see Figure 8.13).Thermostats in each zone control the dampers. They have the same drawback as dual duct systems in that simultaneous heating and cooling can occur. Advantages over dual duct systems include less ductwork, centralization of the mixing dampers, and less noise in the conditioned space. However, multi-zone systems have a high pressure drop through the mixing box dampers which increases fan power requirements.

Variable Air Volume System

Variable air volume (VAV) systems vary the amount of conditioned air delivered to a space based on the heating or cooling load. Reheat, dual duct, and multi-zone systems can be combined with VAV features to improve efficiency. When cooling loads dominate over heating loads (as is the case in many commercial buildings), VAV systems are designed to reduce airflow during periods of lower cooling requirements to the amount necessary to provide adequate cooling with minimal or no simul-taneous heating. They then increase airflow during high cooling loads. The advantage is that fans expend less energy in moving air and avoid or

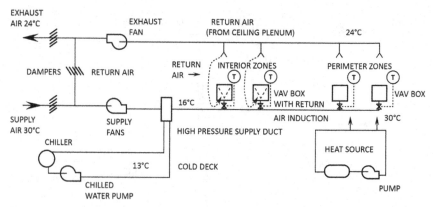

Figure 8.14 Variable air volume system (VAV).

reduce simultaneous heating and cooling during low cooling loads. (The corollary is true when heating loads dominate.)

Airflow reductions are accomplished with dampers to reduce flow or by modulating airflow with variable speed fans. Figure 8.14 shows a system in which a thermostat in each zone controls a valve that lets in a sufficient volume of air to provide needed cooling.

VAV systems are available for small and large buildings and are much more common in newer construction, but constant volume systems in existing buildings can be retrofitted with VAV systems. In large buildings, lighting, equipment, and occupant loads contribute heat to the interior zones and may be adequate for heating purposes. If not, supplemental heat is used. Sometimes the interior zone VAV boxes have built-in venturi action that induces some return air (from the ceiling plenum) into the airflow to help control temperature. Perimeter zones may have an external heat source provided unless they are south-facing.

Passive and Active Beam Systems

Another family of air distribution technologies consists of active and passive beam systems. These systems, which are also referred to as *chilled beam systems*, provide cooling (or heating) to individual zones by use of heat exchangers in "beams" that are embedded in or suspended from ceilings in the zones. Chilled (or heated) water circulates through the heat exchanger tubes thereby providing sensible cooling (or heating) to air that passes across the tubes.

In *passive* beam cooling systems, cooled air sinks toward the floor, while warmer air rises to the ceiling by natural convection. Once the warmer air has been cooled by the tubes, it then falls toward the floor, and so on. This natural flow passively distributes cool air throughout the zone uses only buoyancy forces. Since hot air rises, passive systems are not as effective for heating and may require supplemental heating, for example with baseboard heaters. Passive beam systems also recirculate room air, so do not inherently introduce fresh outside ventilation air. Primary air must be delivered through a decoupled ventilation system to provide adequate outside air ventilation as well as to meet latent space conditioning requirements (humidification or dehumidification).

Active systems incorporate a ducted primary air supply configured in a way to induce flow of the room's air across the tubes by mechanical means. The primary air supply provides outside air to meet ventilation requirements and also satisfies latent requirements, while the induced air flowing across the tubes meets sensible loads. They work for space cooling and heating, but are most effective for cooling. Active systems tend to be more common than passive systems because they result in much larger cooling (or heating) capacity due to the greater airflow across the tubes.

The primary advantages of active and passive beam systems over conventional forced air systems are potentially higher efficiencies, lower noise levels, and lower space requirements. Disadvantages include the need to prevent condensation on the cooling coils by ensuring cooling water temperatures are not too low and/or room humidity is not too high as well as the potential need to provide supplemental heating.

For more information on passive and active beam systems, refer to a design guide recently released by ASHRAE and the Federation of European Heating, Ventilation and Air-Conditioning Associations (REHVA).[18]

ENERGY MANAGEMENT OPPORTUNITIES IN HVAC SYSTEMS

Now that we have established a basis for understanding HVAC fundamentals, we show how to apply the energy management principles listed in Table 8.1. The discussion begins with opportunities to control HVAC systems more efficiently. It then turns to specific measures for the major

---

[18] ASHRAE/REHVA. (2015). *Active and Passive Beam Application Design Guide.* Atlanta, GA: ASHRAE and REHVA. www.ashrae.org, www.rehva.eu.

equipment items: chillers, fans, pumps, cooling towers, and ductwork. It concludes with a system-by-system review of energy management possibilities.

In any energy management program, implementing energy management opportunities that have little or no cost first is generally the best approach. No/low cost opportunities generally consist of operational and maintenance strategies, such as minor control changes. Moderate cost opportunities may include equipment retrofits or system modifications. Major cost items include new equipment or new designs. (Note that costs are relative and application-specific; what seems to be a minor cost item in one building could be a major cost item in another.)

HVAC System Controls

The energy use of an HVAC system is obviously highly dependent on the number of hours the system operates. It also depends on the level of heating, cooling, and ventilation supplied at any given time. Controls provide a means for scheduling operation and controlling the level of supplied space conditioning such that the system only operates when and how it is needed (Principle 1 in Table 8.1). Examples of energy management opportunities for HVAC controls are as follows:

- *Check for Proper Thermostat Operation*: The thermostat is the main interface between building occupants and the HVAC system. Occupants specify a temperature set-point and the HVAC system responds by operating heating or cooling equipment to bring the room air temperature to that of the set-point. Thermostats may also allow the occupants to control other parameters such as on/off, set-back temperatures, or fan operation. In some cases, occupants have no direct control of temperature set-points; instead facility managers or an energy management and control system centrally control the system. If building occupants are allowed to alter thermostat settings, establish policies for operation and make sure they are followed. Increasing awareness of the importance of the policies is a critical step in attaining occupant compliance and maximizing energy efficiency and comfort.
- *Replace Manual Thermostats with Programmable Models*: When building occupants are given the responsibility to manual control thermostats, inefficiencies in HVAC operation inevitably occur. Installing programmable thermostats in place of manual thermostats may address this

problem to a certain degree. When programmed properly, they help prevent needless operation of the HVAC system at night and during other periods when the building is unoccupied (e.g., summer vacation, winter break). They can also help prevent the high and low temperature extremes sometimes experienced when occupants turn the thermostat up excessively during cold periods and turn it down excessively during hot periods. Program temperature set-points to match the building's established guidelines for HVAC operations and settings. If non-maintenance people are allowed to program the thermostats, provide them with the proper training.

- *Turn off HVAC Equipment When Not Needed:* It sounds obvious, but this basic measure is often overlooked. Whether manual or automatic controls are used, make sure the HVAC system is off when the building is unoccupied, except when it is being used to precool or preheat the space.

There are also specific control strategies in addition to the general methods above:

- *Night Cool-down with Outside Air:* During the cooling season, bring in cooler outside air to precool the space overnight or in the morning hours. This measure will reduce the next day's cooling load. One way to do this is with economizers, which we discuss later in the chapter.
- *Night Temperature Setback:* During the heating season, lower the temperature set-point during evening hours when the space is unoccupied to reduce heating energy use.
- *Night Shutdown with Morning Warm-up (or Morning Precool in Hot Climates):* During the heating season, another strategy is to completely shut down the heating system during nighttime (or unoccupied periods) and then turn the system on a couple hours before occupants arrive to warm-up the space. The same concept applies to the cooling system, where the equipment can be shut off at night and then turned on a couple hours before people arrive to precool the space.
- *Minimize Outside Air on Hot or Cold Days:* Codes and standards prescribe minimum allowable amounts of outside ventilation air to safeguard against poor indoor air quality. However, outside air that is much higher in temperature than indoor air during the cooling season, or much lower in temperature during the heating season, puts an additional load on space conditioning equipment. Therefore, at times when *excess* outside air is being pulled into the space, take measures to minimize it to levels required for indoor air quality, unless its

temperature is such that it reduces the building's heating or cooling load. Be sure not to reduce outside air intake to levels that would jeopardize indoor air quality.

- *Improve Temperature Control*: Ways to improve temperature control are addressed more later in the discussion of specific system types. The primary goal is to make the space comfortable while avoiding simultaneously heating and cooling to meet temperature set-points.

When it comes to maintenance, the upkeep, calibration and repair of control systems should receive primary attention. An HVAC system will be unable to function as designed unless controls are properly maintained.

Space Cooling Systems

Controls. As with HVAC systems in general, improved controls (Principle 1) often offer a low-cost opportunity for energy savings with chillers and other types of space cooling systems. Modern systems may already incorporate some of these control strategies. However, many systems in existing buildings still lack sophisticated control capabilities.

A major area for investigation is operation at low loads. Since air conditioning systems are designed on the basis of a maximum heat rejection load, much of their operation is actually at less than full load. Yet, chillers are not as efficient when operated at part load. Careful use of equipment during partial load conditions can lead to significant energy savings. Adjusting controls to permit one unit (if there are several) to run at full load (rather than several at part load), or cycling one unit on and off at full load (rather than running continuously at part load), can save as much as 50% of the energy that would otherwise be used. *Chiller sequencing controls* can automate this process.

For water-cooled systems, lowering the condensing water temperature (*condenser water temperature reset*) is another opportunity for saving energy. In conventional systems, the condensing water temperature is set at a constant value of about 30°C ($\sim$85°F). In systems where the condensing temperature is set at too high of a constant value, or during times when the outdoor wet-bulb temperature decreases, reducing the condenser water set-point improves chiller efficiency and can save energy. A 5°C reduction in temperature could save 5–10% of chiller energy use. Cooler condenser water has the added advantage of a reduced tendency to cause scaling. One limiting factor may be the design of the expansion value, which requires a specified minimum pressure drop to deliver sufficient refrigerant. Another is the fact that lowering the condensing water temperature may come at the expense of increasing cooling tower fan energy use to meet the lower temperatures. There is a balance point at which energy efficiency of the cooling system as a whole is maximized.

Chilled water supply temperature should also be examined. If the temperature of the chilled water leaving the evaporator is set lower than necessary for occupancy comfort or to serve the current load conditions, it may be possible to increase the temperature either manually or with automated controls (*chilled water temperature reset*). A 2–5°C increase in the evaporator temperature could save on the order of 5–15% of chiller energy use. These savings come with a penalty in variable flow chilled water systems because more water must be pumped to meet a given cooling load. There may also be a penalty in variable-air-volume systems since raising the supply water temperature may raise the supply air temperature, making the supply fans draw more power.[19] Typically, there is a net savings with this measure.

Controls can also reduce space cooling energy use by modifying operation for night cool-down (*night precooling*), weekend shut down, or to take advantage of cooling with outside air when conditions are suitable (often accomplished with an economizer).

Proper Sizing. In new designs or new equipment installations, *proper sizing* is of obvious importance (Principle 3). Specify the space cooling system to meet full and part load conditions as efficiently as possible. For reciprocating chillers, this may include using multiple compressors to allow some units to be shut down when the cooling load is low. Or, it may include installing a *variable speed drive (VSD) chiller*. System capacity should also be reassessed and reduced as necessary when other major energy management projects lower cooling loads.

Another commonly encountered problem is use of a large central chiller to air condition a single office on weekends or to provide special conditions for a computer facility or other special installation. In such cases, turn off the central plant and install a small packaged unit sized to meet the need (Principle 3).

Example. An electrical equipment manufacturer operates a product test room to temperature-test electronic components for test cycles of 24 up to 36 h. The test room operates seven days per week. The test

---

[19] Wulfinghoff, D.R. (2000). Chapter 2 – Chiller Plant, *Energy Efficiency Manual*, Energy Institute Press.

room generates its own heat by virtue of the number of electronic components being tested. The maximum allowable temperature in the test room is 43°C (110°F). The facility has a 100-ton air conditioning system to cool the test room and other spaces in the plant. However, on Sundays, the only load serviced by the 100 ton air conditioning system is the test room. The average Sunday usage amounts to 8 h/day, 4 days per month, or 32 h/month. The average chiller load was 110 kW, so the energy usage for Sunday operation totaled 3520 kWh per month. Analysis of the costs of this operation compared with the installation and operating costs of using a thermostatically controlled 5-ton packaged air-conditioning unit for the test room only on Sundays (requiring only 176 kWh/mo) showed a payback of slightly more than three years.

Load Reduction. *Reducing the cooling load* (Principle 4) and *cooling only occupied spaces* (Principle 5) will also reduce space cooling energy use. These principles are most easily applied in new construction, but can also be addressed in retrofit projects. In existing buildings, focus on reducing infiltration and improving insulation to reduce sensible cooling loads. Since humidification and dehumidification loads are also affected by outside air conditions, reducing infiltration and otherwise minimizing exchange with outside air will reduce latent loads. Also consider changes in the air distribution system to prevent cooling unoccupied spaces.

Efficient Processes and Equipment. In cases where system retrofit is being considered, apply Principles 6 and 7 (*efficient processes and equipment*). For example, perhaps there is a source of "free" cold water.

Example. Use of ground water. A paper mill in Oregon found it could run its well water at 13°C (55°F) through the cooling coils in the plant *before* using it to generate steam. This reduced chiller energy use *and* the energy used for boiler feed water heating. This process also make use of Principle 10 (*employ heat recovery*).

It may be feasible to cool directly with the cooling tower if wet bulb temperatures are sufficiently low enough (even for part of the year). If proper filtering is available, the cooling tower water could be connected directly to the chilled water loop. Or, a heat exchanger between the two loops could be used to protect the coils from fouling. Another technique is to turn off the chiller but use its refrigerant to transfer heat between the two loops. This *thermocycle* uses the same principle as heat pipes, but only works on chillers with the proper configuration.

Other chiller retrofit options include replacing a constant speed centrifugal compressor with a VSD compressor to optimize operations at a range of part-loads conditions. This measure should be compared to the other chiller sequencing control strategies mentioned previously to see which approach is most cost-effective and energy efficient.

Whether it be large chillers or smaller-scale packaged DX systems, spending more upfront for high efficiency space cooling equipment is often cost-effective because even small efficiency gains can have a significant impact on annual energy costs. For example, in some buildings, annual energy costs for space cooling are on the order of one-third of a chiller's purchase price, so the incremental cost of purchasing the higher efficiency system can payback very quickly.[20]

Using *heat recovery chillers* designed to extract waste heat from condenser water for hot water and space heating applications is growing in favor for concurrent heating and cooling loads (Principle 10).

Maintenance. Maintenance is another important factor in efficient operation of space cooling equipment (Principle 8). Some maintenance opportunities include the following:

- *Keep Compressors in Good Repair:* Valve malfunction, ring wear, or impeller wear lead to refrigerant leakage or blow-by, and cause excessive energy use.
- *Inspect Chilled Water Pipes:* The insulation on chilled water piping may degrade or be damaged over time, especially if exposed to sunlight. Check chilled water piping regularly to be sure it has adequate insulation to prevent heat gain through the pipes. Also inspect piping for leaks. Repair insulation and pipe leaks immediately.
- *Clean Condenser and Evaporator Heat Exchange Surfaces:* Dirty and fouled coils and other heat exchange surfaces reduce heat transfer (and can block airflow in DX systems), causing the compressor to work harder than necessary to deliver the required cooling. Clean the heat exchange surfaces regularly to increase heat transfer and improve system performance and efficiency.
- *Check Refrigerant:* Contamination of refrigerant with excess oil, water, or air reduces cooling energy efficiency and can damage equipment.

---

[20] EPA. (2008). *ENERGY STAR® Building Upgrade Manual*. Washington, D.C.: Office of Air and Radiation, U.S. Environmental Protection Agency. Chapter 9. Available here: http://www.energystar.gov/sites/default/files/buildings/tools/EPA_BUM_Full.pdf.

Analyze refrigerant to detect contamination and decontaminate or replace refrigerant as needed. Decontamination services are available to help with this process. If the problem persists, consider installing high efficiency purgers to remove oil, water, and air contamination on a regular basis. Newer systems may come with some type of automatic refrigerant purging equipment. Also, inspect refrigerant charge and adjust the quantity according to manufacturer recommendations. Insufficient refrigerant charge will reduce the ability of the system to provide cooling.

• *Shade Roof-Top Refrigerant Coils:* Shade refrigerant coils in roof-top units to minimize heat gain from the environment.

Heat Pumps

Efficient Processes and Equipment. There are several ways to apply Principles 6 and 7 to heat pump systems. One application of the heat pump is a continuous loop of water traveling throughout the building with small heat pumps located in each zone. Each small pump can both heat and cool, depending upon the needs of the zone. This system can be used to transfer heat from the warm side of the building to the cool side. A supplemental cooling tower and boiler may be included in the loop to compensate for net heating or cooling loads.

A double bundle condenser can be used as a retrofit design for a centralized system. This creates the option of pumping the heat either to the cooling tower or into the heating system hot duct. Some chillers can be retrofitted to act as heat pumps. Centrifugal chillers will work much more effectively with the heat source warmer than outside air (exhaust air, for example). The compression of the centrifugal chiller falls off as the evaporator temperature drops.

When installing a new heat pump system, select the highest efficiency alternative that meets cost-effective criteria.

Heat Recovery. Commonly, a convenient source of heat for a heat pump is the building exhaust air (Principle 10). This is a constant source of warm air available throughout the heating season. A typical heat pump design could generate hot water for space heating from this source at around 32–35°C (90–95°F). Heat pumps designed specifically to use building exhaust air (*exhaust air heat pumps*) can reach 66°C (150°F).

Other Space Heating Systems

Many energy management opportunities for space heating systems are similar to those for process heating systems since the basic technologies are the same. Chapter 11 addresses combustion processes, boilers, furnaces, and electric heating technologies in greater detail. Below are some examples of space heating efficiency measures, with an emphasis on low-cost options.[21] Boiler systems offer some of the greatest energy management opportunities of the space heating types, but some of these measures apply to furnaces as well.

- *Practice Careful Load Management of Boilers and Furnaces*: Employ Principle 1 and install controls that load the most efficient boilers and furnaces first; then follow with systems of decreasing efficiency to optimize overall operational efficiency. When possible operate systems on high fire setting and at full load. Turn off boilers or furnaces when not in use.

- *Optimize Boiler Blowdown*: Blowdown is important to remove dissolved solids from the boiler water. Installing blowdown controls that result in frequent or continuous, short blowdowns instead of long, infrequent blowdowns reduces energy losses from the blowdown process (Principles 1 and 6). Recovering heat from boiler blowdown is another energy management opportunity (Principle 10).

- *Control Temperature and Pressure*: When possible (e.g., when outside air temperatures are higher), lower the supply air temperature or pressure to match the system load (Principle 1). Temperature or pressure reset controls like these will reduce boiler steam or hot water supply, thereby saving energy.

- *Monitor Combustion and Boiler Equipment Continuously*: On-going monitoring of combustion parameters will help detect issues. Analyzing equipment is available to measure excess air and carbon monoxide levels. High carbon monoxide levels indicate incomplete combustion, which could be due to a poor air-to-fuel ratio or fouled burners. It is also important to monitor the temperature of stack gases.

- *Adjust Burners on Furnaces and Boilers*: Optimizing the air-to-fuel ratio by adjusting burners will improve combustion efficiency (Principle 8). A small amount of excess air is typically necessary, but the optimal ratio depends on the particular system and fuel type. For example, a forced draft gas boiler may operate well with 5−10% excess air, which

---

[21] Parmenter, K., Arzbaecher, C. (2007). *Simplified Guide to Energy Efficiency in Office Buildings: Low-Cost Measures*. Lafayette, CA: Global Energy Partners.

relates to 1−2% excess oxygen. Controls can be installed to automatically trim excess air based on monitored conditions.

- *Maintain Burners*: Cleaning burners to remove soot and other deposits improves heat transfer and burner efficiency and ensures smooth ignition and proper flame color (Principle 8). It is also important to replace damaged burner tips.
- *Replace Inefficient Burners*: Replacing inefficient burners with new efficient burners has the potential of reducing fuel use in boilers by several percent (Principle 7).
- *Recover Flue Gas Heat:* In some applications, stack economizers are a cost-effective technology for recovering waste heat from the flue gas to preheat boiler feedwater (Principle 10). *Condensing economizers* are particularly effective at recovering heat since they capture both sensible and latent heat from the boiler flue gas.
- *Clean Heat Transfer Surfaces*: Cleaning heat transfer surfaces within boilers and furnaces removes fouling and scale and maximizes heat transfer efficiency (Principle 8).
- *Treat Feedwater*: Better treatment of feedwater minimizes boiler blowdown and fouling on waterside surfaces (Principle 8).
- *Repair Leaks and Insulation*: Reduce heat loss by repairing leaks in hot water and steam pipes, connections, and air ducts, as well as by repairing or replacing poor insulation on boiler jackets, condensate and feedwater tanks, hot water and steam pipes, and air ducts (Principle 8).
- *Inspect Steam Traps*: One of the most important energy management activities for steam systems is regular steam trap maintenance (Principle 8). Malfunctioning steam traps can cause large energy and water losses. The Federal Energy Management Program estimates that space heating systems not proactively maintained lose about 20% of boiler steam through leaky steam traps.[22]
- *Install Efficient Equipment*: When cost-effective, consider replacing existing boilers or furnaces with high efficiency alternatives (Principles 6 and 7). For example, *condensing boilers* are more efficient than conventional boilers and are an attractive alternative in some applications. There also may be an opportunity to lower the capacity of the new installation if the existing system is oversized.

---

[22] DOE. (1999). *Steam Trap Performance Assessment*. Washington, D.C.: Federal Energy Management Program, U.S. Department of Energy. Available here: https://www1.eere.energy.gov/femp/pdfs/FTA_SteamTrap.pdf.

Fans

Energy management opportunities for fans can readily be identified following the discussion above regarding the principles of fan operation. Examples of possibilities include the following:

- *Turn Fans Off*: Apply Principle 1 and use controls to turn fans off when not needed (on weekends, for example). Also consider turning off exhaust fans in unoccupied areas.
- *Optimize Capacity*: Replace oversized fans with correctly sized fans (Principle 3). Often the most efficient way to do this is in conjunction with installing right-sized premium efficiency motors and variable speed drives or energy-efficient synchronous belt drives. If this is not feasible, change pulleys to slow down the fan. (Recall the cubic law of fan power and Figure 8.5.)
- *Reduce Excessive Ventilation:* Consider strategies such as *pressure reset* or *demand-controlled ventilation* to lower airflow to levels needed at any given time (Principles 1, 2, 3, and 5). For example, reduce airflow during periods of low occupancy or during demand response events. It may also be feasible to run HVAC systems for fewer hours per day during seasons with low heating or cooling loads, as long as ventilation is sufficient to meet indoor air quality requirements.
- *Modify Systems to Function as Variable Air Volume Systems*: When feasible and cost-effective, consider converting constant volume systems to variable volume systems (Principles 6 and 7). The Environmental Protection Agency estimates that typical airflow needs for variable air volume systems are 60% those of constant volume systems.[23]
- *Select Efficient Fans and Ducts:* Consider lifecycle costs when sizing systems; reduce pressure drops where feasible (Principle 7).
- *Maintain Fans*: Fans require routine maintenance for optimal operation. Lubricate bearings, adjust or change fan belts, and clean fan blades on an annual basis to help maximize fan efficiency (Principle 8).
- *Replace Air Filters:* Particulate accumulation on air filters reduces airflow and may increase fan energy use and affect air quality. Install new air filters quarterly or at a prescribed increase in pressure drop to improve efficiency and indoor air quality (Principle 8).
- *Use Passive Concepts:* Make use of natural convection if possible (Principle 9).

[23] EPA. (2008). *ENERGY STAR® Building Upgrade Manual*. Washington, D.C.: Office of Air and Radiation, U.S. Environmental Protection Agency. Chapter 8. Available here: http://www.energystar.gov/sites/default/files/buildings/tools/EPA_BUM_Full.pdf.

Pumps

Pumps are important energy users in HVAC systems. Although they frequently receive less consideration than larger equipment (such as chillers, boilers, and air handling systems), they may operate continuously. When the capacity of the larger equipment is reduced, it may also be possible to reduce pumping capacity. For example, with multiple chillers, there may be several chilled water pumps operating in parallel, even when only one chiller is on. Control changes may permit the use of a single pump.

As explained in the section on pump theory, reducing the diameter of the pump impeller is equivalent to slowing it down. This is an economical method for saving energy with certain types of pumps having excessive capacity.

Other approaches to reduce pumping power are as follows:
- Reduce pressure drop and head.
- Use multiple pumps of different capacity, each operated at full load.
- Use variable speed pumps.
- Check chilled water circulation pumps for leaks and repair any leaks immediately.
- Clean water strainers a few days after new construction and after any piping/pump repairs or alterations.

Cooling Towers

We discussed the operating principles of cooling towers previously. Providing adequate water treatment and maintenance for system components is essential for optimum operation. Many of the energy management opportunities already discussed related to fans, pumps, and water treatment also apply to cooling towers. In addition, some of the chiller opportunities overlap with cooling tower opportunities. Examples of general possibilities for energy management include the following:
- Keep associated components such as fill material, fans, valves, and pumps in good working order.
- Ensure water treatment is effectively controlling scaling and biological fouling so heat transfer efficiency is optimized.
- Check for and repair leaks.
- Use two-speed fans for light/heavy load conditions.
- Reduce operation of multiple towers during cool periods.

Example. Energy audits were performed at several naval facilities in Hawaii. During inspection and measurements on several reciprocating chillers, the audit team checked the nearby cooling tower. They found that all of the fill material and the fan had been removed! Cooling was the being provided by a water spray at the top of the tower, with the water falling into the sump. The water spray was produced by a conventional bathroom shower head. This reduced the efficiency of the chiller considerably.

Cooling towers use a considerable amount of water. Therefore, water efficiency is also important to address.

As described in Chapter 6, a detailed energy audit was conducted by the authors and their team at Punahou School in Honolulu, Hawaii. Much of the work focused on various measures to improve HVAC systems, as detailed in the following example.

Example. HVAC energy management opportunities at Punahou School. The audit determined that HVAC accounted for 52% of electricity used on the campus. As part of the audit process, we examined typical weekday load curves as well as weekend load curves. The weekday load curve peaked at a value of 1500 to 1800 kW during the daytime school hours. Due to extracurricular activities in the evening, from about 4 PM to 10 PM, demand gradually declined to around 500 kW. Throughout the night, demand was nearly 1/3 of the peak demand. Some of this was attributed to security lighting, swimming pool filter operation, and miscellaneous equipment, with the balance due to HVAC. On the weekends the demand averaged about 500 KW, peaking at 700 KW at midday. These data suggested that that there was an excessive amount of 24 h/day air conditioning usage.

Based on the building-by-building energy audits and the database we constructed, we determined that there were 18 measures that could be deployed to reduce HVAC energy use. These measures are summarized below (Text Box 8.1):

Each measure was evaluated on a building-by-building basis for applicability. Some illustrative examples are described below.

H1—Install control system to schedule starting and stopping of equipment so it does not run continuously

This measure can greatly reduce energy consumption in buildings. If control is manual, equipment often gets left on accidentally and can run for unnecessarily long periods of time. We recommended controls of this type for every building on campus, to be part of the campus-wide Energy Management and Control System (EMCS). The savings that can be achieved for this measure are difficult to quantify since

TEXT BOX 8.1 HVAC EMOs Recommended for Punahou School

H1 Install control system to schedule starting and stopping of equipment so it does not run continuously.

H2 Replace air conditioning units with high efficiency units

H3 Install thermostats (or other controls) able to set the space temperature up during unoccupied periods

H4 Install thermal energy storage systems so chillers operate during off-peak periods (either completely, or partially)

H5 Replace chillers with high efficiency chillers

H6 Install two-speed motors on cooling tower fans

H7 Install variable frequency drives on the chiller compressor motors to vary refrigeration capacity to match facility requirements

H8 Install VFDs on hot water pumps

H9 Reset supply air temperature higher when cooling demand is low

H10 Convert constant-air volume air-handling systems to variable-air volume (VAV) systems, using variable frequency drives on the supply and return fans to vary airflow according to the requirements of the spaces

H11 Shut down/shut off devices

H12 Improve chiller efficiency

H13 Replace large chillers with small chillers after hours

H14 Insulate pipes or ductwork

H15 Install enthalpy controls

H16 Install regulating dampers on return or supply air ducts of fan coil units

H17 Install regulating valves on chilled water coils of fan coil units

H18 Connect to a central cooling plant

they depend on current operating habits of maintenance personnel. If maintenance personnel diligently turn equipment off when it is unneeded, then energy savings are small (although maintenance savings can be significant). If, however, equipment is left on often, as the load curves seemed to indicate, then energy savings are significant.

H2—Replace air conditioning units with high efficiency units

H5—Replace chillers with high efficiency chillers

There are cooling systems available with efficiencies much greater than older systems. Punahou School had a number of old, inefficient air conditioning systems that would benefit from replacement. Many of the older units were at the end of their useful life. We recommended installation of high efficiency units for all older cooling systems on campus.

H9—Reset supply air temperature higher when cooling demand is low

Manual control of cooling equipment often leads to wide swings of room temperature to keep up with complaints of occupants. A temperature reset controller will regulate the temperature of the supply air. This will reset the supply air temperature higher when the cooling demand is low, and lower the air temperature when the cooling demand is high. The controller should operate by sensing the temperature of the air in the return air duct. This measure was recommended in particular for Building #34 - Bishop Hall. Besides improving the comfort level of the occupants, chiller energy would be saved by reducing the amount of excess cooling that was occurring.

H6—Install two-speed motors on cooling tower fans

H12—Improve chiller efficiency

These two measures were recommended for Building #7 - Cooke Library. The chemical treatment for the cooling tower was disconnected and the tower had a lot of algae growth. This lowered the effectiveness of the tower and made the chiller less efficient. Installing a new cooling tower and two-speed motors on the cooling fan would improve the efficiency of the chiller.

H13—Replace large chillers with small chillers after hours

For buildings that run equipment 24 hours per day to satisfy the needs of one small area in a building, installation of a small chiller dedicated to serve the specific area in place of operating the entire cooling plant can results in substantial savings. This measure was recommended for Building #10 - Dillingham Hall.

H16—Install regulating dampers on return or supply air ducts of fan coil units

H17—Install regulating valves on chilled water coils of fan coil units

These measures were recommended for Building #34 - Bishop Hall to improve air quality. A common complaint of occupants on the ground floor was mustiness or staleness of the air on the first floor. Air quality would be improved by installing regulating dampers on the return-air ducts of the first floor fan coil units and by improving the fresh air supply by enlarging the existing small fresh air inlets and ducts. Maximum benefits would accrue with the installation of regulating valves on the chilled water coils of the fan coil units to shut off or reduce the chilled water supply during the night or unoccupied periods. During the unoccupied periods, the circulation of the supply air and its fresh air content would maintain the quality of the air in the first floor areas without unnecessary cooling or heating.

This is a partial listing of HVAC measures that we recommended. In addition, there were many opportunities for replacing fan motors with more efficient types. For the buildings with major HVAC loads, we estimated the demand and energy savings on a building-by-building basis. We also determined the energy cost savings, cost to implement the change, and the Hawaiian Electric company rebate. In summary, the recommended HVAC measures would reduce demand by 95 kW and energy usage by over 900,000 kWh per year.

Ductwork

Central space conditioning systems use ducts and dampers to distribute air to the spaces requiring heating or cooling. It is important to regularly inspect and maintain this equipment to prevent energy loss.

- *Inspect Ductwork:* Inspect the ducting system for leaks, blockages, and cleanliness (Principle 8). Loss of conditioned air through duct leaks must be made up and reconditioned, resulting in unnecessary energy use. Leaks at joints and connections can be sealed on the outside by using traditional mastic and glass fiber tape. Alternatively the duct can be sealed from the inside with an aerosol sealing system that involves plugging leaks with small droplets deposited around the edge of leaks. This alternative can be performed immediately after testing for leaks by a qualified technician. (Do not spray seal or clean ductwork unless there is an opportunity to ventilate the building for a few days without occupancy.) Also, for packaged systems, inspect the unit's cabinet and duct connections for air leaks.
- *Maintain Dampers:* Inspect the condition of static pressure dampers and make repairs as needed (Principle 8). If they are operating improperly, they may restrict airflow when they should be open or they may leak conditioned air when then should be closed.

Building Envelope

Check buildings for excessive infiltration of outside air or air loss from the building envelope. Either can lead to additional heating or cooling. Equipment is available to pressurize buildings and measure the amount of air leakage. Leaks can be located and sealed by caulking or other means. (See also Chapter 12).

Systems using Preheat or Reheat

Systems that use preheat on air washes for humidification or dehumidification or tempering of air can often be modified to save energy. If it is necessary, the simplest way is to use return air to preheat supply air. Alternatively, another type of heat recovery project can provide preheating (Principle 10).

There are several possible approaches for decreasing the amount of energy used in terminal reheat systems. For example, large buildings seldom need heat in the interior zones because of internal heat generation. Instead, the reheat system can be disconnected and replaced by a variable air volume system operating with the cold deck air (Principle 6).

In addition, the cold deck should always be kept as warm as the hottest room will allow. This can be done by resetting the deck temperature on the basis of the return air from that room or zone (Principle 1). Outside air should be used as well for supplying cold air, whenever temperatures permit.

Strip heaters used for terminal reheat in exterior zones can be replaced with a variety of heat sources. For example, if water is pumped to a cooling tower at 35°C (95°F), this heat might be exchanged with a hot water heating system to provide the reheat (Principle 10). If this won't supply the building's heating needs, a heat pump could be used instead, drawing heat from either the cooling tower water or the building exhaust air (Principles 6, 7, and 10).

Dual Duct Systems

The first modification is to decrease the hot deck temperature and increase the cold deck temperature as much as possible. This can solve some, but not all, of the problems.

Example. In the San Diego City Hall building (discussed in Chapter 1), cold and hot deck temperatures were adjusted manually every hour of the day. Measurements showed that excess heating averaging 1 million Btus per hour was occurring on days with outside air temperatures in the low 70s°F. To compensate for this, an additional 83 tons of air-conditioning was required on average. If the deck temperatures had not been adjusted continually, even more air-conditioning energy would have been wasted.

Hot and cold deck temperatures should be reset automatically with simple electronic controls. If this is already done on the basis of outside air temperature, it can be changed in many cases to return air temperature. This guarantees that heating is only supplied when it is needed in the building, not whenever the temperature drops outside. There may be problems with a particularly hot or cold zone, but these can be alleviated by monitoring the return air from those zones to control the duct temperatures, or adding extra capacity locally. In many cases a heating/cooling lockout can be installed to eliminate the simultaneous operation typical of dual duct systems.

Still another type of beneficial change is the so-called *split flow* modification. Normally, return and outside air are mixed and then sent to the hot and cold decks. Splitting the flows reduces cooling energy during the summer, but the most notable reduction is in heating energy during moderate weather. (Note the split flow modification does not *divide* the plenum, but *channels* the flow. Measurements show it is effective with single fan plenums as well as plenums with dual fans, although the efficiency is better where there are two fans.) Energy savings can be as much as 50% compared to the original condition. Another way to do this is to add a *third duct* that has a separate airstream of outside air which then mixes with either the hot air during heating or the cold air during cooling. That approach avoids mixing of hot and cold conditioned air, but has a higher capital cost for the third deck.

A still more drastic approach (but one that is feasible in certain buildings in mild climates, particularly for the interior or building core zones), is simply to shut off air to the hot duct. This approach relies on the internal heat generation (people, lights, equipment) to provide the required heating. The mixing box then functions as a variable air volume box, modulating cold air according to the demand as indicated by the room thermostat. Airflow must be checked to ensure that minimum requirements are met.

In some buildings, small diameter ducts have been used to reduce space needs and minimize capital costs. These systems operate at higher pressures and require more fan power. HVAC system design standards can influence duct size, and need to be checked for retrofit situations. On a lifecycle cost basis, larger ducts may be more economical.

Multi-zone Systems

Many of the considerations discussed above for dual duct systems also apply to multi-zone systems. An additional approach suitable for multi-zone systems involves monitoring the demand for hot and cold air in each zone. Signals are transmitted to hot and cold deck dampers controlling each zone by mixing hot and cold air to provide the required temperature. The cold air temperature should be just low enough to cool the zone calling for the most cooling. If the cold air were any colder, it would be mixed with hot air to achieve the right temperature. This creates an overlap in heating and cooling, not only for that zone but for all the zones because they would all be mixing in the colder air. If no zone calls for total cooling, then the cold air temperature can be increased gradually until the first zone requires cooling. At this point the minimum cooling necessary for that multi-zone configuration is performed. The same operation can be performed with the hot air temperature until the first zone is calling for heating only. Note that simultaneous heating and cooling is still occurring in the rest of the zones. Therefore, this is not an ideal system, but it is a first step in improving operating efficiency.

Another approach suitable for multi-zone and dual duct systems is to convert them to variable air volume systems. The cost and complexity of doing this varies depending on the design of the system, and is not feasible in all cases.

Economizer Systems and Enthalpy Controllers

Some small HVAC systems and many heating-only ventilation systems have fixed-position outside air openings. However, a large share of ventilation systems use airside economizers to automatically modulate the amount of outside airflow introduced to the mixed air duct. Economizers can yield significant energy savings, particularly in large commercial buildings in moderate climates. Economizers can also be added as a retrofit to existing systems (Principle 7).

Airside economizers allow up to 100% outside air for "free-cooling" during moderate outdoor conditions, but restrict the outside airflow to a minimum setting when it is too cold or hot outside for beneficial use. During cooling, it permits mixing of warm return air with cooler outside air to maintain a preset temperature in the mixed air plenum. When the outside temperatures are mild, 100% outside air is used to provide as much of the cooling as possible. During very hot outside weather, minimum outside air is added to the system.

Typically, the economizer is controlled by the dry-bulb temperature of the outside air rather than its enthalpy (actual heat content). This type of control is adequate most of the time, but can lead to unnecessary cooling or excessive humidity loads when outside air is damp. Using enthalpy controls to measure wet-bulb temperatures can prevent bringing in too much humid outside air, thereby saving energy.

The rules that govern enthalpy controls for cooling-only applications are as follows:

- When outside air enthalpy is greater than that of the return air or when outside air dry bulb temperature is greater than that of return air, use minimum outside air.
- When the outside air enthalpy is below the return air enthalpy and the outside air dry-bulb temperature is below the return air dry-bulb temperature but above the cooling coil control point, use 100% outside air.
- When outside air enthalpy is below the return air enthalpy and the outside air dry-bulb temperature is below the return air dry-bulb temperature and below the cooling coil controller setting, the return and outside air are mixed by modulating dampers according to the cooling set point.

These points are valid for the majority of cases. When mixed air is to be used for heating and cooling, a more intricate optimization plan will be necessary, based on the value of the fuels used for heating and cooling.

A major downfall of economizer systems is poor maintenance. The failure of the motor or dampers may not cause a noticeable comfort change in the building because the system is often capable of handling the additional load. Since the problem is not readily apparent, corrective maintenance may be put off indefinitely. In the meantime, the HVAC system will be working harder than necessary, wasting energy and money. For example, if the outside air damper becomes stuck open, too much outside air may enter the system and the cooling coils can be overloaded. If it is stuck closed, then the opportunity for "free cooling" is lost. Enthalpy controls are also less reliable than their dry-bulb counterparts and may require replacement every few of years depending on their operating conditions. A well thought-out maintenance program is necessary for any economizer installation. It should include examination of damper positions, periodic cleaning and lubrication of all moveable parts, inspection of the actuator movement, and replacement of temperature or enthalpy controls as necessary (Principle 8).

Heat Recovery

We have mentioned a few opportunities for heat recovery (Principle 10) in the proceeding sections. Heat recovery can also be employed in HVAC systems to recover energy from exhaust air. If the reject air has been heated or cooled, it represents an energy loss inasmuch as the makeup air must be modified to meet the interior conditions. The simplest approach to reduce this loss is to maximize recirculated air and minimize exhaust air. If recirculation is not possible (i.e., once-through airflow designs), devices are available to transfer heat between the inlet (outside) air and the exhaust air. These systems are most cost-effective when there is a large volume of exhaust air and when weather conditions are more extreme, thereby offering a greater potential for energy recovery.

Heat recovery systems are broken down into two types: *regenerative* and *recuperative*. Regenerative units use alternating airflow from the hot and cold stream over the same heat storage/transfer medium. This flow may be reversed by dampers or the whole heat exchanger may rotate between streams. Recuperative units involve continuous flow; the emphasis is upon heat transfer from one stream to the other through a medium with little energy storage. (As noted previously, a third approach is to use heat pumps to recover heat from the exhaust airflow.)

Regenerative. The rotary regenerative unit, or *heat wheel*, is one of the most common heat recovery devices for HVAC applications. It contains a corrugated or woven heat storage material that gains heat in the hot stream. This material is then rotated into the cold stream where the heat is given off again. The wheels can be impregnated with a desiccant to transfer latent heat (moisture) as well as sensible heat. Purge sections for HVAC applications can reduce carry-over from the exhaust stream to acceptable limits for most installations. The heat transfer efficiency of heat wheels generally ranges from 60% to 85% depending upon the installation, type of media, and air velocity. For easiest installation, the intake and exhaust duct should be located near each other.

Recuperative. An *air-to-air plate heat exchanger* is a type of recuperative system that can be employed within a convenient duct location. The system is usually lighter, though more voluminous than a heat wheel system. Heat transfer efficiency is typically in the 60% to 75% range. Individual units range from 100 to 11,000 SCFM and can be grouped together for greater capacity. Almost all designs employ counter-flow heat transfer for maximum efficiency.

Another recuperative option to consider for nearly contiguous ducts is the *heat pipe*. This is a unit that uses a boiling refrigerant within a closed pipe to transfer heat. The refrigerant and capillary wick are permanently sealed inside a metal tube, setting up a liquid-to-vapor circulation path. Thermal energy applied to either end of the pipe causes the refrigerant to vaporize. The refrigerant vapor then travels to the other end of the pipe where thermal energy is removed. This causes the vapor to condense into liquid again and the condensed liquid then flows back to the opposite end to the capillary wick. Since the heat of vaporization is utilized, a great deal of heat transfer can take place in a small space. Heat pipes are often used in double wide coils that look very much like two steam coils fastened together. Energy transfer between incoming and outgoing air can be accomplished by banks of these devices. The amount of heat transfer can be varied by tilting the tubes to increase or decrease the flow of liquid through the capillary action. Heat pipes cannot be "turned off" so bypassing ducting is often desirable. Efficiency of heat transfer ranges from 55% to 75%, depending upon the number of pipes, fins per inch, air face velocity, etc.

Runaround systems are also popular recuperative devices for HVAC applications, particularly when the supply and exhaust plenums are not physically close. Runaround systems involve two coils (*air-to-water heat exchangers*) connected by a piping loop of water or glycol solution and a small pump. The glycol solution is necessary if the air temperatures in the inlet coils are below freezing. Standard air-conditioning coils can be used for the runaround systems. Precaution should be used when the exhaust air temperature drops below 0°C (32°F), which would cause freezing of the condensed water on the heat exchanger fins. A 3-way bypass valve will maintain the temperature of the solution entering the coil at just above 0°C (32°F). The heat transfer efficiency of the system ranges from 60% to 75% depending upon installation.

Demand-Controlled Ventilation

To meet outside air ventilation requirements, many buildings use a fixed rate of outside air intake based on an assumed or design occupancy of the space. However, buildings often have peak occupancies below the design value, and occupancies often fluctuate widely throughout the day, reaching the peak occupancy only a fraction of the time. As a result, varying the outside air ventilation rate based on the actual occupancy of the space can yield significant energy savings. This ventilation strategy is referred to as demand-controlled ventilation (DCV). DCV uses CO_2 sensors located throughout the building to relay information on CO_2 concentrations (and, thus, occupancy) to building control systems that then adjust outside air ventilation rates by manipulating dampers in air handling equipment. The result is less ventilation during low occupancy and more ventilation during high occupancy.

DCV (including control algorithms and DCV-related equipment such as CO_2 sensors, variable frequency drives, building automation systems, etc.) can be used as a retrofit ventilation strategy to replace fixed ventilation in a variety of commercial building applications. In addition, DCV equipment and control algorithms can be readily applied in new construction (Principles 1, 2, 3, 5, and 8).

Depending on the application, implementation of a DCV strategy can yield energy savings of up to 60% or more.[24] For example, typical office buildings can expect energy savings of 10−40% (see Figure 8.15), while

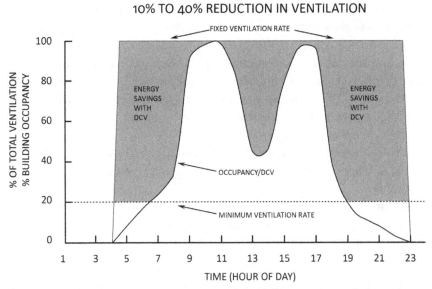

Figure 8.15 Illustration of energy savings potential of demand-controlled ventilation relative to fixed ventilation in an office building.

[24] Parmenter, K. (2006). *Demand-Controlled Ventilation (DCV) for Indoor Air Quality.* Tech Review. Lafayette, CA: Global Energy Partners.

schools and meeting rooms may experience savings of 10—50%, and restaurants and bars may see savings of 30—60%.[25]

Thermal Energy Storage

Thermal energy storage (TES) for space cooling can reduce on-peak electricity demand and cost (Principle 11). This is accomplished by operating the electric chiller predominantly during off-peak hours when electricity rates are the lowest, and storing the thermal energy so that it can be used later during on-peak hours. Thus, TES systems can significantly cut operating costs by cooling with cheaper off-peak energy, and reducing or eliminating on-peak demand charges. Although originally developed to shift electrical demand to off-peak hours, TES applications can also result in lower first costs and/or higher system efficiencies compared to non-TES applications. The lower first cost is due to the fact that the chiller size can be reduced (sometimes to half of the size) with the addition of TES. TES also enables full-load operation of the chiller over longer periods of time, which maximizes chiller efficiency. Moreover, for the case of air-cooled chillers combined with TES, the efficiency of heat rejection to the air increases during the night hours. Therefore, in addition to shifting load, TES systems may also reduce energy use.

Example. Energy simulation modeling was used to generate comparative estimates of energy consumption and demand for four cooling system alternatives, in four climate zones, and for two building types (office and hospital).[26] The cooling systems investigated were as follows:

- WCC = water-cooled centrifugal chiller.
- ACC = air-cooled screw chiller.
- ACC-FS = air-cooled screw chiller with full thermal storage.
- ACC-PS = smaller air-cooled screw chiller with partial thermal storage.

Figure 8.16 shows the demand profiles for each building type for the Miami, Florida climate zone. The HVAC demand profiles for the scenarios without TES are consistent with the hours of operation of the two building types. The hospital operates 24 h/day and the HVAC equipment

[25] DOE. (2004). *Demand-Controlled Ventilation Using CO_2 Sensors*. Federal Technology Alert. Washington, D.C.: Energy Efficiency and Renewable Energy, Federal Energy Management Program, U.S. Department of Energy. DOE/EE-0293.
[26] Parmenter, K., Arzbaecher, C., Prijyanonda, J., Johnson, R., Khattar, M., Krill, W. (2003). *Comparison of Water-Cooled and Air-Cooled Chillers: Impacts of Thermal Energy Storage*. Lafayette, CA: Global Energy Partners.

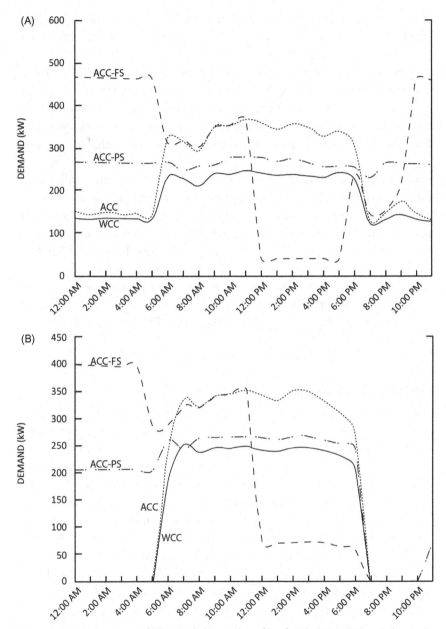

Figure 8.16 Comparison of electrical demand profiles for the Miami Hospital scenarios (a) and Miami Office Building scenarios (b).

also operates around the clock (Figure 8.16a). In contrast, the office building and its HVAC equipment only operate during the hours between 7:00 am and 7:00 pm (Figure 8.16b).

For both buildings, the demand in the WCC and ACC scenarios without TES peaks during the early afternoon hours when the outside ambient temperatures are the warmest, and then drops off to a lower level during the cooler nighttime hours. (Note that the WCC uses less energy than the ACC in both buildings because water-cooled chillers are more efficient.) In the ACC-FS scenarios, the chiller works to charge the thermal storage system during the off-peak hours and then the chiller turns off and lets the TES discharge during the on-peak hours to meet the cooling load between noon and 6:00 pm. In the ACC-PS scenarios, the HVAC demand is fairly constant during the 24-h period for the hospital and for all but the late evening hours for the office building. Therefore, the partial storage scenarios presented here do not yield a significant drop in peak demand, but they do levelize loads by running a smaller air-cooled chiller at night to charge the TES, and then augmenting its operation during the day with discharge from the TES to meet the cooling loads.

Retrocommissioning

Retrocommissioning (RCx) is a term that often comes up in discussions about improving the efficiency of building systems, particularly in regards to HVAC systems. RCx is a systematic process for optimizing the operation and maintenance of energy-using systems and controls in an existing building. The key goal is to achieve the most efficient performance of the systems while meeting the building's most current operational needs. Depending on current needs, RCx may or may not emphasize bringing the building's performance back to its original design intent. RCx is only applicable to buildings that have never been commissioned during design and construction. (Recommissioning is the term used to describe buildings that have undergone prior commissioning.) RCx most often focuses on the dynamic energy-using systems (e.g., HVAC, lighting, controls) with the objectives of identifying and repairing existing problems, increasing energy efficiency, obtaining energy cost savings, and improving occupant comfort. The operation and maintenance energy management opportunities presented in this chapter fall under the umbrella of RCx.

CONCLUSIONS

HVAC systems are major energy users in buildings. System designers now have many options for more efficient designs. There is also a natural reluctance to modify approaches that succeeded in the past. However given the new energy economics and the imperative for green, sustainable building designs, energy managers will find it useful to evaluate the operation of existing HVAC systems and to evaluate carefully proposed new designs.

In many cases major energy savings can be achieved by relatively minor adjustments to existing systems. In other cases major (and expensive) changes may be required. Strategies can take the form of:

- Operation and maintenance strategies applied to heat sources, chillers, fans, pumps, ducts and dampers, and controls.
- Modification and retrofit strategies applied to components or complete systems.
- New system design, utilizing variable air volume or other approaches.
- Heat recovery and economizer systems.

In reviewing HVAC system operations, the energy manager should first review the needed comfort level that has to be provided. Next, an understanding of the principles of operation of an existing system or of a proposed new system is required. In many cases, an analysis or simulation of the operation of the system is helpful, with the objective of determining annual energy use and determining its sensitivity to certain system changes. Modeling of systems is based on fundamental laws of thermodynamics, fluid mechanics, and heat transfer. There are many techniques for analysis ranging from simple hand calculations to computer simulations. Once the system operation is understood, energy-saving opportunities may be apparent. This chapter includes a number of examples of typical system improvements and areas to consider, ranging from control changes to system modification or redesign. Any proposed changes to HVAC systems should be checked to ensure they conform to the latest code and regulatory requirements.

CHAPTER 9

Lighting Management

INTRODUCTION

In this chapter we shall review lighting, one of the major uses for electricity in the residential, commercial, and industrial sectors. In 2012, the U.S. residential and commercial sectors used about 460 million megawatt hours for lighting. More than 50 million megawatt hours were used for lighting in the manufacturing sector (2010 data). Combined, these sectors used over 510 million megawatt hours of electricity, or about 13% of total U.S. electricity use.[1]

GENERAL PRINCIPLES OF ELECTRICAL ENERGY MANAGEMENT FOR LIGHTING

We have identified four general principles for energy management in lighting systems:

Principle 1: Optimize Capacity

The first general principle is to optimize capacity. Excess capacity, which might take the form of excessive illumination or inefficient lighting fixtures, can result in excessive electricity use and greater air conditioning loads, due to the heat dissipation of lighting systems. In the last several decades, great strides have been made in improving the efficiency of light sources. Today, illumination system designers have a wide variety of lamps to consider for lighting design projects.

Principle 2: Optimize Controls

One of the reasons for the rapid growth of electricity use is convenience and ease of control. It is logical therefore to consider optimum controls as the second area to explore in an electrical load energy management program. Timers, occupancy sensors, photocells, or switches to turn lights on only when and where needed are examples.

[1] U.S. Energy Information Administration, www.eia.gov/electricity/.

Energy Management Principles.
DOI: http://dx.doi.org/10.1016/B978-0-12-802506-2.00009-4
189

Principle 3: Use Passive Concepts

The third general principle is to eliminate unnecessary electrical lighting and makes use of daylighting; this strategy is particularly effective in combination with daylighting controls (photocells). Other passive approaches include using light-colored paint on walls and ceilings to increase light reflection.

Principle 4: Improve Operation and Maintenance

The fourth general principle is improved operations and maintenance. In addition to better controls, another way to improve the efficiency of lighting operation is to increase awareness among building occupants of the importance of energy efficiency. When it comes to maintenance, keeping light fixtures clean is essential to maintaining illumination levels.

LIGHTING SYSTEM DESIGN CONSIDERATIONS

Building designers consider lighting from two points of view. The first is that light is one of the elements of design, and is in fact an architectural medium. In this sense, illumination may be combined with building shape, form, color, and other variables to produce an aesthetic and functional structure. Light used to highlight architectural features, showcases, artwork or landscaping is often called *decorative* or *accent* lighting. Decorative lighting is of particular interest to residential structures, museums, monuments, churches, and certain other structures. Though there are fewer opportunities for energy savings with decorative lighting, some possibilities do exist. For example, Figure 9.1 illustrates small LED lanterns illuminating trees in an estate. These lamps can be photocell controlled.

Example. Decorative outdoor landscape lighting with low voltage LED lamps. LED lamps can provide outdoor garden or landscape illumination with lower power usage. Typically 6—7 W LED lamps will provide flood and spot illumination equivalent to a 50 W halogen lamp. With low voltage, there is no electrical shock risk and installation is less expensive. In a typical installation, power is provided by a power supply that converts 120 or 220 VAC to 12 VDC. The power supply has a dusk-to-dawn photo cell control and a digital timer that can be set to energize the lamps for a set period of hours. See Figure 9.2.

Figure 9.1 Decorative landscape lighting.

Figure 9.2 LED garden lighting.

The second viewpoint is to emphasize the pragmatic aspects of illumination (although not neglecting the aesthetic aspects) and to provide illumination required to accomplish visual tasks. This approach combines *ambient* lighting and specific *task-oriented* lighting to provide the required task illumination. Ambient lighting is lighting needed for general purposes, as in hallways or uniform lighting in a room area. Task lighting is the illumination required for work areas or to enable specific tasks (drafting, reading, electronic assembly, etc.) to be accomplished.

To ensure human comfort, illumination must provide more than just light. The factors that enter into acceptable illumination are:

- Illuminance and luminance.
- Glare/reflection.
- Contrast.
- Shadows.
- Color.
- Task difficulty and duration.
- Occupant age.

Light sources admit energy in the ultraviolet (low-frequency), visible, and infrared (high-frequency) portions of the electromagnetic frequency spectrum. The human eye only responds to a narrow frequency band of this energy (the "visible spectrum") and has a peak response at a wavelength of about 550 nm.

Light source strengths are measured in *candelas* (cd), originally the light output of a "standard candle." The amount of light from one candle that falls on a one square foot of area, 1 foot from the candle, is called a *lumen*. Lumens per square foot are also called foot-candles (*fc*), while lumens per square meter are called *lux*. The current definition of the lumen is the amount of light per second from a point source of 1 cd falling on an area equal to r^2, where r is the radius of a sphere with the light source at its center.

While the luminous flux emitted by lights is measured in lumens (lm), the luminous flux per unit area is called the illuminance and is measured in lumens/m^2, which is given the special name lux (lx). The foot-candle (fc) is converted to lux by multiplying by 10.76. The Dlux (dekalux, equal to 10 lux) is roughly equal (7% greater) to the foot-candle (See Appendix B for more complete discussion of units). Lux is the quantity normally measured with a light meter. Specifications for illumination of a task are given in lux. However, visual acuity, which is a measure of one's ability to perceive fine details, depends on *contrast*, which in turn is related to the *luminance* of an object and its immediate background. Luminance is a measure of the luminous flux that passes through or is reflected from an object and has units of cd/m^2. Luminance and illuminance have an important effect on the contrast with which a viewer sees a task or object.

Shadows and colors can aid or hinder vision. Also it takes time to see, so duration and difficulty of a task influences the contrast and illumination required for accurate vision. As people grow older, the lens of the eye hardens and is less able to focus on near objects. Also, with age the size of the pupil decreases, letting less light into the eye, and other changes occur which appear to reduce visual acuity.

Summing up, it is apparent that illumination alone is an insufficient criterion for lighting system design or evaluation. Adequate luminance from the task is also required for good visibility and comfort. On the other hand, excessive luminance (from lights, windows, or other sources) can reduce visibility if at a level 3 to 5 times higher than task luminance, while higher factors could even cause eyestrain, headaches, or discomfort.

Recognizing the limitations inherent in basing lighting system designs purely on illumination, the current trend is to evaluate lighting systems for both *quantity* and *quality* of illumination. The new design processes for various types of buildings and outside areas are described in considerable detail in the publications of the Illuminating Engineering Society of North America (IESNA).[2]

The first step is to choose acceptable illuminance levels according to categories established by IESNA. In the IESNA publications, there are seven illumination categories for building interiors designated A, B, C, D, E, F, and G. Associated with each of these categories is a suggested illuminance value in both lux and foot-candles. These range from category A (public spaces with dark surroundings), with a recommendation of 30 lx, to category G (performance of visual tasks of low contrast and very small size over a prolonged period of time), with a recommendation of 3,000–10,000 lx. As an example, for category E, which involves performance of visual tasks of medium contrast or small size, the recommendation is 500 lx. If this recommendation was applied to a college laboratory, the selected value might be reduced to 400 lx, but for a workshop in a senior center, 600 lx might be more appropriate. In addition to stipulating recommended illuminance categories for hundreds of different types of occupancies, IESNA publications also provide guidelines on lighting characteristics that are "very important," "important," "somewhat important," or "not important or not applicable." These evaluations are

[2] Laura, David D., Houser, Kevin, Mistrick, Richard, and Steffy, Gary, Eds. (2011). *The Lighting Handbook: References and Application.* 10th ed. New York: The Illuminating engineering Society.

applied to parameters that include appearance of space and luminaires, color appearance, direct glare and flicker, light distribution, peripheral detection, reflected glare and shadows, and vertical versus horizontal illuminance.

Thus in applying the illuminance recommendations, one needs to be cognizant of the visual size of the task, night versus day, the age of occupants, and dynamic versus static illumination. Illumination is said to be dynamic when it involves a combination of daylighting and electric lighting to provide the required level.

From the foregoing it is apparent that the previous system—which involved specifying design illuminance levels for hundreds of different types of occupancies—has changed dramatically. The recommendations are now general and more limited, with a much greater dependence on the designer's judgment and evaluation of the illumination requirements. However, to achieve the goal of energy efficiency, standards have now been developed for lighting power densities (also called *lighting power budgets*), expressed in watts per square meter. Lighting power densities range from 5.4 W/m^2 to 21.5 W/m^2 (0.5 up to 2.0 W/ft^2).

LIGHTING REGULATIONS, CODES AND STANDARDS

Because lighting energy use is a significant component of total energy use, many new regulations, codes, and standards have been promulgated to govern the design and operation of lighting systems. As noted above, standards based on lighting power densities stipulate allowable power budgets, but today's designers have more freedom to select lamps, luminaires, lighting controls, and illuminance levels as long as their design does not exceed the power budget. New standards also specify when certain types of controls are mandatory (for example occupancy sensors shall be installed in all classrooms, conference/meeting rooms, employee lunch and break rooms and so on for commercial buildings). The basic codes and standards are National Electrical Codes, the International Energy Conservation Code, or ANSI/ASHRAE/IES Standard 90.1-2013.[3,4,5]

---

[3] www.energy.ca.gov/title24/2013/standards/.

[4] Anon. (2011). 2012 International Energy Conservation Code. Country Club Hills, IL: International Code council, Inc.

[5] ANSI/ASHRAE/IES Standard 90.1-2013: Energy Standard for Buildings Except Low-Rise Residential Buildings and ANSI/ASHRAE/IES Standard 90.2-2007: Energy Standard for Buildings Low-Rise Residential Buildings, http://www.ashrae.org/standards/.

Local building codes also have rules for lighting system design. Local jurisdictions may spell out more restrictive requirements. For example, California has pioneered energy efficiency regulations through its Title 24 series, first adopted in 1978 as part of the California Building Standards Code. These standards have been updated every 2—4 years. The current edition has been modified to reflect California's initiatives to reduce greenhouse gases and global warming, and to encourage green building construction.[3] Sometimes local codes incorporate provisions drafted by the IESNA or by ASHRAE. They may mandate the use of specially labeled lamps or luminaires, such as UL or ANSI approval, or *Energy Star* rated.

OVERVIEW OF LAMP TYPES

Incandescent

The earliest electric lamps were *incandescent*, whereby the electric current heated a carbon filament in a vacuum and the hot filament provided light. These were improved by using a tungsten filament in an evacuated glass bulb containing a trace of inert gas. In a typical incandescent lamp about 5% of the energy is converted to light in the visible spectrum, while the balance is reradiated as infrared energy and heat. The glass bulb of an incandescent 100 W lamp can reach temperatures of 200—250°C during normal operation. Over a period of time, the filament of the lamp gradually evaporates, reducing its diameter and therefore increasing its resistance. This also reduces the light output to about 80% of rated lumens at the end of rated life.

Fluorescent

The next generation of lamps was *fluorescent*, in the form of a glass tube with an electrode at each end. The tube contains mercury vapor, a small amount of inert gas, and fluorescent powders lining the walls of the tube. When a sufficiently high voltage is applied, an arc is produced by current flowing between the electrodes and through the mercury vapor. This discharge generates some visible radiation, but mostly ultraviolet radiation. It is the ultraviolet that causes the phosphors to fluoresce. The phosphors and other additives are incorporated in the lamps to modify the spectrum of emitted light to accentuate certain colors or to improve efficiency in the visible spectrum. These lamps require a ballast, which is basically an inductor to provide the high initial voltage pulse to initiate the discharge.

The efficacy of the lamp increases with the arc length. *Compact fluorescents* solve this problem by wrapping the tube in a spiral shape. The life of a fluorescent lamp is determined by the rate of loss of the electron emitting material on the electrodes. This is influenced by the number of times lamps are started. Typical life ratings are 7,500–12,000 hours (vs. 750–1000 h for incandescent lamps). The average life data for fluorescent lamps is based on the assumption of 3 h of operation per each start. In a typical fluorescent lamp, 21% of the input energy is converted to light, 37% infrared, and 42% dissipated as heat. Early fluorescent lamps were designated as T-12 (diameter is 12/8 inches, or 1.5 inches) and used a magnetic ballast. Beginning around 1980 these have been largely replaced by T-8 lamps (diameter 8/8 inches, or 1 inch) that use an electronic ballast. The electronic ballasts are smaller, run quieter, produce less heat, and increase light output and efficacy. More recently T-5 lamps (diameter 5/8 inch), which have higher light quality and longer life are being used. However, they are more expensive and this needs to be considered.

Mercury Vapor

High intensity discharge (HID) lamps include mercury vapor, metal halide, and sodium lamps. Light is produced by the flow of electric current through a metallic vapor. In *mercury vapor* lamps, there are typically two main electrodes and a starter electrode. When the voltage is applied to the starter electrode, argon gas is ionized and an arc is formed. This arc vaporizes the mercury and eventually an arc occurs through the mercury. Once the arc is extinguished, it cannot be reinitiated until the vapor pressure is lowered to a point suitable for the applied voltage. This typically takes from 3 to 8 minutes. In the mercury vapor ballast, the inductor provides an "inductive kick" to help initiate the discharge and also limits current through the lamp. A capacitor is used to correct the power factor of the inductor.

Metal Halide

Metal halide lamps are similar to mercury vapor lamps except that they contain various metal halides. When the lamp reaches operating temperature, the metal halide disassociates into the metal plus halogen. This has several advantages. First, the efficacy of metal halide lamps is 1.5−2.0 times as great as mercury vapor lamps. Also, the metals permit "white light" to be produced with improved color rendition. Although the construction details and ballasts used for metal halide lamps are different from mercury vapor lamps, the basic concept is the same.

Sodium Lamps

Sodium lamps operate on the principle of an electric current flowing through sodium vapor. In the high-pressure sodium lamp, energy is radiated over a band of wavelengths. In the low-pressure lamp, the light is almost monochromatic, consisting of two lines at 589 and 589.5 nm. The high efficacy of low-pressure sodium lamps has led to their widespread use for street and area lighting. Sodium lamps require special ballasts, capable of providing high voltage for initiating the arc. However, the basic principles are similar to those described above.

Light Emitting Diodes

Light Emitting Diodes (LEDs) resemble conventional diodes, which are designed to allow current to flow in one direction and not in the opposite direction. In a diode, free electrons are driven across the diode interface (from the negative to positive side of the junction) by a potential difference. When an electron fills a vacancy (called a "hole") on the positive side, it drops to a lower energy state by emitting a photon. In conventional diodes, the energy change is small and no visible light is produced. In an LED, the energy drop is greater and a higher energy photon is emitted. By selecting the materials used, different frequency photons (and thus different colors of light) can be produced.

LEDs have a number of advantages compared to other light sources. The first advantage is that they have a higher light output in lumens compared to incandescent or compact fluorescent lamps and they are starting to rival the efficacy of HID lighting technologies. Projections indicate they will soon surpass all lighting technologies in terms of efficacy. For example, depending on light output, an LED will produce the same amount of light as an incandescent lamp, but uses only 1/5 to 1/10 the amount of power required by the incandescent lamp. Compact fluorescents, for a given light output, use anywhere from 1.5 to 3 times the amount of power as an LED. Other advantages are the fact that LEDs are very durable, they emit little heat, they can be cycled on and off frequently with no effect, and they are extremely long-lived, 25,000–50,000 h. The main disadvantage to date has been cost; LED lamps cost from 5 to 20 times more than a compact fluorescent or incandescent. However, the costs have been declining rapidly as the types of LED lamps increase.

Today LEDs are available in a wide variety of forms and are finding much broader applications. In addition to the wide range of styles to replace screw-in incandescent and compact fluorescent lamps, LED reflected flood lamps in the PAR format are available. They require 7.5–17 W, produce much less heat, and last 10 times longer than compact fluorescents. Today there are LED lamps available as replacements for 4-ft T-8 fluorescent lamps. The LED lamps are more efficient (typically 15–22 W vs. 32–40 W for the T-8) and have longer lives, typically 50,000 h. They also contribute less heat to the air conditioning load. The disadvantages are that they are more expensive and have lower light outputs (1500–1900 lumens vs. 3000). Light distribution may be a problem depending on the type of troffer. However, improvements are occurring rapidly and they should be considered. LED lights are also replacing HID products. For example, today traffic signals increasingly use LED lamps for the familiar red, orange, and green lights we see at traffic intersections. Besides requiring less power, LED traffic lights have longer lives and require less frequent maintenance and replacement.

Efficacy Comparison

The various lamps types described above have efficacies in the range of 5–10 lm/W to upwards of 150 lm/W. Typical efficacy ranges are as follows, but some products may have higher or lower values (units of lm/W):
- Incandescent, 5–20.
- Mercury vapor, 20–50.
- Compact fluorescent, 55–70.
- LEDs, 60–100.
- Conventional fluorescent, 30–70.
- High efficiency fluorescent, 85–100.
- Metal halide, 45–95.
- High pressure sodium, 45–110.
- Low pressure sodium, 100–150.

ENERGY MANAGEMENT OPPORTUNITIES IN LIGHTING SYSTEMS

Following are specific techniques for meeting lighting energy budgets in existing or new facilities. Even in facilities where energy management programs have been implemented, additional savings are most likely possible, due to advances in efficient lighting technology. There are at least a dozen practical techniques for reducing energy use in new and existing facilities (See Table 9.1).

Lighting Survey

The first step in reviewing lighting electricity use is to perform a lighting survey. An inexpensive hand-held light meter can be used as a first approximation. However, the raw measured levels of light intensities (lux or fc) are only part of the story. Illumination quality is another important element. The lighting survey will give a good first indication of areas where light levels are potentially excessive or inadequate.

Delamping

This energy management opportunity (EMO) is an example of Principle 1 and is appropriate in installations where excessive illumination occurs. Excessive illumination could be the result of several causes:

- Overdesign.
- Changed building purpose or occupancy.

Table 9.1 Energy management opportunities and lighting systems

1. Delamp (reduce lighting levels if excessive).
2. Disconnect ballasts where lamps have been removed.
3. Relamp (use a lower wattage lamp).
4. Use more efficient lamps: replace incandescent with fluorescent; replace fluorescent with metal halide or sodium; substitute LED lamps for any of the preceding types.
5. Improve controls with time clocks, photocell controls, or occupancy sensors.
6. Use zone switching or install more switches.
7. Use task lighting.
8. Make effective use of day lighting where possible.
9. Use more efficient lenses to reduce losses.
10. Clean luminaires to increase illumination.
11. Improve color and reflectivity of walls, ceilings, floors to reduce lighting energy needs.
12. Train and motivate personnel to turn lights off when not needed.

- Changed illumination requirements.
- Changed illumination standards.

Current illumination standards should be reviewed, since the older standards typically called for higher illumination levels. A reduction of illumination levels without regard to production or employee morale is not recommended. A loss of productivity will more than wipe out savings in energy. One study of computer operators indicated that a loss of 12% of production output resulted from reducing the lighting from 150 fc to 50 fc. Since the cost per square meter per year of working people is several orders of magnitude greater than the cost of working lights per square meter per year, far more could be lost than gained. However, in many areas such as unoccupied spaces or storage rooms, perhaps lighting levels can be reduced. The key is to use common sense!

Disconnect Ballasts

Ballasts dissipate energy in much the same manner as a transformer. The energy is lost in the form of heat given off by the windings and metal of the ballast. Ballasts can account for 10–20% of total energy use in a lamp. When the lamp is switched off, the ballast does not use any energy. However, it is frequently the case that some lamps will be removed from the circuit during a delamping program, while others will remain in the circuit. In this situation, when the circuit is energized, the ballast will still dissipate energy even though the lamps have been removed. To obtain further savings, disconnect the ballasts by cutting the connecting wires and then insulating them (Principle 1).

Low Wattage Lamps

In some cases a reduction in lighting energy use can be achieved simply by switching to lamps that require less energy, using the existing fixtures. This is another example of Principle 1. Along with delamping and disconnecting unused ballasts, switching to lower wattage lamps in existing fixtures is generally one of the easiest, lowest cost, and quickest improvements to implement.

For example, retrofitting incandescent lighting with compact fluorescents offers significant cost savings to the user. Money saved through reduced energy use, fewer lamp replacements, longer lamp life, and related maintenance savings can quickly recoup initial investment and provide continuing operating cost savings. Additionally, utility rebates often partially offset initial retrofit costs.

The average compact fluorescent lamp consumes only one-quarter to one-third as much energy as its incandescent counterpart and will last up to ten times longer. Incandescent lamps typically operate 1000–2000 h; compact fluorescent lamps last approximately 10,000 h.

Compact fluorescent lamps have good color rendering properties as measured by their Color Rendering Index (CRI). CRI is a measure of how well lamps illuminate colors. They are also available in a variety of sizes, shapes, and wattages.

Compact fluorescent lamps are actually lighting systems consisting of a lamp (often with a starter integrated into the base), a lamp holder, and a ballast. Sometimes, a screw-in socket adapter is incorporated into the package. Generally, there are two different types of compact fluorescent lamp-ballast systems:

- Integrated systems are self-ballasted packages and are made up of a one-piece, disposable socket adapter, ballast, and lamp combination.
- Modular systems are also self-ballasted packages, consisting of a screw-based incandescent socket adapter, ballast, lamp holder, and replaceable lamp.

When replacing lamps without integral ballasts (e.g., linear fluorescent lamps), make sure replacement lamps are compatible with existing ballasts.

Today many buildings will have undergone lamp replacement, meaning that the "low hanging fruit" of easy changes has already occurred. Further improvements will require more extensive, and expensive, lighting system changes. These may require redesign and replacement of fixtures. In this situation economic cost comparisons become important determinants of the decision to make changes. Including the cost of maintenance and lamp replacement is also important.

Relamping

Previously, building designers had limited choices when it came to selecting lamps. Today there are hundreds of different light sources available for various uses. In new construction, use of newer lamps and luminaires with improved ballasts is cost effective because they are more efficient, have longer lives, and produce less heat. When relamping in existing buildings, it is important to give consideration not only to the cost efficiency of the new lamp and luminaire, but also to the cost of installation and the savings achievable.

Example. Retrofitting T12 fluorescent lamps and ballasts with more efficient T8 lamps and electronic ballasts decreases energy usage and improves light quality. T8 lamps have a smaller diameter and use better coatings to produce greater light output with lower wattage input than T12 lamps. T8 systems also provide the advantage of a greater CRI over standard systems. T8 lamps have a CRI of 75, while warm white and cool white T12 lamps have CRI values of 53 and 67, respectively. Thus, T8 lamps and electronic ballast produce light that is more energy efficient, and that makes colors appear more pleasing.

Example. Retrofit with high pressure sodium lamps. Consider a 100 foot by 50 foot by 20 foot laboratory area which is currently being lit to 30 fc by 500 W incandescent lamps. There are currently 27 units with a total power requirement of 13,500 W. An equal or greater light intensity could be realized by the retrofit of only ten 250 W high-pressure sodium lamps. Their total wattage (including ballast) would be 2900 W. At 24 h use per day, this represents an annual energy savings of 92,856 kWh or US$9,286 per year at an energy cost of US$0.10 per kilowatt hour.

An extended service 500 W incandescent lamp has an average life of 2500 h at an average use of 24 h/day. This represents 3.5 lamp changes per year. The 250 W high-pressure sodium lamp has an average life of 24,000 h or 0.365 lamp changes per year. The labor cost to change a lamp is US$20 per fixture. Table 9.2 is the calculation of the annual savings just for reduced maintenance due to longer lamp life. Total savings, energy plus maintenance, is US$11,503/year. Assuming an installed cost of US$500 per fixture, this change would have a simple payback of less than 6 months.

Table 9.2 Calculation of maintenance labor savings for sodium lighting example (US$)

BEFORE (incandescent lamps)	
27 fixtures × 3.5 changes × $20 labor	= $1890.00
27 fixtures × 3.5 changes × $5 lamp cost	= $473.00
Total annual maintenance cost	**= $2363.00**
AFTER (high-pressure sodium lamps)	
10 fixtures × 0.365 changes × $20 labor	= $73.00
27 fixtures × 0.365 changes × $20 lamp cost	= $73.00
Total annual maintenance cost	**= $146.00**
SAVINGS (sodium vs. incandescent)	
$2363−$146	= $2217.00/year savings

Use the Most Efficient Light Source

The light intensity in lux at a given work surface is a function of several variables. First is the light output from the lamp. Next is the efficiency of the lighting fixture, luminaire, or troffer, since this determines what fraction of the light output is directed toward the work surface and what fraction is reflected or absorbed elsewhere. Other factors include the reflectance of ceiling, walls, and floor in the room as well as their colors. Selecting the most efficient light source involves more than just lamp choice; these other variables can have an important effect on the result. Using the most efficient light source, with consideration to quantity *and quality* is another example of Principle 1.

One of the most common lamps world-wide is the 60 W incandescent A-19 lamp. However, many governments have initiated regulations to stop production and phase out these lamps. Today, there are many replacements available that not only save energy but have reduced life cycle costs compared to the venerable 60 W incandescent lamp. (see Text Box 9.1).

Example. LED lamp replacement. High-efficiency LED lamps are available to replace any traditional incandescent lamp. While the LED lamps have a higher initial cost, they have the advantage of much longer lifetimes, lower power requirements for a given lumen output, and produce virtually no heat. A comparison follows for R-20 indoor flood lamps, 6.35 cm (2.35 inch) in diameter:

Type	Watts	Lumens	Life (years)	Cost (US$)	Cost/year
Incandescent	45	385	2.2	$3.50	$4.93
Halogen	45	490	2.7	$5.50	$4.93
LED	12	830	22.8	$16.00	$1.31

This calculation assumes 3 h/day and energy cost of US$0.10/kWh.

From these data it is apparent that the 22.8 year lifecycle cost, including lamp replacement cost, is US$46 for the LED lamp, US$149 for the incandescent lamp and US$159 for the halogen lamp. During its lifetime, the LED lamp will use 824 kWh less (at 3 h/day) when compared to the incandescent or halogen lamp, while producing twice as much light. In new construction, the added benefit is that half as many light fixtures would be required for equivalent illumination. Even more striking, over its lifetime this simple lamp change will save more than *half a ton* of greenhouse gas emissions compared to its incandescent

TEXT BOX 9.1 60-W Lamp Life Cycle Cost Comparison

Type (A-19 lamps)	Watts	Equivalent Watts	Initial Lumens	Life Hours	Lamp Cost	Energy Cost $/yr	25,000 hr kWh	25,000 hr cost	Lamp Re place cost	Life Cycle Cost*
Incandescent	60	60	840	1000	1.62	$6.57	1500	$150.00	$40.50	$190.50
High efficiency incandescent	43	60	750	1000	2.5	$4.71	1075	$107.50	$62.50	$170.00
Halogen	43	60	750	2000	1.51	$4.71	1075	$107.50	$18.88	$126.38
Compact fluorescent	14	60	850	8000	6.45	$1.53	350	$35.00	$20.16	$55.16
LED	9.5	60	800	25000	7.87	$1.04	237.5	$23.75	$7.87	$31.62

NOTES: Energy cost is US$0.10/kWh; 3 hours use/day, excludes lamp replacement labor cost.

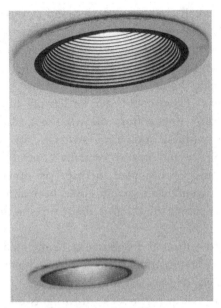

Figure 9.3 Halogen (top) and LED (bottom) R-20 lamps.

predecessor. Figure 9.3 shows the halogen lamp on the top and the LED lamp on the bottom. Due to the angle of the photograph, the halogen lamp appears brighter, but this is not the case.

Improve Lighting Controls

Application of Principle 2 consists of improving lighting controls by various means. The purpose of lighting controls is to turn on, turn off, or dim lamps in accordance with the need. This can be done manually, on a schedule, by sensing daylight, by sensing the presence of occupants, or a combination of these. The simplest and least cost measure is improving existing manual controls. However, advanced automated controls often offer greater potential for energy savings.

A system of on-off switches on individual or specific groups of light fixtures is called "selective switching." This enables an individual to turn off specific light fixtures when no one is using that particular area. This is an essential element in lighting energy management. Of course, the switches are useless unless people use them. Therefore, put switches in convenient, easily accessible locations such as near the doorway or entrance to the particular area they control. Carefully design the particular fixtures that each switch controls to correspond to specific task areas that may not be used for periods of time within a larger area.

Another important use of selective switching is in conjunction with natural light through windows and skylights. An office, for example, with windows on one side, should have separate switches for the light fixtures nearest the windows. These could then be left off during the day while using only the fixtures distant from the windows. Many older buildings were constructed without individual switches for light fixtures. In these cases, a single circuit breaker typically controls groups of rooms. When only one office is occupied, perhaps six others may be lit but empty. Obviously, a significant savings could be realized with individual switches, as long as people are diligent about turning the lights off when they leave.

Another way to ensure the lighting is being used only when it is required is by using automatic switching. This is typically either by *time clocks*, *occupancy sensors*, or *photo cell* control.

Simple time clocks can be preset to turn lights on and off at specific times of the day and for certain days of the week. They are a low cost and effective solution for spaces with predictable occupancy, but lack the sophistication and customization of more advanced controls. These types of on/off devices are applicable in indoor and outdoor applications. In a parking lot, for example, where the final shift leaves at midnight, timers are an efficient solution. This type of scheduling can also be programmed into energy management and control systems for building wide lighting control.

Better yet, *occupancy sensors* automatically turn off lights when a room is unoccupied for a certain period of time, which saves even more energy than time clocks. According to the U.S. Environmental Protection Agency, savings of up to 50% are typical in office areas. These devices operate by sensing either the presence or absence of a person and then sending a signal to a relay to close or open the light circuit accordingly. They are best in spaces with intermittent or unpredictable occupancy, such as restrooms, hallways and stairwells, closets, private offices, conference rooms, lunchrooms, classrooms, etc. Sensors come in many different styles. For the sensors to work properly, they must be mounted in a location where they will accurately detect occupancy. Typical styles include both ceiling and switch mounted units. Sensors normally feature adjustment settings for activation time delay as well as override control.

Figure 9.4 Combination light and occupancy sensing light switch.

Photocell controls are light-sensitive devices that turn fixtures on and off according to the intensity of existing natural light. They can be used indoors in conjunction with on/off switching or dimming controls to take advantage of natural light near windows or skylights. This approach is called *daylight harvesting* or *daylighting* and is discussed in greater detail in the daylighting section below. Another possibility is a combined occupancy sensor/photocell control switch. (Figure 9.4).

Photocells are also well-suited to outdoor, garage, and security lighting and have some advantages over timers in these applications. For example, seasonal changes in the time of sunset and sunrise require periodic readjustments of clock timers; this is unnecessary with photocell controls. Security lighting (outdoor lighting, parking lots, etc.) can be a major energy user. A single extra unneeded hour of security lighting because of non-adjusted clock timers can lead to hundreds of unneeded kilowatt hours in a large facility. Energy audits often reveal outdoor lighting that is energized during daylight hours.

Dimmers, either manually operated or used in conjunction with light level sensors, can also lead to energy savings and they also help align lighting levels with human needs. Dimming controls reduce the output of lighting systems in either a step-wise or continuous fashion. Though dimming controls are available for many types of lighting, some dimming controls require dimmable ballasts. Other dimming controls may rely on electronic packages installed in the panel board or elsewhere within the lighting system.

In our experience, advanced lighting control systems that incorporated daylighting, occupancy sensors, and automatic dimming, typically reduce lighting energy use by 30—40%.

Figure 9.5 Mezzanine storage area lamp.

Night audits are also a useful tool for reviewing lighting energy use. A night audit, or on the weekend, (when a building is not occupied or not in production) will reveal lights that are left on at night in unoccupied areas. For example, inspection of a manufacturing plant on the weekend showed a lamp that was operating 24 h/day in an unoccupied mezzanine storage area. During the work day when the plant was illuminated, it was difficult to tell that this lamp was on. (See Figure 9.5)

Finally, the rapid expansion of the internet, Wi-Fi, tablet computers and "smart phones," has created entirely new channels for lighting control. Today an individual on vacation can remotely turn off lights that were accidentally left on, or upon returning, can activate security lighting before approaching or entering a building. Beyond a doubt we will see many new control strategies during the coming years.

When considering advanced lighting controls, each case must be examined separately to determine cost effectiveness as well as applicable codes and standards. As mentioned previously, new standards in some jurisdictions specify when certain types of controls are mandatory. Therefore, occupancy or other controls may be required in major renovations or new construction.

Use Task Lighting

The major determinant of luminaire efficiency is its location in relation to what needs to be seen. Artificial light intensity decreases exponentially as the distance from the light source to the task increases. To achieve optimum efficiency keep the light sources as close as reasonably possible to the work plane. This is especially true with work requiring very high illumination levels such as technical drafting, bookkeeping, or fine assembly work. Fortunately, these tasks require relatively small working areas and specific *task lighting* is appropriate. Localized light sources such as a flexible arm lamp can be used at the work, provided the required higher illumination levels for surrounding areas, lit by ceiling fixtures, are at lower levels. Energy auditors have encountered entire rooms that were lit at levels of 215 Dlux (200 fc) to provide intense light for a single desk. These and many other less dramatic examples would benefit from a task lighting approach (Principle 1).

There are some specialized areas that do require a uniform light level over an entire room, such as some offices where a large surface area is being used simultaneously. Generally, however, nonuniform illumination is appropriate in offices. For example, relocate ceiling fixtures to provide high illumination over desks and work areas and significantly less illumination over walkways, filing areas, etc. In many instances simply removing lamps and fixtures from certain areas of the room (i.e., delamping, as discussed above) and leaving others intact optimizes task lighting while minimizing energy use.

Another approach applicable in some cases is to change the type of fixture, reflector, or diffuser to more carefully control the focus of the light and, again, only put the light where it is required. Because this increases efficiency, it may significantly decrease the number of fixtures or wattage of lamps.

The effectiveness of task lighting can be equally true in the case of outdoor lighting. For example, for security fences, select a fixture that emits a narrow beam appropriate to the fenced area requiring illumination. This beam shape, coupled with brighter, more efficient lamps, may reduce the number of fixtures required by as much as one half.

Task-oriented illumination is non-modular producing a variable lighting fixture pattern with fewer luminaires. This characteristic can be employed to create an interesting environment, especially when augmented with supplementary highlighting of interesting spaces (meeting places, sculptures, and display areas). The common notion that task-oriented lighting is inflexible for changing furniture arrangements can be refuted by considering the following commercially available techniques:

- Flexible above ceiling distribution systems.
- Multiple lamp types.

- Furniture with integrated lighting.
- Interchangeable ceiling lighting patterns.

Task-oriented lighting has a further advantage in that it creates an awareness of light used by the user and provides an automatic stimulus to extinguish lighting when the task is complete.

Make Effective use of Daylighting

For centuries buildings were designed to permit the use of natural light and therefore applied Principal 3 (*use passive concepts*). The convenience of artificial light has slowly modified this practice, but today we see a trend back in the other direction. Daylighting can be incorporated in office buildings easily in the perimeter areas. In the core areas, it requires special building designs, but is feasible. One example is the so-called "light tube" or tubular skylight. They consist of a transparent lens on the roof that is attached to a tube or duct that leads to the interior space. The tube may be lined with highly reflective material. It directs daylight into the space and a diffuser spreads the light into the room. In single-story buildings, skylights can be used. Daylighting is also practical in many industrial facilities. Chapter 12 discusses passive design techniques for optimizing the availability of natural light.

Buildings lit by daylight use substantially less energy while the occupants reap the benefits of natural lighting. A well-designed daylighting application may realize annual energy cost savings of 20–30% compared to buildings without daylighting design or controls. Daylighting coupled with controls can maximize peak electrical demand savings from lighting systems operating at reduced power and the associated reduction in the building's cooling load. For many businesses, the demand savings can be significant since the hours of greatest daylight availability often coincide with the peak electricity demand period.

In addition to energy savings, there are many other benefits of daylighting. Studies have shown that there is a direct relationship to sunshine and human well-being and comfort levels. Natural light provides the best CRI. The sun produces light that makes colors appear more pleasing. New technologies have emerged that increase the opportunities to apply daylighting technologies cost-effectively. Sensors and electronic dimming ballasts provide distributed control that operates independently to detect conditions in small areas and controls the light from the light fixtures in that area.

In a daylighting design for fluorescent lighting, a photosensor reads the ambient light. If the daylight is sufficient to light the space, the connected electronic dimming ballast is signaled to reduce light output to its minimum value. As weather conditions change, the photosensor output will signal the dimming ballasts to increase the amount of light from the fluorescent fixtures to supplement the daylight.

Multiple dimming ballasts connect directly to each sensor. This control system produces a fade feature which ensures that lighting is not altered too abruptly. The dimming ballasts can dim down to 10%. The effectiveness of daylighting controls depends on the careful place-ment of photosensors and the proper calibration of controls during the commissioning phase of the installation.

Areas utilizing HID type lighting can also successfully achieve energy reductions with daylighting controls. Lighting can be dimmed either by a high-low step dimming system or a continuously variable system. The lower limit of the control system is determined by the lamp type. Metal halide lamps are not usually dimmed lower than 50% to avoid unstable lamp operation.

Example. Maximizing the use of natural light in an industrial facility. Using natural light led to substantial energy savings in a major industrial facility. A new, large equipment fabrication shop (30 m wide, 80 m long, with a 12 m high ceiling) was constructed. A comparison was made between conventional lighting and conven-tional plus daylighting. Daylighting saved 38% of the lighting energy compared to the conventional approach by using the following strategies:

- Large fiber glass skylights were installed in the roof. Two rows of these ran the length of the building.
- Natural light was also admitted by providing a 7.3 m high opening the entire length of the south wall of the shop.
- Electric lighting was controlled by photocells located outside the building.

The shop was designed for illumination of 21.5 DLux (20 fc), which was provided by 50 overhead metal halide lamps, each rated at 426 W (lamps plus ballasts). On the average, daylight provided adequate illumination for six out of 16 working hours (38%), saving approximately 50 fixtures $\times$ 426 W $\times$ 250 days per year $\times$ 6 hours per day $\times$ US$.10 per kilowatt hour = US$3195 per year.

Use More Efficient Lenses to Reduce Losses

Lamps are typically shielded by shades, louvers, or lenses to distribute or to direct the light and to reduce glare. These devices absorb some fraction of the light output; in extreme cases, more than 50% of the light can be stopped by the lens. The light transmission of the lens when new varies with the type. For example, a clear plastic lens will transmit 45–70% of the light; polarizer or diffuser-type lenses transmit 40–60%; white metal louvers, 35–45%; dark metal louvers, 25–40%. The performance of the lens degrades with time due to aging and dirt buildup.

Plastic lenses are susceptible to yellowing or darkening that eventually reduces the light output substantially. If this situation is observed, it may be economical to replace the lenses, which will increase the light output, and then reduce the number of lamps. Cleaning or replacing the lens is an example of Principle 4.

Clean Luminaires to Increase Illumination

Dirt and dust collect on the luminaires over a period of time, causing attenuation of light transmission. Depending on the type of luminaire, light output can be decreased under typical operating conditions to as little as 60% of the clean value in the course of 12 months. Several benefits can be obtained, simply by cleaning light fixtures:

- More light delivered: removing dirt increases light output. This may increase productivity, and it will bring illumination back to design levels. If illumination is excessive, cleaning means fewer lamps to do the job and some lamps can be eliminated.
- Better appearance: clean lighting systems improve the appearance of a work area or display area. This is conducive to improved employee morale and better housekeeping.

In summary, clean fixture lenses or diffusers at 12-month intervals or even more frequently in high dust or dirt environments (Principle 4).

Improve Room Color and Reflectivity

Since darker colors absorb light, the actual illumination levels in the room are affected by the color and reflectivity of walls, ceilings, and floors. This can be tested by the simple technique of taking a light source into rooms of different colors and measuring illumination levels at similar distances from the source. Illumination can often be improved in by cleaning the walls or repainting them with lighter colors. Incorporating light-colored, reflective surfaces in building and interior designs is an example of Principle 3.

Create an Energy Awareness Program

Many of the devices and energy management techniques recommended in this section are automatic, utilizing switches, photocells, or other controls. However, another equally important aspect of controlling energy usage is the human element (Principle 4). This begins with a basic awareness on the part of each individual of the uses of energy. Within buildings, energy is used for heating, cooling, and ventilating; however, lighting is one of the most visible and obvious energy uses.

Lighting is an area in particular where energy savings can be realized by conscientiousness on the part of individuals. User carelessness can negate the energy efficiency of the most sophisticated lighting system. Ideally, artificial light would focus only on the area where a specific task is being performed and would be energized only while it is being performed. Through the use of carefully designed and placed fixtures and readily accessible selective switching, a situation approaching this ideal can be realized. The key element, however, is the individual using the system. A conscientious effort on the part of the user to use light only when and where it is required can result in important savings.

A carefully planned energy awarenes program should be considered for any business with a large lighting load and many employees. It should stress the importance of energy conscientiousness in day-to-day activities. Conspicuous posters and signs on light switches and machine and appliance controls have proven to be effective. Also, consider the following:

- Distribute pamphlets on energy management and the importance of lighting.
- Give short seminars for managers and supervisors and ask them to pass information to their respective employees.
- Study and implement suggestions and recommendations from employees themselves.

The point can also be made that energy is equivalent to money and jobs. Finally, a major point of any awareness program should be to dispel the notion that energy use by lighting systems is insignificant. See Chapter 14 for more information on training and motivating personnel to achieve energy management objectives.

Heat and Light

Lighting systems deposit heat in buildings. In most cases, all of the heat goes into the building. The exception is where special cooling is provided to remove some of the heat. The heat from lights will be partially deposited in conditioned spaces, and partially deposited in unconditioned spaces (such as an attic or ceiling plenum), depending on the design of the system. In cases where lights are on walls or ceilings and all the heat is released to the conditioned space, the heat gain is:

$$E_L = Q_L t \qquad\qquad [9.1]$$

where

E_L = Heat added by lights, J
Q_L = Operating lighting load, W
t = Time lamps are in operation, s

This additional heat is a benefit during the heating season and a detriment during the cooling season. Credit may be taken for reducing the heating load. The additional cooling load may be calculated approximately by:

$$E_c = E_L/COP \qquad\qquad [9.2]$$

where

E_c = Cooling load caused by lights, J
E_L = Heat added by lights, J
COP = Average coefficient of performance of the air-conditioning system

Under typical conditions the COP is approximately 2.0. This signifies that every watt of lighting in air-conditioned space adds the requirement for 0.5 W of cooling. Under full load conditions and with newer, more efficient air conditioning systems, the additional requirement may be closer to 0.3 W of cooling.

Building simulations of various prototypical commercial and industrial facilities have shown that installing energy efficient lighting often results in a net annual reduction in HVAC energy use because of the greater share of space cooling in these types of buildings relative to space heating. However, the net impact depends on climate, building type and function, and lighting and HVAC system design. For example, simulations of several types of commercial and industrial facilities with lighting efficiency improvements were used to estimate the interactive effects of the efficient lighting with HVAC systems for IECC Climate Zone 5. The results showed that the incremental impact on energy savings from the efficient lighting projects due to interaction with the HVAC system ranged from −9% to 14%.[6] That is, in buildings where space heating represented a greater share of HVAC energy use (e.g., in schools that were closed in the summer), the HVAC interaction lowered energy savings from the lighting project by 9% compared to a case of no interaction. However, for facilities like office buildings and manufacturing plants with year-round operation, the HVAC interaction had a positive effect on energy savings, typically increasing saving for the lighting project by 10% (and up to 14%) relative to a case of no interaction.

Example. Lighting energy management opportunities at Punahou School. We refer again to the energy audit conducted by the authors and their team at Punahou School that is described in Chapter 6. The second greatest use of energy at Punahou School is due to lighting. Based on the building-by-building energy audits and the database constructed by the audit team, we determined that there were 11 measures that could be deployed to reduce lighting energy use. These measures are summarized below (Text Box 9.2):

Each measure was evaluated on a building-by-building basis for applicability. Some illustrative examples are described below.

TEXT BOX 9.2 Lighting EMOs for Punahou School

L1   Replace combinations of T12 lamps and magnetic ballasts with T8 lamps and electronic ballasts
L2   Replace small incandescent lamps (40–150 W) with compact fluorescents
L3   Install LED retrofit kits in exit signs currently containing incandescent lamps
L4   Install occupancy sensors in offices, bathrooms, conference rooms, storage rooms, and other areas that may be lit continuously but are intermittently
L5   Replace large mercury vapor lamps with metal halide lamps
L6   Modify fixtures or reflectors
L7   Reduce number of lamps
L8   Convert to task oriented lighting
L9   Lighting controls (timers, etc.)
L10 Take advantage of natural daylighting with photocell controlled dimmable ballasts
L11 Install high pressure sodium or metal halide lamps with 2 stage ballast

---

[6] Prijyanonda, J. (2010). "Building simulation analysis for Global Energy Partners as part of an evaluation of the 2009 Energy Conscious Blueprint Program for the Connecticut Energy Efficiency Board." Walnut Creek, CA.: Global Energy Partners LLC.

TEXT BOX 9.3 Note on Lamp Substitution

At the time of the audit, we recommended using this technology where incandescent lamps currently existed. Today, our recommendations would of course be different, as now there are even more efficient LED lamps available as replacements for virtually all incandescent lamps. As noted elsewhere in this chapter, LED lights have much longer lives, use far less electricity, and produce virtually no heat.

L1—Replace combinations of T12 lamps and magnetic ballasts with T8 lamps and electronic ballasts

The majority of fluorescent fixtures at Punahou School consisted of T12 lamps and standard magnetic ballasts. We recommended that all existing T12 lamps and magnetic ballasts be replaced with T8 lamps and electronic ballasts.

L2—Replace small incandescent lamps with compact fluorescents

The majority of the campus incandescent lamps were assumed to be 100 W. We recommended that they be replaced with compact fluorescent lamps with wattages ranging from 7 to 26, depending on necessary light output. For analysis, we assumed a 13 W compact fluorescent with reusable ballast (line wattage = 15) represented an average replacement (Text Box 9.3).

L3—Light Emitting Diode (LED) exit signs and kits

Retrofitting exit fixtures that contain incandescent lamps with LED kits, or replacing the entire fixture with an LED fixture reduces operating and maintenance costs. LED exit signs use approximately 3–7 W and have a life expectancy of approximately 20 years. During their life, LED exit signs should require no maintenance, unlike incandescent exit sign that have a lamp life expectancy of approximately 2000 h. Also, conversion to LED is extremely simple; existing exit signs can be converted with LED retrofit kits in minutes. We recommended that Punahou School convert all incandescent exit signs to LED exit signs with LED retrofit kits.

L4—Occupancy sensors

Occupancy sensors are typically installed in private offices, conference rooms, bathrooms, copy machine areas, hallways and stairwells. We recommended the use of occupancy sensors throughout the campus in the areas mentioned above and in classrooms.

Determination of the exact number of light fixtures and rooms applicable for installing occupancy sensors was beyond the scope of our work. Therefore we calculated occupancy sensor savings by conservatively assuming that 50% of the lights in each building would benefit from sensor installation. In addition, although savings can approach 50% as stated above, Punahou staff and students were diligent about turning off lights as they left the room. Consequently, a savings factor of 30% was used in energy savings computations (see Equation 9.3).

Occupancy sensor savings = [building total energy consumption after
relamping and ballast replacement]   [9.3]
× 50% × 30%

L5—Replace large mercury vapor lamps with metal halide lamps

We recommended that the 400 W mercury vapor lamps in the locker rooms (and the exterior lights near the swimming pool) be replaced with more efficient 325 W metal halide lamps.

L7—Reduce number of lamps

As noted previously, reducing the number of lamps in areas that have more than sufficient lighting is an obvious way to reduce energy consumption. We recommended this measure for all bathrooms on campus. Existing bathrooms had an average of 200 W of lighting. This wattage would be reduced significantly by replacing lights with two 32 W Circline compact fluorescent fixtures. Total demand reduction would be 136 W for each bathroom. The number of bathrooms was determined for some buildings by counting, and for others by estimation.

L10—Daylighting

We recommended installing daylighting controls on fluorescent lights in the following buildings: Building #3, Bingham Hall, Building #11, Dole Hall, Building #31, Ing Learning Center, and Building #32, Castle Hall. In addition we recommended daylighting controls on metal halide lights in Building #13, Hemmeter Field House and Building #38, Alexander Pavilion. These areas were selected based on the amount of sunlight penetrating into the rooms during the field visit, and the existence of southerly facing windows.

Determination of the exact number of light fixtures applicable for installing daylighting controls was beyond the scope of our work. Therefore daylighting savings were calculated by conservatively assuming that 15% of the lights in each of the above buildings would benefit from sensor installation. This is based on 50% of the buildings' lights being on the south side, and 30% of these lights being next to windows. In typical applications there is enough daylight to turn lights off for 6 h/day. Since lights are operated for roughly 11 h, the energy savings exceed 50%. To be conservative, we assumed a savings factor of 40% (See Equation 9.4).

Daylighting savings = [building total energy consumption after relamping and ballast replacement] × 15% × 40%

$$[9.4]$$

In summary, the recommended lighting measures would reduce demand by 158 kW and energy usage by over 725,000 kWh per year. These projects were eligible for over US$50,000 in Hawaiian Electric Company rebates.

CONCLUSIONS

Based on our experience, it has *always* been possible to find cost-effective energy management opportunities in lighting systems. Several decades ago, we could walk into a commercial office building, look around for five minutes, and guarantee 10% energy reduction. (It was always more in practice). Today the situation is more complex, because many of the easy improvements have already been made, but with the advances in lamp types and greater efficacies, as well as many new control options, savings are still possible. The U.S. Department of Energy funded a study comparing the savings that would be realized in sixteen typical building types by the implementation of ANSI/ASHRAE/IES Standard 90.1-2013 compared to ANSI/ASHRAE/IES Standard 90.1-2010. The results indicated an average reduction of 8.5% in source energy use and 8.7% savings in energy cost for new building construction.[7] The California Energy Commission estimates that its energy standards have saved Californians more than US$74 billion in reduced energy bills since 1977 Much of this is due to lighting, with most of the balance from more efficient HVAC.

[7] Halverson, M. et. al. (2014). *ANSI/ASHRAE/IES Standard 90.1-2013. Preliminary Determination: Quantitative Analysis*, PNNL-23236, Richland, Washington: Pacific Northwest National Laboratory.

CHAPTER 10

Transportation

INTRODUCTION

Transportation is a major energy use in the global economy. Worldwide, there are approximately 1,000,000,000 passenger cars in use, and this number is expected to double in the next two decades, especially in Asia and non-OECD countries, where growth will be approximately 2–3% per year, or 10 times that of the OECD countries. In 2012, transportation was 27.9% of global energy use, 33.1% of total OECD energy use, and 27.8% of U.S. energy use.[1] In the global economy, nearly 55% of world petroleum and liquid fuels were used by the transportation sector. In the U.S., this percentage was 72%, while in OECD Europe it was 58%.[2] The balance of transportation energy is a small percentage fueled by natural gas or electricity, and a miniscule amount by fuel cells or alternative fuels such as vegetable oil.

From these data, it is apparent that transportation is a major contributor to greenhouse gas emissions on a global scale. In 2012, the U.S. emitted 6.5 billion metric tons of greenhouse gases. Transportation produced 28% of the greenhouse gas emissions, second only to electric power generation.[3] While at first glance transportation might seem to be out of the purview of the energy manager, it is an area where decisions made at the individual or corporate level can make a difference in the long run.

HISTORY

The earliest powered transportation vehicles (or vessels) were powered by animals (or wind). James Watt developed the steam engine in 1784, but the first practical train was not developed until 1804 in the United Kingdom. Diesel locomotives gradually supplanted the early steam

---

[1] Sources: World and OECD data. IEA, 2014 *Key World Energy Statistics*; U.S. data: U.S. Energy Information Administration (2015). *Monthly Energy Review*. Refer to Table 2.3.

[2] DOE/EIA-0484 (2014) September 2014 *International Energy Outlook 2014*.

[3] Center for Climate and Energy Solutions. (2014). *Greenhouse Gas Emissions by Sector*. www.c2es.org/facts-figures/us-emissions/sector).

Energy Management Principles.
DOI: http://dx.doi.org/10.1016/B978-0-12-802506-2.00010-0
219

engines, which first burned wood or coal. Meanwhile, in the late 1800s, electrically powered locomotives were developed and saw service. Electric powered trolleys and trains were of interest because they avoided the smoke problem, which was undesirable around cities and especially in tunnels. The earliest trains used direct current (DC), but later versions used alternating current (AC). In the last half of the 1900s, electric trains entered high-speed passenger service, particularly in the case of the Japanese "Bullet" trains and the French TGV system. In some mountainous countries, such as Switzerland and Japan, electric traction found widespread use. Today India hauls over 80% of passengers and freight by electric locomotives, mostly AC powered. Electrification is much more widespread in Europe, Japan, and India than in the U.S. The infrastructure for electric trains is more costly, so where there are greater distances to be covered, diesel locomotives are selected because of the lower infrastructure costs.

Around the same time as the advent of steam powered trains, a few impractical steam powered automobiles were built. It was not until 1886 that Karl Benz patented the predecessor of the modern automobile. Then it was another 22 years (1908) until Henry Ford's model T launched the passenger car as a replacement for the horse and buggy. During the war years, there were also some vehicles fueled with wood or other alternate fuels.

Within transportation, passenger cars and light trucks are major energy users, followed by public transportation (buses, trains, airlines and ships) and freight. In describing important issues for energy management research, the first edition of *Energy Management Principles* stated that "Improvements and major new developments in transportation, (for example, in the area of electric vehicles) are required."[4] Other than that brief comment and a few sentences regarding the need for improving vehicle fuel efficiency, encouraging van and carpooling, and promotion of the use of bicycles, transportation was not considered.

At that time, a visible symbol of the importance of energy availability was the long lines of motorists attempting to purchase gasoline during the 1973 oil embargo. When supplies resumed, public concern decreased to a low level. Within six or seven years, the price had tripled, but people continued to purchase gasoline because they were dependent on it for their lifestyle. This is all the more remarkable because in the four decades

---

[4] Smith, Craig B. (1980) p. 8, *Energy Management Principles*, NY: Pergamon Press.

that have elapsed since that fateful year, gasoline prices increased by a factor of more than ten, from around \$0.35/gallon (regular grade) to as much as \$4.00 per gallon in the U.S., and higher in Europe and other areas. People are still dependent on gasoline for their lifestyles, but energy costs, environmental regulations, and consumer demand have been driving forces behind increased vehicle fuel efficiency.

In the intervening thirty-five years since the first edition was published, we have seen remarkable changes in passenger vehicle transportation energy use.

- In the U.S., Department of Transportation standards for the fuel efficiency of passenger automobiles has increased from an average of 18 mpg in 1978 to 38 mpg in 2014 (smaller passenger vehicles). Different standards apply to larger passenger vehicles and light trucks.
- Alternate-fueled vehicles (using natural gas) were introduced during WWI in Europe, when vehicles with gas stored in bladders replaced gasoline. In the 1960s, vehicles and buses using compressed natural gas started appearing.
- The first truly successful hybrid (Toyota Prius) was introduced in 1997 in Japan, and in 2000 in the rest of the world. By 2013 Toyota had sold 3 million of these vehicles with a fuel efficiency of 48 to 51 miles per gallon (20 to 22 kilometers/liter).
- The first commercially successful all electric automobile capable of a range greater than 200 miles (the Tesla Roadster) was introduced in 2008.
- The Chevrolet Volt was introduced in 2010, initially with an all-electric range of 38 miles, which later increased to 50 miles. Fuel economy was reported to be 41 mpg as a hybrid.

Along the way, there have been many experimental vehicles, including solar powered, fuel cell powered, bio-fueled and hydrogen-fueled.

RECENT TRENDS IN FUEL EFFICIENCY

During the last four decades the average fuel efficiency of passenger vehicles has nearly doubled. From 1975 through 1988 there was a steady improvement in mileage for new gasoline and diesel powered vehicles as measured by the U.S. Environmental Protection Agency. As mileage improved, there was a corresponding drop in CO_2 emissions per mile. (See Figure 10.1). This trend flattened out and then began to reverse in

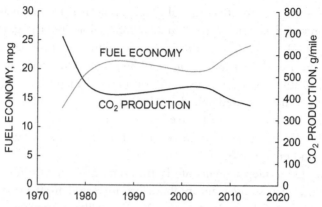

Figure 10.1 Adjusted fuel economy and CO_2 production.

1990 until 2005, when once again vehicle efficiency improved and emissions dropped to new lows.

Similar improvements have occurred internationally. In the European Union, fuel consumption of new cars dropped from 7.9 liter/100 km (29.75 mpg) in 2000 to 7.1 liter/100 km (33.1 mpg) in 2010. The specific CO_2 emission of new cars decreased by 20% during the same period, reaching 140 g CO_2/km (225 g/mile).[5]

Example. Effect of new fuel economy standards. A 2003 Mercury Sable 4-door passenger car had an EPA mileage rating of 21 mpg (combined city and highway driving) and produced 518 g/mile of greenhouse gas emissions. This was replaced in 2008 with a Toyota Prius that had an EPA rating of 46 mpg and 236 g/mile of greenhouse gas emissions. Based on actual measurements, the Prius has averaged 50 mpg over the last 7 years (75,000 miles of southern California driving.) That is a savings of 2586 gallons of gasoline. During the period from 2008 to 2014, local gasoline prices ranged from US$ 2.50 a gallon to more than US$4.00, with the average around US$3.00 per gallon, for a savings on fuel cost of US$7,758. Vehicular emission reduction was 21,150 kg, or about 21 metric tons of avoided pollution.

Improvements in vehicle fuel efficiency have come about by the use of composites and nonmetallic components in automobile bodies to reduce overall weight. Engine weights have been reduced. Power is required

---

[5] European Union (2012) *Energy Efficiency Trends in the Transport Sector in the EU: Lessons from the ODYSSEE MURE Project.* http://www.odyssee-mure.eu/publications/br/trans-port-energy-efficiency-trends.html.

to overcome rolling friction, climb grades, overcome aerodynamic drag, and for acceleration. New design approaches have also reduced aerodynamic drag. Engine weight and horsepower declined in the U.S. during the period 1975 to 1983. Then, with the increased demand for sport utility vehicles and trucks, horsepower once again increased.

For electric vehicles, the critical factor is the battery. Lead-acid batteries have a specific energy storage capability of 30 Wh/kg. To have enough energy for a reasonable trip (say 30 to 50 miles) requires a heavy battery load. Another drawback is the short life of lead acid batteries (about 3 years). The Toyota Prius uses a 53 kg nickel-metal-hydride battery with storage capability of 1.31 kWh that has demonstrated lifetimes in excess of ten years. The Tesla 85 kWh lithium–ion battery has a storage capability of 150 Wh/kg, which enables the vehicle to have a range in excess of 200 miles. Further electric vehicle development will depend on continuing to improve battery performance and reduce cost.

GREENHOUSE GAS EMISSIONS FROM VEHICLES

Carbon dioxide is the principal greenhouse gas, although methane (CH_4) and nitrous oxide (N_2O) also contribute. Carbon monoxide (CO) is emitted but converts to CO_2 in the atmosphere. The amount of CO_2 produced depends on the amount of carbon in the fuel. Since combustion is basically combining carbon and oxygen, virtually all of the carbon in the fuel is converted to CO_2 unless the engine is running "rough" and incomplete combustion is occurring. The U.S. Environmental Protection Agency uses average carbon content values to estimate vehicle CO_2 emissions factors:

Ef_g = CO_2 emissions from gasoline: 8,887 g CO_2/gallon
Ef_d = CO_2 emissions from diesel: 10,180 g CO_2/gallon

To determine annual greenhouse gas emissions for a specific vehicle, the following equation can be employed:

$$CO_2 \text{ emissions/yr} = (Ef/mpg) \times D \times 10^{-6} \quad (mt/yr) \qquad [10.1]$$

Where:

Ef = the emission factor (gasoline or diesel)
mpg = the vehicle fuel efficiency rating, miles per gallon
D = miles driven per year

Using Equation (10.1), a gasoline powered vehicle with fuel efficiency of 25 mpg, driven an average of 12,000 miles per year, will emit 4.3 metric tons of CO_2 per year.

HOW VEHICLE ELECTRIFICATION CAN HELP

A study by the Electric Power Research Institute and the National Resources Defense Council examined future scenarios that combined new plug-in hybrid electric vehicles and improved low emission electric power generating stations. The study considered combinations of nuclear, wind, solar, clean coal, and combined cycle natural gas generating stations, along with three different types of plug-in electric vehicles, each with different range capability with a single battery charge. The simulation included the effect of retiring older, less efficient, generating stations and replacing them with more efficient, less polluting new types. The study indicated that by 2050 transportation sector greenhouse gas emissions could potentially be reduced by as much as 612 million metric tons annually (9.4% of 2012 U.S. emissions.)[6]

The International Energy Agency has organized an Electric Vehicles Initiative with fifteen participating member countries. The objective is to stimulate public/private initiatives in vehicle electrification. In 2012, the member countries had approximately 90% of the world's inventory of electric vehicles (200,000, or 0.02% of the world's 1 billion passenger cars). The goal of the project is to increase the number of electric vehicles to 20 million (equal to 2% of all passenger vehicles) among the member countries by the year 2020. In parallel with this effort there would be a goal to expand the network of electric vehicle charging stations in the member countries.[7]

ENERGY MANAGEMENT OPPORTUNITIES

For the energy manager, the incentive to improve vehicle efficiency is likely to be economic rather than consideration of greenhouse gas

[6] Electric Power Research Institute (2007). *Environmental Assessment of Plug-In Hybrid Electric Vehicles (volume 1: Nationwide Greenhouse Gas Emissions)*, Electric Power Research Institute and the National Resources Defense Council, http://mydocs.epri.com/docs/CorporateDocuments/SectorPages/Portfolio/PDM/PHEV-ExecSum-vol1.pdf).
[7] International Energy Agency. (2013) *Electric Vehicles Initiative, "Global EV Outlook: Understanding the Electrical Vehicle Landscape to 2020."* http://www.iea.org/publications/freepublications/publication/GlobalEVOutlook_2013.pdf.

emissions. The incentive could also stem from a corporate mandate or goal for environmental stewardship. Any improvement in vehicle fuel efficiency will have accompanying environmental benefits.

There are number of measures that energy managers can take to improve transportation energy use. The following suggestions apply primarily to an energy manager working for a corporation or municipality.

Low-Cost Measures

The easiest and most obvious thing to do is to pay careful attention to proper maintenance of existing vehicles. Engine tune-ups and proper tire inflation will help achieve optimum fuel efficiency. For deliveries, there are many new GPS-based techniques for optimizing travel routes and avoiding traffic delays that cost fuel.

Commuting

Policies can be established to encourage ridesharing, carpooling and use of public transportation rather than individual automobiles. Many companies today offer van pooling to transport employees to and from the workplace. Some companies provide bus passes or offer financial incentives for employees to use public transportation. Another approach is to alter the work week to 4-ten hour days, eliminating 20% of mileage and time spent commuting. Still another approach that is being used is telecommuting, or allowing certain job classifications to work at home without requiring regular attendance at the office.

Vehicles

Converting vehicle fleets to improved designs with better mileage is an obvious first step, particularly if the vehicle fleet has older vehicles. Conversion can reduce fuel costs by as much as 50%, going from average fuel economy of 25 miles per gallon to as much as 50 miles per gallon. Historically it has been noted that sales of better mileage vehicles increase as fuel prices increase. At the time this book went to press gasoline prices were see-sawing, first going down, then back up, due to falling crude oil prices. This should be viewed as a temporary condition and should not be taken as a reason to not select more fuel-efficient vehicles.

Conversion to compressed natural gas or liquefied natural gas (CNG or LNG) is another option that offers a reduction of 6–11%

in greenhouse gas emissions compared to gasoline or diesel engines.[8] Natural gas is being used in many countries for passenger vehicles, buses, forklifts, municipal trash trucks, and other types. Currently, natural gas offers economies compared to diesel or gasoline fuels. Disadvantages or limitations include the fact that fuel storage tanks take up more room and fueling stations are not widely distributed.

Around the globe all automobile manufacturers are working on hybrid electric vehicle, and many are starting to work on plug-in electric vehicles. The benefit of electric vehicles could be increased if there was a significant effort to expand the number of *solar powered* electric vehicle charging stations, in order to take electric vehicles totally off the grid.

Example. Efficiency of electric vehicles needs to consider the source of electricity. The Tesla model S has a claimed range 265 miles with an 85 kWh battery storage capacity at 65 mph average speed, depending on temperature, load and other factors. You would not want to run the battery to zero, so say you retain 10% charge. Then the trip would require 76.5 kWh. Based on 115,000 Btu/gallon of gasoline, (the equivalent to 33.7 kWh on a straight energy basis), you would use the equivalent of 2.27 gallons of gasoline or 116 mpg. (EPA says 90 mpg equivalent using their test methods).This would be okay if you charged the car with solar or wind power. Plugged into the utility grid, assume a power plant heat rate of 10,000 Btu/kWh for an oil-fired generating station, with fuel oil at 140,000 Btu/gallon. Then to produce 76.5 kWh requires 765,000 Btu or 5.46 gallons of fuel, so the mileage is now 48.5 mpg, about the same as a Prius, and with a corresponding amount of greenhouse gas emission.

Depending on range requirements, electric vehicles should be considered for vehicle fleet upgrades. For local deliveries and commuting requiring less than 20 to 40 miles driving per day, electric vehicles can be charged overnight during off-peak hours using low-cost time-of-use rates. In addition, the energy manager can use corporate or municipal buying power to purchase better mileage vehicles, thereby encouraging production of more efficient models. Companies can consider providing free charging stations for electric vehicles or provide preferred parking for employees utilizing electric vehicles.

---

[8] U.S. Department of Energy. (2014). *Natural Gas Vehicle Fuel Emissions*, www.afdc.energy. gov/vehicles/natural_gas_emissions.html. (accessed March 16, 2015).

CONCLUSIONS

Transportation accounts for a large percentage of global energy use and of global output of greenhouse gases. This important sector of the energy economy is starting to benefit from efficiency improvements in conventional gasoline and diesel engines, and also from new fuels (LNG, NGL, and others, including bio-derived fuels). The rapid advances in hybrid and electric vehicles are encouraging. Comparisons based on "well-to-wheel" studies (raw fuel to distance traveled) include the inefficiencies of fuel refining and electricity generation. Such studies demonstrate that to take full advantage of electric or hybrid vehicles, solar powered charging stations should be deployed.

CHAPTER 11

Management of Process Energy

INTRODUCTION

Process energy is defined as energy required to process materials in order to provide goods or services. Examples readily come to mind: the heat required to cast or forge steel, electricity that powers welders and machine tools, and the fuels that operate furnaces and ovens. It consists of energy used in specialized production equipment, as well as in common support systems such as air compressors, motors, boilers, and refrigeration systems. Process energy is generally associated with the industrial sector and specifically with industrial manufacturing facilities.

About two-thirds of industrial energy use is in the form of process heat (roughly half as steam and the balance as direct heat); about one-fourth is in the form of electric energy (primarily electric drives); the balance of industrial energy use consists of miscellaneous uses and fuel used as feedstocks. It is apparent that process heat and steam are important considerations in industrial energy management. Electricity is important as well, as it is the fastest growing segment of industrial energy use.

For the global economy as a whole, industrial energy amounted to about half of the total in 2011 (refer to Table 2.3). Though industry still accounts for one-third of total energy use in the U.S. (31% in 2011), the situation is changing in some countries. In the United Kingdom, industry's share of total energy use declined from 40% in 1970 to 16% in 2013. This decline is partly due to sectorial changes, but most importantly, energy intensity in industry fell by 70% during the same period, due to efficiency improvements.[1] Contrast this trend to China, where industrial energy use accounts for about 70% of total energy usage, even though energy intensity has decreased by about 46% from 1996 to 2010.[2] India represents another important player in global energy use.

[1] Khan, S. and Wilkes, E. (2014). "Energy Consumption in the UK (2014): Overall Energy Consumption in the UK Since 1970." London: UK Department of Energy and Climate Change.

[2] Ke, J., et al. (2012). "China's Industrial Energy Consumption Trends and Impacts of the Top-1000 Enterprises Energy-Saving Program and the Ten Key Energy-Saving Projects," *Energy Policy*, vol. 50 pp. 562–569, Nov. 2012.

Energy Management Principles.
DOI: http://dx.doi.org/10.1016/B978-0-12-802506-2.00011-2
229

With "business as usual," Indian industrial energy use is likely to double or triple over the next several decades.[3]

Industrial energy use, always important in the global economy, today has even greater importance as a key contributor to climate change. The five most energy-intensive industrial sectors (iron and steel; cement; chemicals and petrochemicals; pulp and paper; and aluminum) account for an estimated 77% of industrial direct CO_2 emissions.[4] While increased efficiency alone will not achieve the global reduction of greenhouse gas emissions considered necessary to address climate change, it is a vital first step.

In light of these statistics, it is clear that improving industrial energy management—specifically management of process energy—is increasingly important from a global perspective. However, it also has myriad benefits at the facility or corporation level, including reducing operating costs, increasing industrial productivity, and helping companies demonstrate environmental stewardship, all of which improve industrial competiveness. It can also help companies comply with carbon policies.

SCOPE OF THIS CHAPTER

Though process energy is often associated with industrial manufacturing facilities, that is too restrictive a definition. For example, commercial laundries and restaurants use large amounts of energy in specialized processes, as do the agriculture and water and wastewater industries. Even though these market segments are not considered manufacturing industries, they do share the common thread that they process materials to provide a product or service. Therefore, the processes and associated energy management opportunities (EMOs) described in this chapter extend beyond the confines of industrial manufacturing plants.

The structure of this chapter follows the relative significance of process equipment energy use. In view of the importance of heat as a process energy form, we have placed a major emphasis on fuels, combustion, steam generation and distribution, direct- and indirect-fired furnaces and ovens, heat recovery, and electric heat. The next major area addressed is

---

[3] Trudeau, N. et al. (2011) "Energy Transition for Industry: India and the Global Context," *IEA Information Paper*, Paris: International Energy Agency.

[4] Taylor, P. et al. (2010). "IEA Energy Technology Perspectives 2010," p. 162. Paris: International Energy Agency.

other electric process loads, namely transformers, electric motors and drives, pumps and fans, refrigeration and process cooling, and finally electrolytic processes including welding, plating and anodizing, and electrochemical machining. This chapter concludes with a discussion of a variety of other process energy forms, such as compressed air, machine tools and manufacturing processes, paint spraying, and energy storage and process controls.

GENERAL PRINCIPLES FOR PROCESS ENERGY MANAGEMENT

The general principles of energy management presented in Chapter 3 can be applied to the management of process energy use in a variety of ways. Table 11.1 lists the basic principles in approximate order of importance relative to process energy use and provides typical examples of the application of each principle. These principles can be applied to fuel and electricity. For example, process heating and cooling systems should be properly sized, avoiding excess capacity. Likewise, excess capacity in electric motors leads to inefficiencies of two types. First, motors are less efficient at partial load. Second, power factor decreases at low loads, causing the electrical distribution system to incur greater losses.

Approach

In the discussion that follows, we assume that the preliminary steps outlined in Chapter 4 have been accomplished, including the initiation and planning phase and the historical review. For the process under consideration, the basic facts are presumed known, either by energy audits, calculations, or measurements. The basic items required are the various energy forms (fuel, electricity, steam, compressed air, etc.), quantities (liters, kilowatt hours, kilograms per hour), and states (residual oil, 480 V, three phase, 235°C at 1.5 MN/m^2 pressure, etc.), both entering and leaving the process. The process itself must be understood so the need for these various energy forms can be evaluated.

With this as background, our goal now is to determine how to manage process energy in the most efficient way and still get the job done.

The field of process energy use is so broad that it would take volumes to cover it in a comprehensive manner. In fact, entire books have been written about specific processes, e.g., energy use in foundries or plating operations. Therefore, our approach will be to review the fundamental

Table 11.1 Examples of principles for process energy management

Principle	Operational and maintenance strategies	Retrofit or modification strategies	New design strategies
Energy confinement (reduce energy losses)	Reduce infiltration in buildings; repair leaking valves or steam traps; reduce compressed air leaks; optimize excess air to combustion processes; clean lamps and filters	Insulate buildings, steam pipes, furnaces, ovens, etc. Reduce radiation losses from high temperature processes or heaters. Decrease i^2R losses, apply power factor corrections	Properly orient buildings on sites; use passive design techniques for new buildings; consider sustainable design
Heat or power recovery	Use less than 100% makeup air for HVAC purposes by returning heated or cooled air; use waste heat to preheat combustion air. Provide proper ratings for motors, pumps, fans. (Avoid excess capacity.)	Recover waste heat from cooling towers, stacks, air compressors, processes, equipment, building exhausts; return condensate to boilers; recover power from pressurized liquids	Add topping or bottoming cycles, waste heat boilers, recuperators, or economizers. Recover heat from motors, transformers, or equipment
More efficient equipment	Operate fans, pumps, and motors at rated load; follow proper maintenance procedures for all equipment; observe lubrication and filter change-out schedules	Use cooling towers instead of chillers; optimize air compressor types for the job; replace inefficient lamps with more efficient ones, use high-efficiency motors	Use high-efficiency lamps; install heat pumps; select high-efficiency motors
More efficient systems or processes	Reduce boiler stack temperatures to minimum; lower hot water temperature settings and compressed air pressures; remove unnecessary lamps	Use outside air for cooling; use retrofit equipment or processes; use microwave of dielectric heating rather than direct-fired heating. Convert pneumatic powered equipment to electric	Select new processes that use less energy (e.g., painting rather than plating, electric motors rather than compressed air drives); use cold forging rather than hot forging; use induction heating rather than carburizing

Integration of energy uses	Use condensate or cooling water to preheat materials or feedstocks; where multiple air conditioning or heating units are used, be sure they do not "buck" each other; avoid simultaneous heating and cooling	Review electrical system needs and design for the highest voltage practical; eliminate transformers	Use integrated utility systems for towns or building complexes; install district heating systems or other loads to use power plant waste heat
Aggregation of energy uses	Combine metallurgical processes to eliminate cooling and reheat of process materials	Combine heating and cooling operation so heat rejected from one can be used in the other or vice versa (e.g., use refrigeration system reject heat to warm buildings)	Design processes that degrade process heat by the smallest steps possible; e.g., "continuous" steel rolling mills; use cogeneration or combined cycles to meet multiple energy needs
Select appropriate materials	Reduce scrap; use materials or feedstocks with higher impurity content if this can be tolerated	Modify production processes to use materials with lower melting point, reduced hardness, or better machineability	Use water-based paints rather than solvent-based paints to reduce drying energy
Materials recovery	Recycle scrap or waste products; reduce corrosion	Investigate new processes or improved processes for extracting materials for ores or feedstocks	Use cast parts rather than machined parts; design to reduce obsolescence and permit recovery of energy-intensive materials or components
Integration of material use	Use scrap from one process as feedstock for another	Investigate material shaping or forming operations that perform several steps or handle multiple materials in one operation; use scrap or wastes as a substitute for another material	Use municipal or industrial wastes as a fuel for municipal or industrial power generation

(Continued)

Table 11.1 (Continued)

Principle	Operational and maintenance strategies	Retrofit or modification strategies	New design strategies
Aggregation of material use	Relocate stock rooms to reduce material handling; use one material rather than several to get better utilization of scrap	Modify production lines to reduce conveyor or transportation energy use	Design systems to employ materials sequentially in their highest forms gradually degrading to the lowest; e.g., wood–lumber–paper–fuel
Improve controls	Modify controls to permit separate switching of lighting and other electrical loads; provide time clocks or photocell controls where appropriate; consider enthalpy controls	Improve lighting controls; install peak demand limiting controls for electrical loads. Use microprocessor controls and variable speed motor drives	Use computers for process simulation and large-scale process control. Use demand control techniques to shift loads off-peak
Improve monitoring or metering	Read meters and provide periodic reports to department managers; add excess air monitoring equipment to boilers	Install submetering, demand meters, steam meters, air meters	Provide centralized data acquisition and analysis facilities for monitoring plants or facilities on the process or subprocess level

aspects of some major areas, providing illustrative examples, and note sources of additional information where appropriate.

Potential Savings and Benchmarking

There have been a number of international studies to evaluate "best practices" in major segments of industrial energy use.[5,6] These studies examine specific energy use in various industrial applications (such as glass manufacture, petroleum refining, and iron and steel mills) at different plants in different countries. From these data, the most efficient plants are used to establish a "best practice," or best available technology for each end use. Then, by comparing the best practice to the average specific energy required for that end use, an estimate of the savings possible by deploying the best practice can be made. Compilations of such data provide an "efficiency benchmark" for individual companies to compare their process performance against others in the same industry.

These studies indicate that savings in the range of 20–26% of current industrial energy use are possible.

Today there are many tools available to the energy manager to analyze or evaluate specific industrial processes. An example is the U.S. Department of Energy's Advanced Manufacturing Office (AMO), formerly the Office of Industrial Technology. This office provides many information sources, case studies, and tools for improving industrial process energy efficiency. Some of the specific types of equipment addressed are boilers and steam systems, air compressors, motors, fans, pumps, and process heat technologies. The website is www.energy.gov/eere/amo/information-resources. Besides free tools that help energy managers assess the efficiency of various industrial processes, the website has links to other sources of information. Another example is a website hosted by the European Community at http://iet.jrc.ec.europa.eu/energyefficiency. Technical guides, training courses, manuals, and other documents can be accessed at this site in various languages.

Caution must be exercised in applying any of the principles that follow on a general basis. Process energy use tends to be site- and

[5] Saygin, D. et al. (2010) "Global Industrial Energy Efficiency Benchmarking: An Energy Policy Tool Working Paper," Vienna: United Nations Industrial Development Organization.
[6] Pellegrino, J.L. et al. (2004) "Energy Use, Loss, and Opportunities Analysis: U.S. Manufacturing and Mining," Washington, D.C.: U.S. Department of Energy, Office of Energy Efficiency and Renewable Energy, Industrial Technologies Program.

process-specific. What is appropriate for one process and one plant may not be appropriate for a similar process in another plant, depending on fuel types and costs, weather patterns, variations in material properties, and a host of other variables. Judgment, careful analysis, and sound engineering are prerequisite for any successful energy management program.

PROCESS HEAT

Process heating applications range from the simple functions of heating water or generating steam to industry-specific applications such as drying, melting, curing, annealing, cooking, distilling, softening, and fusing various products and materials. Process heating technologies are also quite varied. The most common types are boilers, furnaces, ovens, and kilns, but there are also many specialty process heating technologies for specific applications. Some of the technologies for process heating overlap with those used for space heating (see Chapter 8), such as boilers, furnaces, and heat pumps. Both process heating and space heating systems are significant energy users.

The discussion below first describes the mechanisms of heat transfer and loss in process heating systems. It then provides an overview of energy use and management in five key areas of process heat: combustion, steam, hot water, furnaces and ovens, and electric heat.

Heat Transfer Mechanisms

To reduce heat losses, it is important to recognize that heat is transferred (and lost) through three mechanisms: conduction, convection, and radiation. Refer to Figure 11.1 in connection with the following equations:

Conduction. Consider a section of material thickness L, area A, thermal conductivity k, and exposed to temperature T_1 at one side and T_2 at the other, such that: $\Delta T_{1-2} = T_1 - T_2$. The heat flow *conducted* through this material is

$$\dot{Q} = kA[dT/dx] \quad \text{W} \qquad [11.1]$$

where
$\dot{Q}$ = rate of heat flow, W
A = area, m^2
k = thermal conductivity, W/m · K
T = temperature difference, K
x = distance into material, m
(see Table 12.2 for typical values of thermal conductivity, k.)

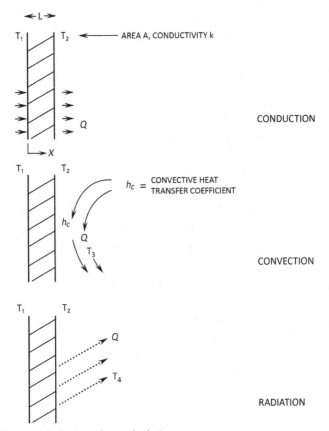

Figure 11.1 Models for heat loss calculation.

Equation 11.1 can be rewritten as

$$\dot{Q} = kA\, \Delta T_{1-2}/L \quad \text{W} \qquad [11.2]$$

Thus, all other parameters being equal, the driving force for conduction heat losses is thermal conductivity and ΔT.

Convection. *Convective* heat losses are described by a similar equation:

$$\dot{Q} = h_c A\, \Delta T_{2-3} \quad \text{W} \qquad [11.3]$$

where

h_c = convection heat transfer coefficient (also called film coefficient), $\text{W/m}^2 \cdot \text{K}$

A = heat transfer area, m^2

ΔT_{2-3} = temperature difference between the surface and convection fluid, K

Radiation. The third method for heat loss is *radiation*, given by this equation for a relatively small body in a large enclosure or space:

$$\dot{Q} \sim A\epsilon\sigma(T_2^4 - T_4^4) \quad W \qquad [11.4]$$

where

σ = Stefan-Boltzmann constant, 5.67×10^{-8} W/m$^2 \cdot$ K^4

T_4 = temperature of surroundings, K

T_2 = temperature of radiating surface, K

A = area of body, m^2

ϵ = emissivity of body, dimensionless

Reducing heat loss. With Equations 11.1–11.4, it is possible to see how to reduce heat losses and to evaluate the benefits from doing so.

For a given situation, the driving force for conduction heat loss is the thermal conductivity of an insulating material and the temperature difference across its thickness. In the case of convection, the loss depends on the temperature difference between the surface of a heated material and the ambient temperature, as well as the transfer area (in m^2) and the convection heat transfer coefficient (in W/m^2 K). Radiation heat loss is influenced by the area radiating heat, its emissivity, its temperature, and the temperature of its surroundings, where the temperatures are in Kelvin raised to the fourth power. At low temperatures, conduction and/or convection heat losses typically will dominate, while at high temperatures radiation becomes important.

To reduce conduction losses, either more or better insulation is required to increase insulation thickness and decrease thermal conductivity. To reduce convection losses, e.g., from a tank containing a hot fluid, an insulating cover should be added to reduce the convection heat transfer coefficient and the temperature difference. This would also reduce radiation losses. Radiation losses can also be reduced by adding radiation shields.

Combustion

Combustion is the starting point for all forms of process heat except nuclear or solar, neither of which is currently of great importance for manufacturing processes. In direct-fired operations, process heating is caused by radiant energy or is transferred by the products of combustion. In indirectly heated operations, the heat of combustion is first transferred to a working fluid (such as steam) that does the heating or that produces the electricity that does the heating. Thus, combustion is the first area to examine when considering energy management options.

Fuels can be rated according to their *heating value*, which is a measure of the available work they possess (see Appendices for typical values). In theory, nearly all the fuel's potential to do work should be recoverable; however, due to practical limitations, there are losses due to the thermodynamics of combustion. The only current prospect for avoiding these losses is through the use of fuel cells that extract fuel energy without combustion. Size, economic, and lifetime constraints limit fuel cell applications for process energy use at present.

The next loss of available work in the combustion process results from the decrease in temperature from the flame temperature (on the order of 2000°C) to the process temperature. In the case of a steam cycle, the process temperature might be in the range of 200−500°C. In concept, efficiency could be improved if hot combustion products could be utilized at as high a temperature as possible, subject to the limitations imposed by useful engineering materials. Turbine technology permits temperatures of about 1100°C; this suggests that one improvement would be to use a topping cycle involving a direct-fired turbine to extract work. The turbine exhaust (typically at 600−650°C) could then be applied to a waste heat boiler to generate steam. The usual approach, to generate steam directly from fuel combustion, requires a drop in temperature from 2000°C to about 550°C due to material and corrosion limitations. This leads to a decrease in available work before the energy is even applied to a process.

While a combined cycle probably will not be feasible in an existing plant, there are several other EMOs that can be considered to improve combustion efficiency:

- Monitor fuel quality and improve if required.
- Provide correct amounts of excess air.
- Add or improve automatic controls or monitoring equipment.
- Establish and follow burner maintenance procedures.
- Use waste heat to preheat combustion air.
- Consider mixing combustible waste with fuel (solid waste with coal, waste lubricating oils with fuel oil, etc.).

We discuss these opportunities below as well as in Chapter 8.

Fuel Quality and Excess Air

Fuel should be analyzed periodically to determine that it still meets requirements. For example, excess fines (small coal particles) in coal could lead to improper operation, including too much excess air, which in turn leads to energy losses up the stack. Excess air is best controlled by monitoring flue gas oxygen content, either by periodic measurements, or better yet, by permanently installed instrumentation. The proper setting for the type of fuel can be established by testing. Fuel savings of 2–19% are possible from correcting excess air. Reviewing the combustion process helps explains the science behind the excess air issue.

Combustion of carbonaceous fuels requires oxygen. Oxygen is normally supplied from atmospheric air, which has the following standard properties:

Volumetric composition: N_2: 78.1%; O_2: 21%.

Mass composition: N_2: 75.5%; O_2: 23.2%.

Molecular weight: air: 29; N_2: 28.02; O_2: 32.0.

Mole ratio: 3.76 moles N_2 per mole O_2.

Moisture content: 60% RH, 0.013 kg water vapor per kilogram of dry air (26.7°C dry bulb temperature).

The combustion process may be outlined schematically as shown in Figure 11.2.

The following notation applies:

m = mass, kg.

x = excess air fraction, %,

Subscripts have the following meanings:

a	= dry air	S	= sulfur
w	= water vapor	T	= trace elements
C	= carbon	dg	= dry gases
H	= hydrogen	A	= ash
O	= oxygen		

If the theoretical amount of oxygen required for complete combustion is calculated, one obtains (for any fuel):

$$w_O = 2.67w_C + 8(w_H + w_O/8) + w_S \quad \text{kg oxygen/kg fuel} \qquad [11.5]$$

where

w equals weight fraction, % and

subscripts are as defined above.

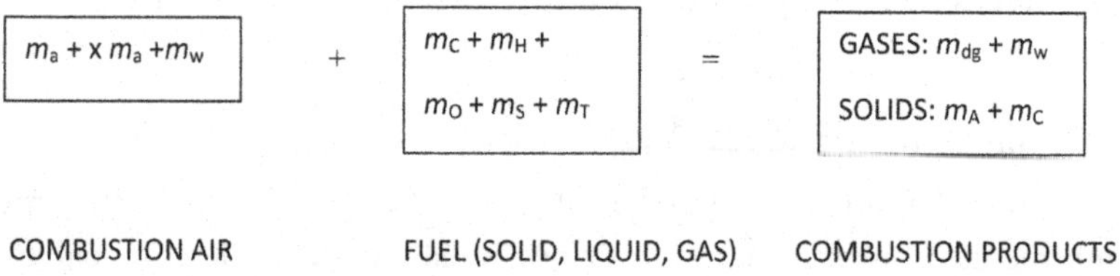

$m_a + x\, m_a + m_w$	$+$	$m_C + m_H +$ $m_O + m_S + m_T$	$=$	GASES: $m_{dg} + m_w$ SOLIDS: $m_A + m_C$

COMBUSTION AIR	FUEL (SOLID, LIQUID, GAS)	COMBUSTION PRODUCTS

Figure 11.2 Combustion process schematic.

The weight fractions for the various constituents in the fuel would normally be provided by the supplier or obtained by ultimate fuel analysis.

Using the properties of air, Equation 11.5 can be converted to an air basis:

$$w_a = 11.5w_C + 34.5(w_H + w_O/8) + 4.31w_S \quad \text{kg air/kg fuel} \quad [11.6]$$

This is obtained by recognizing that O_2 is 23.2% (by weight) of air.

Combustion air calculations can be made on a mass basis, volume basis, or molar basis. The data in Table 11.2 are useful for such calculations.

Using the data in Table 11.2 and Equation 11.6 will provide the minimum amount of air (or oxygen) required for complete combustion of a given fuel. In reality, more air than this must be supplied, since some excess air is required for complete combustion. The amount of excess air depends on the particular system and fuel type. For natural gas, an efficient boiler might use about 15% excess air or roughly 19 kg per kg of fuel. Since this is 75.5% by weight N_2, it means that for every kilogram of fuel, there is nearly 15 kg of N_2 flowing through the boiler, being heated from ambient temperature to combustion temperature, and eventually carrying heat up the stack without bringing any benefit to the process!

Table 11.2 Properties of fuel materials

Material	Chemical formula	Molecular mass ratio	Density, 15.6°C, 1.0 atmosphere (lb/ft³)	Density, 15.6°C, 1.0 atmosphere (kg/m³)	Heating value (Btu/lb)	Heating value (MJ/kg)
Carbon	C	12.00	—	—	14,544	33.8
Hydrogen	H_2	2.02	0.0053	0.085	62,028	144
Sulfur	S	32.7	—	—	4050	9.4
Carbon monoxide	CO	28.00	0.0739	1.18	4380	10.2
Methane	CH_4	16.03	0.0423	0.68	23,670	55
Oxygen	O_2	32.02	0.0844	1.35		
Nitrogen	N_2	28.02	0.0739	1.18		
Air	—	29	0.0765	1.23		
Carbon dioxide	CO_2	44.00	0.1145	1.83		
Sulfur dioxide	SO_2	64.07	0.1692	2.71		
Water vapor	H_2O	18.02	—	—		

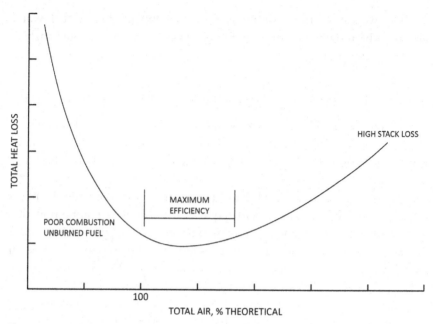

Figure 11.3 Relationship between boiler heat loss and excess air.

Complete combustion requires that the following conditions be satisfied:

- An adequate air (oxygen) supply.
- Adequate fuel/air mixture.
- Appropriate combustor temperature for fuel ignition.
- Adequate combustor residence time for complete combustion.

Too much airflow (excess air) leads to fuel waste by increasing the energy loss up the stack, while not enough air will lead to incomplete combustion.

Figure 11.3 schematically illustrates the relationship between heat loss and excess air. As the figure indicates, when the amount of excess air increases, the stack losses increase. For typical industrial type boilers, the correct amount of excess air is best determined by analyzing flue gas oxygen or carbon dioxide concentrations. Table 11.3 shows typical data for a natural gas-fired boiler.

On well-designed natural gas-fired systems, an excess air level of 10−15% is attainable. Higher amounts (e.g., 45%) will reduce boiler efficiency by 1−4%, depending on the stack temperature (Table 11.3).

Table 11.3 Combustion efficiency for natural gas

		Combustion efficiency				
Excess %		Flue gas temperature minus combustion air temperature (°F)				
Air	**Oxygen**	**200°F**	**300°F**	**400°F**	**500°F**	**600°F**
9.5	2.0	85.4	83.1	80.8	78.4	76.0
15.0	3.0	85.2	82.8	80.4	77.9	75.4
28.1	5.0	84.7	82.1	79.5	76.7	74.0
44.9	7.0	84.1	81.2	78.2	75.2	72.1
81.6	10.0	82.8	79.3	75.6	71.9	68.2

Assumes complete combustion with no water vapor in the combustion air.
Source: U.S. DOE, Energy Efficiency and Renewable Energy program, Steam Tip Sheet #4.

Combustion Monitoring Equipment

On-going monitoring of combustion parameters will help identify and correct issues with the air-to-fuel ratio and will allow for assessment of the combustion efficiency. Measurements include the amount of O_2, CO_2, CO, and unburned hydrocarbons in the flue gas and combustion air and flue gas temperatures.

The O_2 measurements indicate levels of excess air and the CO measurements indicate incomplete combustion. The proper ratio of air and fuel can be maintained manually by an operator if care is taken. However, electronic combustion analyzers and controls can automatically maintain efficient operation. If not already installed, such a system would likely pay for itself in a very short period of time, even by reducing fuel consumption by as little as 5%.

The combustion products (O_2 and/or CO_2) and net exhaust gas temperature can be used to assess the combustion efficiency for a given type of fuel and burner using manufacturer's literature or combustion efficiency curves and tables in handbooks. The net exhaust gas temperature is defined as the flue gas temperature minus the combustion air temperature (ambient temperature) if no air preheater or economizer is installed. If there is an air preheater or economizer, determine the temperature at its outlet. The combustion air temperature should be measured at the forced draft fan inlet.

Burner Maintenance Procedures

Proper maintenance and operation of burners, stokers, and other combustion equipment is important. For example, deposits on burners can reduce burner efficiency and, in oil-fired systems, the temperature at which oil is delivered to the burners contributes to proper atomization and combustion.

Waste Heat for Preheating Combustion Air and Feedwater

Efficiency also improves with increasing combustion air and feedwater temperature, since less energy has to be supplied to heat them. In some instances, waste heat from stacks can be recovered and used to heat the incoming air. Preheating combustion air from ambient temperatures up to several hundred degrees centigrade can save as much as 5−10% of fuel consumption. It may be feasible to use heat recovery to preheat feedwater.

Steam Generation and Distribution

Improvements in steam systems fall into two broad categories. The first applies to the steam system itself, while the second applies to the end uses of steam. For steam systems, consider:

- Steam leaks from lines and valves.
- Defective steam traps.
- Proper sizing and maintenance of distribution systems, including insulation.
- Proper management of condensate return.
- Proper maintenance of steam tracing systems.

For steam end uses such as process heating, operating steam driven equipment, or heating buildings, some EMOs include the following:

- Supply steam at the lowest pressure possible.
- Review steam uses to see if more efficient alternatives exist.
- Apply the cascade principle to steam uses.

The subsections below summarize these boiler and steam system EMOs. See Chapter 8 for additional opportunities.

Steam leaks and defective steam traps. Small steam leaks resulting from defective traps or valves can lead to surprisingly large energy waste, as shown in Figure 11.4. These types of leaks may also cause significant loss of water, which is an increasingly expensive and scarce commodity particularly in drier climates.

Sizing and maintenance of distribution systems. As steam loads change over time, the distribution system may be used for purposes other than those for which it was originally designed. If the lines are too small,

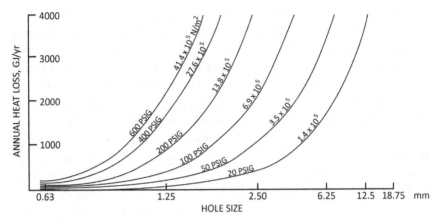

Figure 11.4 Heat loss from steam leaks.

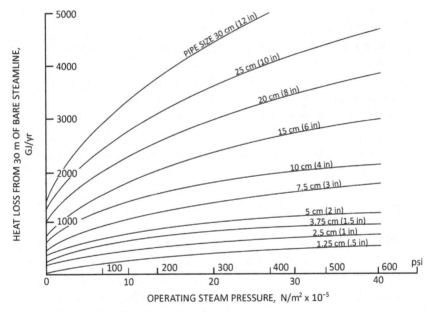

Figure 11.5 Heat loss from bare steam lines.

pressure drops may be excessive. If they are too large—i.e., supplying small loads—the losses may be disproportionately large. In addition, the importance of proper steam line insulation cannot be overemphasized. Figure 11.5 illustrates the magnitude of heat loss from uninsulated lines.

Management of condensate return. In many processes, after use, steam is condensed and the resulting fluid is returned to the feedwater supply to the boiler. Otherwise, lost feedwater must be replaced and heated to provide a constant supply of steam. Condensate return saves energy in a number of ways. Not only is less energy needed to heat feedwater, but also less energy will be expended in pumping and chemically treating makeup water. Condensate return also saves water.

Example. Benefit of condensate return. Consider a boiler producing 25,000 kg/h steam. The feedwater temperature is 15°C and the condensate temperature is 50°C. With no condensate return, the energy input to heat feedwater is 8.9 GJ/h. With 100% condensate return, the energy input to heat feedwater is 5.2 GJ/h or a savings of 3.7 GJ/h. Suppose the fuel is oil rated at 5.8 GJ/bbl and the boiler thermal efficiency is 70%. The theoretical savings is nearly one barrel of fuel per hour. In reality, this is an upper limit, since 100% condensate return is unlikely.

In addition to using condensate return to recover energy, if there is some other source of waste heat, a heat exchanger can be provided to preheat feedwater to the extent possible.

Maintenance of steam tracing systems. Steam tracing systems (used to heat pipes, tanks, etc.) can waste energy if not maintained properly. An obvious caution is to turn them off when not needed.

Steam supply pressure. Steam usually will be supplied at the pressure of the highest load. In industrial operations, this may be $1-3 \text{ MN/m}^2$ if steam turbine driven equipment is used. Steam is also used to transfer fluids by means of steam jets. For heating purposes, pressures are more typically in the range of $0.1-0.3 \text{ MN/m}^2$. If most of the loads are at lower pressure, steam should not be supplied at high pressure. Instead, a substitute energy source could be found for high-pressure loads.

Efficient alternatives to steam systems. In some cases, electric motors or some alternative drive system will be more efficient than steam turbines. This would be true in cases where the steam loads are small or are distant from the steam plant. Electricity might even be better for heating in a particular case, if line losses and the effect of more precise temperature control are included.

Cascading of steam use. Sometimes steam must be provided at high pressure. Instead of using pressure-reducing valves, look for opportunities to cascade steam use, within acceptable limits on steam pressure, temperature, and quality. For example, high-pressure steam can first be expanded through a noncondensing turbine to do useful work, and then the turbine exhaust steam can be used for process or building heating.

Hot Water and Water Pumping

The heating and transport of water and other fluids requires utilization of energy to raise the water temperature, to make up heat losses from pipelines and storage tanks, to pressurize fluids, and to overcome the resistance to fluid flow of pipelines. There are EMOs associated with each of these energy requirements.

Water heating. The energy input necessary to raise the temperature of water (or another liquid) is given by

$$E_{in} = mc_p(T_f - T_i)/\eta \quad \text{J} \qquad [11.7]$$

where

E_{in} = energy input, J
m = mass, kg
c_p = specific heat at constant pressure, J/kg K
T_f = final temperature, K
T_i = initial temperature, K
η = efficiency of the heating system

This can also be expressed as a rate in J/s by using a mass flow rate in kg/s.

Therefore, for a given quantity of water, the main opportunities to save energy in the water heating process are to increase the efficiency of the heating technology or lower the required temperature rise. There are several types of technology advancements that heat water more efficiently, including condensing boilers, heat pump water heaters, solar water heaters, or tankless water heaters. There may also be opportunities to lower the temperature rise by using waste heat recovery to preheat the feedwater or to only heat water to the temperature absolutely necessary for the given application.

Water heaters range from small 50 gallon storage units found in residences to large boiler-fired systems found in commercial establishments and industry. Today, for residential and small commercial applications, *tankless water heaters* are available with flow rates of up to 10 gpm. Tankless water heaters have an "instant on" feature that avoids standby losses typical of storage tank units (see Figure 11.6).

Example. A seafood restaurant used hot water for cleaning, rest rooms, and to preheat water used in large 60 and 80 gallon kettles. (Preheated water would come to boiling temperature more quickly—important during the rush hour.) The conventional 150 gallon water heater proved to be unreliable so the restaurant installed three tankless units. Two served the restaurant loads and one was dedicated to the large kettles in the kitchen.

Figure 11.6 A tankless water heater.

Other restaurant applications have reported gas savings with tankless water heaters of as much as 40–50% compared to storage units.

When considering the hot water system as a whole, it is also important to reduce losses. Losses due to heat transfer to the environment can be reduced by the following steps:

- Reduce thermostat settings (which also reduces the energy required to heat the water).
- Cover open tanks.
- Insulate tanks and pipes.

A major source of loss in hot water systems is the standby losses that occur when tanks are continuously maintained at elevated temperatures or they are not well insulated. Water heaters should be maintained at the minimum required temperature. Lowering the water temperature setting

5–6°C will reduce annual fuel cost by 4–5%. For smaller applications, consider investing in high-quality units with extra insulation and anode protection to prevent tank corrosion. Water heater "jackets" or "blankets" are also available to reduce heat loss in existing systems. In addition, insulate all hot water pipes to prevent heat loss and install insulating lids on any open tanks containing hot water (or other liquids) to prevent both heat loss and liquid losses. Additional benefits of preventing these types of heat losses in conditioned spaces are reduced load on HVAC equipment and improved occupant comfort and safety.

As noted previously, heat recovery is another useful technique that can be used in industry. Waste process heat (e.g., refrigeration compressor cooling water) can often be reclaimed to heat or preheat water.

Water pumping. The power required to move an incompressible fluid through a piping system is given by

Pumping power = (mass flow) (work done per unit mass).

This may be written as

$$pp = \dot{m}\left(\Delta P / \rho_i\right) \quad \text{W} \qquad [11.8]$$

where

pp = pumping power, W
$\dot{m}$ = mass flow rate, kg/s
ΔP = system pressure drop, N/m^2
ρ_i = fluid density at pump inlet, kg/m^3

Or, in more convenient form, the power expended in pumping a fluid is related to the volume pumped:

$$pp = \dot{V}\Delta P \quad \text{W} \qquad [11.9]$$

where now

$\dot{V}$ = volume pumped, m^3/s.

We also derived Equation 11.9 in Chapter 8 using the concept of *head* developed by a pump (see Equation 8.4).

The pressure drop depends on the system (pipes, channels, orifices, bends, etc.) and must be determined for each case. In the case of round pipes, it is given by the following equation:

$$\Delta P = \frac{f\bar{\rho}v^2 L}{2D_e} \quad N/m^2 \qquad [11.10]$$

where

f = friction factor, dimensionless
$\bar{\rho}$ = average density, kg/m^3

v = velocity, m/s
L = length of pipe, m
D_e = pipe diameter, m

The losses through bends, enlargements and contractions, and valves and fittings are similarly proportional to velocity squared.

For incompressible fluids, the volume flow rate $\dot{V}$ is related to the velocity v by

$$\dot{V} = Av \quad m^3/s \tag{11.11}$$

where

A = area, m^2

solving

$$v^2 = \dot{V}^2/A^2 \quad m^2/s^2 \tag{11.12}$$

Substituting these results in the equation for pumping power, we obtain

$$pp = \frac{\dot{V}^3 f \rho L}{2A^2 D_e} \quad W \tag{11.13}$$

This is a very important and fundamental result. It indicates that once a pipe size is fixed, the power required for pumping increases as the cube of the volumetric flow rate. The practical significance is that if the flow rate is *reduced 20%, the pumping power is cut in half.* Conversely, for a fixed flow rate, the pumping power decreases in proportion to the fifth power of the diameter.

For a wide range of Reynolds numbers, $f = 0.022$ for clean, commercial steel pipe. This holds true in either American/British or SI units, since f is a dimensionless ratio having units of length/length. If American/British units are used, a conversion factor ($g = 32.2$ ft/s^2) must be inserted in the denominator of the pumping power equation.

Pressure drops for complete piping systems require determination of losses through all components. This analysis is most easily performed with computer simulation models. Alternatively, handbooks provide data that convert flow resistance of typical pipe fittings to equivalent lengths of pipe.

Major pumping system EMOs can be summarized as follows:
- Reduce system pressure.
- Reduce friction losses (increase pipe size, eliminate pressure-reducing valves).

- Stop leaks.
- Use storage tanks or accumulator so pumps can be shut down. part-time or operated off-peak.
- Operate pumps at full load when possible.
- Consider installing variable speed pumps.
- Maintain pump systems.
- Recycle or reuse water.

It is a common practice to provide water at the pressure required to meet the highest pressure load. An alternate approach that will sometimes save energy is to provide water at the pressure needed by most of the load and provide booster pumps for high-pressure loads.

For additional energy management resources for pump systems, refer to the U.S. Department of Energy's webpage: http://www.energy.gov/eere/amo/pump-systems.

Direct- and Indirect-Fired Furnaces and Ovens

Furnaces, ovens, dryers, and kilns have many similar attributes in terms of their basic heating functions as well as in their opportunities for energy management. For simplicity, we refer to this family of technologies collectively as furnaces and ovens. Direct-fired furnaces and ovens rely on heating directly by the products of combustion (fuel-fired) or by electric heating elements. Indirect-fired furnaces and ovens involve some type of heat exchanger for transferring heat from the heat source (e.g., steam or hot water) to the process. The majority of process heat is from combustible sources, although the share of electric process heat systems is likely to increase because of advantages related to ease of control, no emissions at the point of use, safety, smaller footprint, and widespread applicability.[7,8]

There are three basic approaches for managing efficient use of process heat (Table 11.4):

- Reduce heat losses.
- Use more efficient equipment and processes, including electric process heat alternatives.
- Recover heat.

[7] Gellings, C., Parmenter, K.E., and Hurtado, P. (2002). *Efficient Use of Fossil Fuels in Process Operation*, in "Efficient Use and Conservation of Energy," edited by C. Gellings, in Encyclopedia of Life Support Systems (EOLSS), Oxford, UK: EOLSS Publishers.

[8] Gellings, C., Parmenter, K.E., and Hurtado, P. (2002). *Efficient Use of Electricity in Process Operation*, in "Efficient Use and Conservation of Energy," edited by C. Gellings, in Encyclopedia of Life Support Systems (EOLSS), Oxford, UK: EOLSS Publishers.

Table 11.4 EMOs with process heat

Reduce heat losses

- Insulate furnace walls, ducts, piping
- Put covers over open tanks or vats
- Reduce time doors are open
- Avoid cooling time for heated products
- Shutdown heating systems on tanks and ovens when not in use, or at least lower temperatures (reduce standby losses)

More efficient equipment or processes

- Use alternative processes (microwave, dielectric rather than fuel-fired)
- Employ recuperators, regenerators, or preheaters
- Use direct-fired rather than indirect-fired systems
- Use less energy-intensive materials and processes
- Use heat pumps for low temperature process heat
- Reduce moisture content mechanically in materials used in drying processes
- Use lower temperature processes (cold rinses, etc.)

Recover heat

- There are multiple sources: stacks, processes, building exhaust streams, cooling towers, compressors, etc.
- Recovered heat can be used for space heating, water heating, process preheating, cogeneration, etc.
- Many types of heat recovery systems are commercially available (heat wheels, run-around systems, heat pipes, heat exchangers, heat pumps, etc.)

The discussion below explains each type of opportunity in greater detail. (Refer also to Chapter 8 for the special case of heat recovery in HVAC systems.)

Reduce heat losses. Earlier in this chapter, we described three modes of heat transfer—conduction, convection, and radiation—and explained general ways to minimize heat loss through these mechanisms. Those general approaches apply to furnaces and ovens. Reduction of losses can be accomplished by insulation, reflective shielding, use of curtains, other methods to contain heat and isolate it from conditioned spaces, reduced temperatures, maintenance of heat transfer services, equipment shutdown when not in use, or improved designs. Generally, the economically optimal amount of insulation or other heat containment methods for furnaces and ovens (and associated piping or ductwork) depends on the temperature range and fuel costs, so there is no simple rule of thumb. Each case will usually have to be analyzed on its own merits.

Example. Improve insulation. A useful approach is to consider an analysis on a unit area basis. Consider an electrically heated oven or tank that loses 3.5 kW/m^2 from its uninsulated walls under normal operating conditions. Analysis indicates that this could be reduced to 0.7 kW/m^2 by adding 2.5 cm of Rockwool insulation. Energy costs are \$28/GJ and annual operation is 2000 hours per year. Management requires a 1-year payback. The analysis indicates:

Unit energy loss without insulation:

$$(3.5 \text{ kW/m}^2)(2000 \text{ h/yr})(3.6 \times 10^6 \text{J/kWh}) = 25.2 \text{ GJ/m}^2 \text{ yr}$$

Unit energy loss with insulation:

$$(0.7 \text{ kW/m}^2)(2000 \text{ h/yr})(3.6 \times 10^6 \text{J/kWh}) = 5.04 \text{ GJ/m}^2 \text{ yr}$$

Energy savings:

$$25.2 - 5.04 = 20.16 \text{ GJ/m}^2 \text{ yr}$$

Cost savings:

$$(20.16 \text{ GJ/m}^2 \text{ yr})(\$28/\text{GJ}) = \$564/\text{m}^2 \text{ yr}$$

Therefore, if the installed cost of the insulation is equal to or less than \$564 per square meter, a 1-year payback will be obtained.

Other forms of heat losses that should be evaluated include:

- Heat absorbed by the work or product and lost as the work cools.
- Heat absorbed by auxiliary equipment (conveyors, trays, etc.).
- Heat lost up stacks or outdoors.

More efficient equipment and processes. Some process heating applications may benefit from equipment retrofits or replacements to improve efficiency. The literature in this area is extensive and growing. Examples include improved controls for better regulating process temperatures and other parameters, and new more efficient equipment designs. More efficient equipment designs generally require improved heat transfer capability, although sometimes installing an entirely different process will lead to savings.

One thing to consider in improving a process is how to approach the theoretical maximum efficiency for accomplishing a task. Consider the drying process. Much of the energy used in drying is to remove water. Given that the specific heat of water is $4,180 \text{ J/kg} \cdot \text{K}$, and since the heat of vaporization is 2.26 MJ/kg, the minimum theoretical energy input to evaporate water originally at room temperature (20°C) is

2594 kJ/kg. Real processes will not do as well, since some energy is lost, is used to heat other materials, or heats the container or dryer. However, since a great many heating processes involve drying, distillation, or related end uses, more efficient ways of accomplishing these tasks are great interest.

There are so many potential possibilities for new designs that it is impossible to enumerate them here. Instead we provide few examples. The general approach should be to carefully examine the specific needs of the process and then attempt to provide heat only when and where its use is essential.

Example. Jet impingement heating. A unique approach is *jet impingement heating*, used in the metals industry.[9] This allows the heat to directly impinge on the object being heated, thereby penetrating the surface film barrier and increasing heat transfer efficiency by as much as a factor of 3. Incidentally, the same technique can be used for cooling, as in gas turbines.

Process heat savings also result from converting batch-type processes to continuous operation. This conserves fuel by eliminating or reducing heat-up and cool-down periods. It often increases productivity as well.

Example. Continuous operation. One company replaced a group of batch-type kilns with a gas-fired walking beam kiln, and found the new kiln operated with 20% less fuel, and had 2.5 times the product throughput. Labor savings also resulted, since only two personnel (rather than ten) were required to operate the new unit.

Another process improvement relates to excess air. Excess air is provided to certain types of ovens for diluting the exhaust air. For example, in solvent drying ovens, air is introduced to create an air–gas mixture that is below the lower explosive limit (LEL). Many ovens use excessive amounts of dilution air, thereby wasting heat and fuel.

Typical industry practice is to operate in the range of up to 25% LEL concentrations. With automatic controls, operation up to 50% LEL limits is possible. Many ovens operate below 25% LEL and may be operating as low as 5 or 10% LEL. This is equivalent to about four times the excess air actually needed, or about twice the energy use actually required.

Refer to the "Electric Heat" section for examples of some promising electric process heat alternatives.

---

[9] Zuckerman, N. and Lior, N. (2006). "Jet Impingement Heat Transfer: Physics, Correlations, and Numerical Modeling" p. 565 *Advances in Heat Transfer*, vol. 39, Oxford, UK: Elsevier.

Heat recovery. As noted throughout this book, heat recovery is an important tool for the energy manager. Following the 1970s oil embargo, great emphasis was placed on identifying new methods to reclaim heat. Long practiced in industry, heat recovery is now being extended to agriculture and commerce. The technology can be applied to many waste energy streams, including hot gases, hot air, hot water, hot chemicals, and other hot process streams. The variables that determine the feasibility of heat recovery include the value of the heat, when it is available, the cost of installation, and the uses to which it can be put.

Typical uses for recovered heat include process heating and drying, preheating fuels or materials, space heating, and hot water. Sometimes heat recovery leads to other economies, such as reducing fan or pump operation, or permitting cooling tower operation to be reduced. The potential list of applications for this technique for process heating is too long to include here, but representative examples included throughout this book will illustrate the possibilities. A few examples include the following:

- Recover heat from building exhaust air using run-around systems.
- Recover heat from milk and dairy operations using heat exchangers or heat pumps.
- Recover heat from air compressor cooling water and preheat hot water with it.
- Recover heat from steam condensate or boiler blowdown and use it to preheat feedwater.
- Recover heat from chillers and use for hot water or space heating.
- Recover heat from flue gases to preheat combustion air.
- Recover heat from hot process streams, exhaust, or stacks.

In evaluating heat recovery possibilities, the first step is to evaluate the temperature of and the quantity of heat available. If it is available at low temperatures (50−100°C) there is generally a use for it, if only for heating water and spaces. In the moderate temperature range of 100−200°C, there are many possibilities, including raising steam. Above 200°C, cogeneration and other high temperature applications begin to look attractive.

Example. Assume that 1.0 GJ of heat costs $10 in the form of steam. Thus for each 1.0 GJ of heat recovered per year, one could justify an investment of say $10 if a 1-year payback was required. Depending on local tax codes and credits, utility pricing and incentives, fuel cost escalation, and the cost of money, the allowable investment might be higher, say $15 or $20. After this determination is made, the next step would be to determine the cost of purchasing and installing heat recovery equipment capable of recovering 1.0 GJ/yr (nominal). From this analysis, a decision could be made as to whether or not the investment was justified.

Often the heat exchange equipment is a critical item, depending on the temperature range and corrosiveness of the hot effluent streams. The selection of the optimum heat recovery system size involves a balancing of the cost of the equipment compared to the benefit of recovering still more heat.

See Chapter 8 for more information on heat recovery applications and technologies for recovering heat, such as regenerative and recuperative systems as well as heat pumps.

Electric Heat Applications

Due to its relatively higher cost, electricity is not used extensively for process heat. However, there are some types of applications where electricity offers advantages for heating. Electric heat can take several forms:

- Resistance heating.
- Induction heating.
- Dielectric heating.
- Electric arc heating.
- Microwave heating.
- Infrared heating.
- Heat pumps.

Resistance heating. Resistance heating makes use of the i^2R law; i.e., power dissipated is proportional to the square of the current times resistance. An example of this is a conventional residential electric water heater, which has two resistance heating elements, nominally rated at 3800 W and 240 V, single phase. This form of resistance heating is efficient because all the heat is transferred to the material being heated; i.e., the water. Losses result from conduction through the tank walls and distribution piping.

Induction heating. Induction heating is similar to resistance heating in that the actual heating is caused by current flowing through resistance. However, in the induction heater, the heating current is induced in the work piece. An example is the heating of transformers, cores, and motor windings. Even though they are laminated to produce high resistance to the flow of such currents, transformers are in effect inductance heaters.

In an induction furnace, a coil surrounds the work piece, which must be a conductor. A variable frequency power source (oscillator) is connected to this coil, inducing eddy currents that in turn heat the work piece. The eddy currents exhibit a "screening" effect; i.e., the current density at the surface of the work piece is maximum and decreases exponentially with depth. A "penetration" depth can be defined, wherein the current has decreased to about 37% of the surface value. Approximately 90% of the heating occurs within the penetration depth. Since the penetration depth is inversely proportional to frequency, a low frequency would be used for heating a large piece and a high frequency for a smaller size.

Example. A forge heater. Billets of steel are brought by a conveying system into a water-cooled copper coil. The frequency is in the range of 1−10 kHz; specific power is about 300 kWh/ton. Advantages of induction heating include excellent temperature control and no surface decarburization. The disadvantage is a low power factor (typically 0.1−0.5), which can be corrected with capacitors.

Dielectric heating. Dielectric heating refers to the heating of nonconducting materials by an electric field. Basically, this is similar to the heating that occurs in the dielectric of a capacitor on which a high-frequency voltage is impressed. The electromagnetic fields excite the molecular makeup of material, thereby generating heat within the material. As a result, the heat is distributed uniformly throughout the work piece. Dielectric heating can be applied to wood, paper, food, ceramics, rubber, glues, and resins. The heating effect is proportional to the dielectric loss factor, the applied frequency, and the electric field strength.

Dielectric systems can be divided into two types: RF (radio frequency) and microwave. RF systems operate in the 1−100 MHz range, and microwave systems operate in the 100−10,000 MHz range. RF systems are less expensive and are capable of larger penetration depths because of their lower frequencies and longer wavelengths than microwave systems, but they are not as well suited for materials or products with irregular shapes. Both types of dielectric processes are good for applications in which the surface to volume ratio is small. In these cases, heating processes that rely on conductive, radiative, and convective heat transfer are less efficient.

Most industrial applications are below 200 MHz. Depending on the frequency, care must be given to design the system to avoid interference with telecommunication.

Electric arc furnace. The *electric arc furnace* has three electrodes connected to the secondary windings of a three-phase transformer. The principle is the same as in electric arc welding. When an arc is struck, the nearby gas is raised to such a high temperature (in excess of 5000°C) that it becomes highly ionized. In this state, it is a sufficiently good conductor to be maintained at high temperature by the resistive heating produced by the current. The high temperature of the plasma permits very efficient heat transfer. Arc furnaces with capacities in the range of a few tons to hundreds of tons are in use. The primary application of electric arc furnaces is for melting and processing recycled steel.

Microwave heating. *Microwave heating* (a form of dielectric heating) is a highly efficient technique for heating by high-frequency electromagnetic radiation. Typically, frequency bands are 896 or 915 MHz and 2450 MHz, corresponding to wavelengths of about 0.33 and 0.12 m. Energy is deposited in the work piece according to the same principles as the dielectric heater described above.

Furnaces can be designed to be resonant or nonresonant. The microwave oven found in many homes is an example of a resonant cavity device. Resonant systems have efficiencies generally in excess of 50%. Again, because the heat is deposited in the work piece, losses are minimized.

Infrared heating. *Infrared heating* is generated by i^2R losses in heating lamps or devices, and this is a special case of resistance heating. The difference, however, is that infrared energy can be generated in a narrow bandwidth. This can be applied more efficiently in some cases than combustion energy that spans a broader bandwidth. To be most efficient, infrared heaters should concentrate their output at the peak of the absorption spectrum for the material being heated. For water, this corresponds to a wavelength of about 2.8×10^{-6} m. There are applications in papermaking, drying paints and enamels, and production of chemicals and drugs. Intensities in the range of $10-40$ kW/m^2 are possible.

Heat pump. The *heat pump* is basically a refrigerator operating in reverse. An evaporator receives heat from a low temperature heat source (the air, waste heat, ground, water, etc.). This causes evaporation of the working fluid; the vapor is then compressed by the compressor. In the condenser, it gives up the heat collected at the evaporator as well as the heat of compression. As this heat is delivered, the vapor condenses, and the hot condensate passes through the expansion valve.

Heat pumps fall into the several categories, depending on the type of heating and the purpose. Those used for residential HVAC and water heating are primarily air-source or ground-source heat pumps, meaning they extract heat either from the air or from underground pipes. Therefore they use air-to-air or liquid-to-air heat transfer. Larger units for commercial and industrial applications employ liquid-to-liquid heat transfer.

Example. Potential for reduction in CO_2 emissions. We conducted a study for EPRI on the potential to reduce greenhouse gas emissions in the U.S. by 2030 by expanding end-use applications of electricity in the residential, commercial, and industrial sectors.[10] For the industrial sector, we identified five electric process heating technologies capable of yielding net reductions in CO_2 emissions relative to their fossil-fueled alternatives, even considering losses due to electricity generation, transmission, and distribution. The technologies in order of highest to lowest potential for energy savings and CO_2 reductions are

- Heat pumps.
- Electric arc furnaces.
- Electrolytic reduction.
- Electric induction melting.
- Plasma melting.

The study timeframe was 2005−2030 and was based on the actual and projected generation mix in the U.S. for the four census regions (West, South, Midwest, and Northeast), as forecasted in the Energy Information Administration's 2008 Annual Energy Outlook's reference case scenario.

The results yielded a *realistic* potential of achieving CO_2 reductions of 45.1 million metric tons per year by steadily expanding use of these five technologies by 2030 across the U.S. The estimated *technical potential* was 73.1 million metric tons per year; however, the technical potential does not take into account cost-effectiveness or market acceptance, while the realistic potential attempts to factor in those key market barriers. The South and Midwest census regions had the highest potential for reductions because their generation mixes are more carbon intensive than the West and Northeast census regions.

This example helps highlight some of the benefits of electric process heating alternatives. In addition to electricity's advantages in delivering heat when and when it is needed, as well as in its ability to extract heat from lower temperature sources (i.e., using heat pumps), some electric

---

[10] EPRI (2009). *The Potential to Reduce CO_2 Emissions by Expanding End-Use Applications of Electricity*, Palo Alto, CA: EPRI. 1018871. Available for free download at http://www.epri.com.

process heat technologies save considerable energy in certain applications compared to fossil-fueled counterparts—so much so that they lead to lower overall greenhouse gas emissions, even when generation, transmission, and distribution losses are factored in.

TRANSFORMERS AND ELECTRICAL DISTRIBUTION SYSTEMS
Three-Phase AC Circuits

Virtually all heavy industrial power is provided via three-phase circuits. The reason for this is one of size and economy. Larger blocks of power can be transmitted more efficiently and machinery can be more compact with three-phase systems.

Since copper and other metals have an important effect on capital costs of large equipment, there is also a trend to increase voltage, permitting lower currents, smaller size conductors, and lower i^2R losses as well.

Three-phase voltages are generated with a generator having three sets of windings. Transformers likewise have three sets of windings. In a three-phase circuit, there are two types of connections: the wye and the delta. For either configuration, three-phase power is given by

$$\dot{W} = \sqrt{3}\ v_{line} i_{line} pf \quad \text{W} \qquad [11.14]$$

where

$\dot{W}$ = power, W
v_{line} = line voltage, V
i_{line} = line current, A
pf = power factor, dimensionless

Common nominal line-to-line three-phase voltages are 208, 230, 460, 575, and 2300 V.

Electric System Losses

Motors, transformers, and other electromotive and conversion devices make use of magnetic and electric fields for energy conversion. Mechanical energy is produced because of the fact that a conductor of length l carrying a current i in a magnetic field B experiences a force equal to

$$F = Bli \quad \text{N} \qquad [11.15]$$

where

F = force, N
B = magnetic field intensity, weber/m^2

l = conductor length, m

i = conductor current, A

Transformers, motors, and other all electrical mechanical devices lose energy through various mechanisms, including

- Internal resistance.
- Hysteresis.
- Eddy currents.

Internal resistance may be characterized as the losses caused by conductors (i^2R losses). Hysteresis losses are those that are inherent to magnetic fields. When the magnetizing force is removed, not all the stored energy is recovered. An empirical relation for hysteresis losses is given by

$$\dot{W} = k_h f B^n \quad \text{W} \qquad [11.16]$$

where

$\dot{W}$ = hysteresis loss, W

k_h = a material constant

f = alternating magnetic field frequency, Hz

B^n = magnetic field maximum intensity weber/m^2

n = an empirical exponent, typically 1.5–2.0

Eddy currents are losses that occur because the magnetic core material itself consists of materials that conduct electricity; as voltages are induced in this material by alternating magnetic fields, currents called *eddy currents* are produced. An approximate expression for these losses is given by

$$\dot{W} = k_e f^2 B^2 \quad \text{W} \qquad [11.17]$$

where

$\dot{W}$ = power, W

k_e = a material constant

f = alternating magnetic field frequency, Hz

B = magnetic field intensity, weber/m^2

When all types of losses are considered, the large transformers used in industrial processes typically have efficiencies at the high end of the 90 + percent range.

The energy lost in electrical distribution systems at voltages typical of industrial processes is predominantly in the form of i^2R losses. These losses can be determined by knowing the length of the conductor, the conductor material and cross-sectional area, and its resistivity per unit length. With these conductor properties known, the losses can be calculated as follows:

$$\dot{W} = i^2 \rho l \quad \text{W} \tag{11.18}$$

where

$\dot{W}$ = power lost, W
i = line current, A
ρ = line resistivity, ohms/m
l = line length, m

At very high voltages (experienced in transmission lines, but generally not in industrial applications), there are two additional sources of losses: *radiation* and *dielectric heating*. Radiation losses occur when the voltage is sufficiently high that the line acts as an antenna and radiates energy. Dielectric heating occurs when voltages are induced in insulating materials.

A secondary effect of transmission and distribution systems is voltage drop. Conductors need to be sized to minimize the voltage drop. There is a trade-off between the capital costs of larger conductors and reduced performance of equipment installed at the end of a long distribution line, where voltage drop causes the equipment to operate at a lower than specified voltage.

ELECTRIC MOTORS AND DRIVES

Electric motors and drives account for a substantial share of electricity use in the commercial and industrial sectors. In fact, in many industries they are the largest end-use of electricity, followed by nonprocess uses (such as lighting and HVAC, which also uses motors) and electric process heat. Therefore, improving energy management has the potential for significant energy savings.

The same factors that cause energy loss in transformers act in motors as well, contributing to losses in the stator, rotor, and windings. There will be $i^2 R$ losses in both the stator and rotor windings. Hysteresis and eddy current losses will occur in the stator and rotor core materials. Furthermore, there are at least two other sources of losses: friction and windage losses caused by the bearings and the motor fan, and stray losses, due to electrical harmonics and stray currents.

There is a great variety of motor types, including DC motors (less common in industry) and induction and synchronous AC motors. Induction motors can be further subdivided into squirrel cage and wound rotor types. Each of these exists in both single-phase and three-phase motors.

Although larger motors are inherently efficient, they are often used inefficiently. Oversized motors operating at less than full load have lower efficiencies and in addition have a lower power factor. Good motor maintenance (lubrication, proper cooling, etc.) are also essential to efficient operation. The selection of the motor type can have an impact on energy efficiency. Efficiency in motors basically depends on good design and use of high-quality materials. There is a wide variation in motor efficiency, ranging from 70% to 94% or higher for large (>200 hp) industrial motors. The National Electrical Manufacturers Association (NEMA) publishes specifications for electric motor efficiency. The latest is the "NEMA Premium."[11] It outlines the minimum requirements for three-phase AC induction motors applied to municipal and industrial applications for operation on voltages 5000 V or less, rated 2500 hp or less, operating more than 2000 hours per year at >75% of full load. In addition to the type and size, efficiency is also affected by motor load. Electric motors are most efficient when operated at or near full load. Refer to Table 7.1 for typical motor efficiencies.

Conversion of Electrical Energy to Mechanical Energy

Electric motors are generally found as components of larger systems, e.g., pumps, fans, process cooling systems, air compressors, conveyors, elevators, and assorted equipment for crushing, grinding, stamping, trimming, mixing, cutting, and milling operations. In each of these systems, there are other losses in addition to those in the motor. These may be broadly summarized as follows:

- Prime mover (motor).
- Power coupling (clutch).
- Power transmission (gears, belts, change shafts).
- Mechanical device (shaping, forming, transporting).

The efficiencies of mechanical components such as clutches, gears, belts, and chains are typically high in the range of 80−98% if used properly. However, when there are several of these components in the machine, the combined efficiency can be low. The wasted energy

[11] The National Electrical Manufacturers Association (NEMA) (December 2014) *General Specification for Consultants, Industrial and Municipal: NEMA Premium—Efficiency Electric Motors (600 V or Less)* Arlington, VA: NEMA.

is primarily expended overcoming frictional losses and is eventually dissipated as heat. In some circumstances, this is useful. In air-conditioned spaces, this heat will add to the total air conditioning load.

Variable Speed Drives

Variable speed drives (VSDs), also called variable frequency drives, are a valuable tool for the energy manager. Typically the VSD system consists of a three-phase AC induction motor and a variable frequency power supply. The variable frequency power supply uses solid state components to produce a pulse-width modulated current that varies the power and frequency supplied to the motor. This enables accurate control of the motor speed over a broad range. VSDs are used in connection with pump and fan applications to vary the pump or fan speed according to demand, often with large savings in energy use.

EMOs with Motors

Table 11.5 summarizes EMOs with motors. The table lists a series of actions that users can employ to reduce energy use. (Note that many electric utilities offer incentives for installing efficient motors and VSDs.)

Table 11.5 EMOs with electric motors

Operational improvements
• Supply rated voltage, properly balanced between phases
• Improve controls; turn motors off when not in use
• Schedule regular maintenance of motors
• Provide regular lubrication
• Provide adequate cooling
• Reduce peak demand by rescheduling motor operation

Retrofit improvements
• Improve power factor
• Improve cooling
• Employ heat recovery from large motors
• Replace old inefficient motors with efficient ones
• Properly size motors to run at full load

New installations or designs
• Purchase more efficient motors; evaluate on a lifecycle basis
• Consider using variable speed motors if loads vary
• Use higher voltages for motor drives
• Use three-phase motors rather than single-phase motors

It is clear that there is not much one can do with an existing motor, so long as it is operated efficiently. The inherent i^2R losses in the stator and rotor cannot be reduced without rewinding the motor or purchasing a new one. However, by understanding how these losses occur, the user can optimize motor use. Since running at no load wastes energy and results in a low power factor, it should be avoided. Operating motors as close to the fully loaded condition as possible will improve efficiency and power factor. If the power factor is very low (say less than 80% for a large motor or group of motors), it may be economically advantageous to install capacitors to correct it.

In the limited space available in this book, it is not possible to discuss all of the many applications of electric motors, so here (or in Chapter 8) we have focused on pumps and fans, refrigeration and process cooling, and compressed air systems. Additional details may be found in the references cited in this chapter.

PUMPS AND FANS

Pumps are of three types: centrifugal, axial-flow, or positive displacement. Optimum pump performance requires that it operate at its point of maximum efficiency. As noted above (see Equation 11.13), pumping power is related to the cube of the volumetric flow rate. Thus small changes in flow have an important effect on power requirement. The same is true of fans. Pumps and their EMOs are discussed in more detail in an earlier section of this chapter and in Chapter 8. Fans are covered in Chapter 8.

REFRIGERATION AND PROCESS COOLING

Commercial and industrial refrigeration systems and process cooling technologies function in much the same way as the space cooling systems described in Chapter 8. The primary difference is the application. Instead of providing space heating for occupant comfort, they serve functions such as the following:

- Refrigerating food and other products with commercial refrigeration systems.
- Providing cooling to food and beverage processing applications.
- Cooling refrigerated warehouses.
- Cooling data centers.
- Cooling materials and machinery.
- Regulating temperatures in industrial processes.

Great advances have been made in the energy efficiency of refrigeration and process cooling systems during the last several decades. (For an example that we can all relate to, today's refrigerators use only about 25% as much energy as those sold when the first edition of this book was published!) As another example, ammonia refrigeration systems have increased in popularity for industrial applications due to restrictions and phase out of chlorofluorocarbon, hydrochlorofluorocarbon, and hydrofluorocarbon refrigerants because they damage the ozone layer. Ammonia is a very effective refrigerant and is not ozone-depleting, nor does it directly contribute to greenhouse gas emissions.

The list below contains examples of EMOs with refrigeration and process cooling systems:

- Set the operating temperature in the proper range.
- For refrigerators or cold storage areas, avoid excessive opening and shutting of the doors.
- Install strip curtains for better containment.
- Improve insulation.
- Place cooling equipment in a "cool" location, i.e., not next to a source of heat.
- Optimize set points and controls.
  - Floating head pressure controls.
  - Fan scheduling or cycling.
  - Demand-based evaporator defrost schedules.
  - Raise suction pressure.
  - Sequence compressors.
- Maintain condensers (remove scale; maintain water quality).
- Reduce lighting and other building loads to minimize load on cooling system.
- Install more efficient fan motors.
- Install VSDs on compressors, condenser fans, evaporator fans, pumps, as applicable.
- Consider two-stage versus single-stage compression for low temperature refrigeration.
- Consider applications of thermal energy storage or heat recovery.

Refer to the literature for detailed descriptions of these and other opportunities.[12]

---

[12] Reindl, D., Jekel, T, and Elleson, J. (2005). *Industrial Refrigeration Energy Efficiency Guidebook*. Madison, WI: Industrial Refrigeration Consortium.

Example. Efficient process cooling and heating in a cheese and whey manufacturing facility. A cheese manufacturer in the northwestern U.S. uses heat recovery between milk and whey process lines to reduce heating and cooling energy requirements. The process works as follows. Milk collected from dairies is trucked to the plant 24 hours per day at 7°C (45°F) and stored in refrigerated silos. During processing, the milk is pumped from the silos and converted into cheese for 21 hours per day; the cheese making equipment is cleaned-in-place (CIP) during the other 3 hours per day. The first step in the cheese making process is pasteurization, where the milk is raised to a temperature of 74°C (165°F) for 15 s. Subsequent cheese making steps produce hot whey which must be cooled prior to further processing. Whey processing occurs 24 hours per day; during the 21 h of overlap between the milk and whey processing, the cold milk from the silos is routed through a plate and frame heat exchanger to precool the whey and preheat the milk. The manufacturing facility uses a refrigerated glycol loop to provide the additional cooling necessary to reach a whey processing temperature of about 16°C (60°F). The heat recovery that occurs in the milk-whey heat exchanger optimizes energy use in two ways: (i) precooling the whey with the cold milk lowers refrigeration (electricity) requirements and (ii) preheating the milk with the hot whey reduces steam (natural gas) requirements for milk pasteurization.

In a recent project, the facility explored energy-efficient and cost-effective options to increase the capacity of its whey cooling system. One concern with the previous system was that the glycol loop lacked sufficient capacity to meet all cooling needs of the whey stream during the 3 hours of CIP (when the milk was not available for precooling). The facility installed a fluid cooler to augment the existing cooling system during those 3 h instead of increasing the capacity of the glycol-cooled refrigeration system. The fluid cooler is essentially a cooling tower that uses noncontact cooling water to cool the whey, whereas the glycol-cooled refrigeration system relies on a more energy-intensive refrigeration cycle. The facility still uses the glycol system to get the whey to the correct temperature, but instead of using the glycol system to bring the whey from about 74°C to 16°C (165−60°F) during those 3 h, the plant uses it to bring the whey from about 32°C to 16°C (90−60°F).

According to energy analysis estimates, installation of the fluid cooler saves the facility over 900,000 kWh per year. This level of savings equates to energy cost savings of roughly US$46,000 annually. The facility received an incentive of over US$110,000 for the project from the local utility, which covered approximately one-fourth of the fluid cooler cost. Other improvements made to steam use on the milk pasteurization side of the process save an estimated 6770 GJ (6420 million Btu) per year in addition to the fluid cooler electricity savings.

ELECTROLYTIC SYSTEMS

Electrolysis involves movement of positively- or negatively-charged ions within an electrolyte between an anode (positively-charged electrode) and a cathode (negatively-charged electrode). These familiar processes involve electrolysis:

- Storage batteries.
- Welding.
- Corrosion.
- Electro winning (refining of metals such as aluminum).
- Plating and anodizing.
- Electroforming, electrochemical machining, and etching.
- Fuel cells.

Example. A lead–acid storage battery. In this battery, a plate of metallic lead is attached to the negative terminal. The positive terminal is connected to a plate consisting primarily of porous lead dioxide paste in a matrix. In between the plates is a solution of concentrated sulfuric acid. As the battery discharges, the reaction produces positive ions that drift away from the negative terminal, leaving behind electrons. During discharge, acid is consumed and water is formed, decreasing the specific gravity of the fluid in the battery. When recharging, water is consumed and acid is produced. Under fully charged conditions, the potential of a single cell is slightly more than 2 V. Six cells are required to produce a 12 V battery.

Faraday's laws

Faraday's laws form the basis for electrolysis. Briefly stated, Faraday discovered that 1.0 Faraday (26.8 ampere-hrs or 96,494 coulombs) would liberate one gram-equivalent of an element at an electrode during electrolysis.

Using Faraday's law, it can be shown that only a small fraction of total mass of the storage battery reacts during charging/discharging (about 4 g equivalents for a typical automobile battery or about 0.4 kg).

Corrosion

Corrosion occurs as a result of oxidation—reduction reactants between a metal or alloy and a corroding agent. Corrosion can occur as a result of chemical reactions, which usually require high temperatures and a corrosive environment, or due to electrochemical reactions, which are more common. Note that corrosion is an important indirect use of energy.

The electrochemical reactions resemble the processes that take place in a battery. They can arise when dissimilar metals occur in the presence of an electrolyte or in the presence of external electric currents. A common electrolyte is water with trace amounts of dissolved salts, acids, or alkalis. The rates of corrosion reactions are dependent on the concentration of salts, acids, or alkalis in the electrolyte, and on the surface, temperature, and chemical constituents of the corroding metal.

Welding

Where possible, AC welders are preferred as they offer a better power factor and more economical operation. Automated systems reduce standby power losses compared to manual welding because they place the weld bead more consistently (less start/stop).

Electro winning

An important use of electrolysis is the refining of metals such as aluminum. Basically the original process involved the electrolysis of a solution of aluminum oxide in molten cryolite, using carbon anodes and electrodes. In the electrolyte solution, aluminum oxide disassociates into aluminum and oxygen ions. As currents on the order of 10^5 amperes pass through the cells (at potentials of 5.0—5.4 V), the aluminum ions migrate to the cell lining (cathode) where they are reduced to metallic aluminum. This process required 15—20 kWh per kg of electricity. New processes have been developed that reduce the amount of electricity required.

Plating and Anodizing

An electric current flows in a tank where the object to be plated or anodized serves as one of the electrodes. In plating, the plated object serves as the cathode and the anode has the material to be electrodeposited. Alternatively, the anode may be nonconsumable carbon and the plating material may be drawn from the bath. In anodizing (typical for aluminum), the object to be anodized is the anode and a direct current produces a buildup of aluminum oxide on the surface. By use of various organic acids, colored finishes can be produced.

Electroforming, Electrochemical Machining

Electroforming is a process whereby a thin layer of metal is deposited on an object to be coated or on a mold that is later removed. The classic example is copper plated baby shoes! Electrochemical machining is the reverse of plating; a high current is passed between an electrolyte and the part, removing metal. This process is used for fine, intricate parts or hard, difficult-to-machine metals.

EMOs in Electrolytic Processes

Table 11.6 summarizes typical energy management possibilities for electrolytic processes. The greatest users of energy in this field (aside from the large indirect use caused by corrosion) are in primary metals production, particularly aluminum and magnesium.

Table 11.6 EMOs in electrolysis

Corrosion protection

- Use protective films, paints, epoxy
- Provide cathodic protection (sacrificial anodes)
- Cathodic protection with an applied voltage
- Electroplating and anodizing
- Use chemical water treatment (corrosion inhibitors)
- Avoid contact of dissimilar materials (dielectric unions)

Storage batteries

- Provide adequate maintenance (replace electrolytes, clean terminals, etc.)
- Use efficient charging techniques, charge at proper rates
- Avoid overheating, provide adequate ventilation

Electrolytic processes

- Insulate plating tanks
- Provide proper maintenance of electrodes and rectifiers
- Recover waste heat
- Use more efficient rectifiers (semiconductor vs. mercury arc)
- Use more efficient controls
- Develop improved electrode design and materials to increase efficiency

COMPRESSED AIR

Compressed air is a major form of process energy and is used in a variety of applications, such as to drive power tools and machinery, for conveyance, for spraying and painting applications, to operate controls, in injection molding, and in drying and cleaning processes. Compressors are either fuel-fired (engines) or electric motor driven.

Example. Replacing an air compressor with a variable frequency drive (VFD) air compressor. In a candy factory on the East Coast of the U.S., process air was supplied by a 150-hp air compressor that was considered oversized for the load. The air compressor operated for 24 h, 5 weekdays per week, off on Sunday, and on for part day Saturday. Measurements indicated that the original air compressor used 515,269 kWh per year. The decision was to replace it with a "right-sized" 100 hp unit with a VFD. Figure 11.7 shows the measured performance of the new unit during a typical weekday. Based on the load measurements, the VFD compressor used 174,101 kWh per year, so the project energy savings were estimated to be 341,168 kWh per year with coincident peak demand savings of 17 kW (summer) and 28 kW (winter). The project cost was US$116,000 with a utility rebate of US$45,000. The average electricity cost (demand and energy charges combined) was US$0.16, so the savings was US$54,587 per year. This savings, when added to the rebate, resulted in an attractive simple payback of 14 months for this project.

Whether fuel-fired or electric motor driven, the energy required depends on the efficiency of the machine and the work necessary to

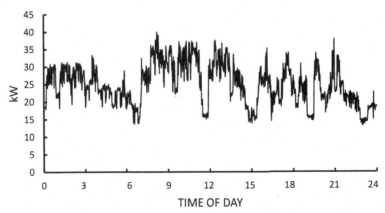

Figure 11.7 Demand profile for a 100-hp air compressor in candy packaging plant.

compress the air. EMOs are related to reducing the amount of compressive work, recovering heat, reducing losses of potential energy stored in the compressed air, and avoiding misuse of the compressed air. Efficient operation is influenced by

- Selecting appropriate type and size (rotary, reciprocal, etc.).
- Providing for intake air as cool as possible.
- Minimizing system pressure.
- Maintaining filters and coolers.
- Eliminating leaks.
- Recovering heat.
- Replacing inappropriate uses of compressed air.
- Replacing air-powered tools with electric tools when appropriate.
- Optimizing load with unloading controls, automatic shutdown, compressor sequencing controls.
- Using air receivers to minimize loading/unloading cycles.
- Installing dew point demand controls on air dryers instead of timed purge.
- Using higher efficiency motors.
- Installing VSDs on compressors to eliminate idling.

The optimum type of compressor depends on the capacity, pressure, and duty cycle. Selection of the best type for the job can lead to 10–20% improvement in efficiency. Intake air should be as cool as possible, as can be seen in the following equation:

$$PV = mRT/M \quad \text{J} \tag{11.19}$$

where

P = pressure, N/m^2
V = volume, m^3
m = mass, kg
R = universal gas constant, $8314.4\,\text{J/kmol} \cdot \text{K}$
T = temperature, K
M = molar mass (equals 29 kg/kmol for air), kg/kmol

At atmospheric pressure, the volume of air to be compressed will be directly proportional to its temperature. The compressor will have to do more work to compress hot air. Table 11.7 shows a savings resulting from moving the compressor air intake to a cooler location.

The pressure of a compressed air system is dictated by the pressure requirements of equipment on the system. Significant savings, in the range of 10–20%, can be obtained by reducing the operating pressure of the

Table 11.7 Power requirements and compressor air inlet temperatures

Temperature of air intake		Intake volume in m³ required to deliver 1000 m³ of free air at 21°C (70°F)	% KW savings or increase relative to 21°C (70°F) intake
°C	°F		
−1	30	925	7.5% Savings
5	40	943	5.7% Savings
10	50	962	3.8% Savings
16	60	981	1.9% Savings
21	70	1000	0
27	80	1020	1.9% Increase
32	90	1040	3.8% Increase
37	100	1060	5.7% Increase
43	110	1080	7.6% Increase
49	120	1100	9.5% Increase

Figure 11.8 Air compressor.

compressed air system. Leaks reduce system pressure and result in longer compressor operating hours, so repairing leaks will permit system pressure to be maintained with less compressor power. In a large plant, various types of equipment will have different pressure requirements. The typical approach is to provide a compressor that can meet the highest pressure required (see Figure 11.8 for a photograph of a compressor).

An alternative approach is to segregate the load by maximum pressure. The low pressure loads can be served by one compressor, while the high-pressure loads are served by another.

In multicompressor systems, control strategies can go a long way in improving efficiency. Sequencing controls optimize operation of compressors to match loads most efficiency. VSDs can also help compressors meet variable loads more efficiently, especially compared to fixed speed rotary screw compressors that operate in either full load or idle mode. Compressors still use energy when idling even though they aren't serving any loads!

Obviously, leaks represent a real and often substantial loss of energy. It is common for leakage rates in a compressed air system to be in the range of 2−20% of system capacity. Compressed air is an expensive energy form. Efforts should be undertaken to eliminate all leaks. Compressed air leaks can occur in piping, valves, fittings, flanges, hose, traps, and filters. A good test is to shut down all equipment that uses compressed air (e.g., on the weekend). Then measure the amount of compressor operation and search for leaks. Air lost through leaks represents energy expended to maintain the system pressure. Tag the leaks and repair them immediately.

Example. Leak repair. Repairing a 9.5-mm- (3/8-in) diameter hole in a 793-kPa (115-psi) system operated for 8520 hours per year would avoid an energy loss of about 392,000 kWh per year. A leak of that size corresponds to a leak rate of about 7.4 m^3/min.

It is also important to minimize or eliminate unnecessary uses of compressed air. Where feasible, use blowers or air knifes since they use less energy. Train staff to use compressed air only when needed, and not for tasks such as cleaning off work surfaces, floors, or clothing.

Heat recovery is another possibility with compressed air systems. Compressors are either air-cooled or water-cooled. Heat recovery may not be economical with small air-cooled units. However, larger units, especially water-cooled ones, should be reviewed for heat recovery opportunities, particularly if a cooling tower is being used to reject waste heat.

Example. Compressed air heat recovery. A large aircraft equipment manufacturing plant used electricity, natural gas, and compressed air as major energy forms. Hot water was provided by three natural gas-fired package boilers, one rated at 300 boiler hp and 200 psi, and two rated at 150 boiler hp and 165 psi. (*Note*: in U.S. and British usage, 1 boiler hp = 9.81 kW.) The plant also had five large air compressors totaling 16,000 hp, three of which were rated at 4600 hp while the other two were rated at 1500 hp and 700 hp. Either the 300 hp boiler or the two 150 hp boilers with a combination of the compressors were operated 24 hours per day, providing heat for the cafeteria, various processes and heat exchangers, and for building heating. The 24-h average heat load of the boilers was 4.7 GJ/h. A study was made to determine if heat recovery from the air compressors could reduce natural gas usage in the boilers.

The three large three-stage reciprocating compressors were considered for this purpose. These compressors had water-cooled interstage heat exchangers. Cooling water was conducted to a cooling tower where the heat was discharged to the atmosphere. Typically, two out of three of these compressors operated, with the load varying between 75% and 100% of full load. Under these conditions, measurements indicated that 3.7 GJ/h of heat could be recovered at temperatures above 65°C (150°F). On average, the recovered heat from the compressors could provide about 80% of the boiler heat. If one compressor operated at full load, all the heat could be provided. The principal difficulty was the fact that compressor operations do not always coincide with the peak heating requirement. Thus, in reality it would be difficult to meet all the needs without adding some form of energy storage. A more reasonable proposition would be to supply 50% of the heat required based on the observed compressor operation. This could be accomplished by adding several heat exchangers and a small amount piping. In this facility, it was convenient that the air compressors and boilers were adjacent to each other in the same building. The payback for these modifications (based on the natural gas savings at current costs) was estimated to be less than 1 year. The estimate did not include additional savings that would accrue from modifying cooling tower operation. The plant had four cooling towers, each with 40 hp fans and 9 cooling water pumps totaling 450 hp. Eliminating the use of one cooling tower would result in additional savings that were not considered in the analysis.

The U.S. Department of Energy has a variety of resources for analyzing and improving compressed air efficiency (http://www.energy.gov/eere/amo/compressed-air-systems).

MANUFACTURING PROCESSES[13]

In manufacturing processes, there are two forms of energy to consider. The first is the direct energy required to carry out the process. The second is the indirect energy embodied in the material being processed.

[13] See also: Smith, C.B., Capehart, B.L., and Rohrer, W.M. Jr.(2007) "Industrial Energy Efficiency and Energy Management," Chap 14 in Kreith, Frank and Goswami, D. Yogi, eds. *Handbook of Energy Efficiency and Renewable Energy*, Boca Raton, FL: CRC Press.

Figure 11.9 Computer numerical controlled (CNC) lathe.

Consider a chemical processing plant producing a product involving several stages of processing. Spillage or waste of raw materials at an intermediate processing step constitutes an energy waste in a very real sense of the word. Even discharged cooling water is a waste not only because of any heat it may carry but also because of the energy expended to pump, filter, and chemically treat it.

The following typify the broad range of EMOs to be encountered in manufacturing operations:

- In machining, eliminate unnecessary operations and reduce scrap.
- Use computer numerically controlled machine tools (Figure 11.9).
- Substitute powder metallurgy for machining operations (use induction furnace).
- Size the tool, the feed rate, the depth of cut, and the motor to the job.
- Carefully review hardening requirements. Carburizing is energy intensive; induction heating less so, plating, metalizing, flame spraying or cladding still less. Select the method that does an adequate job most efficiently.
- Substitute cold forging for hot if possible.
- Avoid overdesign that requires excessive use of energy-intensive materials.
- Reuse or recycle scrap.
- Design products to permit recovery of energy-intensive materials or components.

- Control air usage in paint spray booths. Exhaust air carries away heat.
- Evaluate paints and coatings. Solvent-based paints are less energy intensive than plating; water-based paints still less. Powder coating has the advantage of not requiring a liquid solvent.
- Improve plating efficiency by reducing heat losses from tanks, and by improving rectifier performance, checking cables and conductors, and modifying controls.

Agriculture and Food Processes

Modern agriculture is both a user of energy and a producer of fuels. In the post-World War II period, there has been a sharp decline in the number of family-owned and operated farms in the industrialized nations. Commensurately, the use of on-farm energy has increased dramatically. Sharp increases in farm yields per hectare (acre) have resulted. Today, many farms are both energy users and fuel producers. In addition, energy use in agriculture is both direct and indirect.

Direct Uses

The major direct uses are
- Buildings, barns, greenhouses, livestock and poultry enclosures.
- Pumping and irrigation.
- Mechanized equipment, tractors, engine driven pumps, harvesters, etc.

Electricity and natural gas are the primary energy forms for the first two categories, while diesel and gasoline fuels are the primary energy sources for mechanized equipment.

Farm building energy uses are primarily heating, cooling, and ventilation, lighting, and processes. On-farm energy use is ubiquitous, ranging from electrically powered milking machines to heating of poultry brooders to maintaining illumination levels that stimulate egg laying. Refrigeration is also important, especially in dairy operations. Many of the energy management principles discussed in Chapters 8 and 9 and elsewhere in the current chapter are equally applicable in the agricultural sector. Some of the principles discussed in Chapter 10 can be considered for on-farm vehicles as well as for vehicles used to transport materials and products to and from farms.

Pumping and irrigation are important in arid areas where rainfall is inadequate for crop production. Many new technologies have emerged to improve pumping efficiency and to reduce water loss by evaporation, including drip irrigation rather than sprinklers.

In responding to higher fuel prices, studies show that farmers keep farm equipment engines serviced and tuned to improve fuel economy. In addition, they reduce the number of trips over fields for tilling, planting, and harvesting through the use advanced technology (global positioning satellites and mapping). These techniques, along with soil testing and surveys by aircraft or drones, enable more precise irrigation and fertilization of crop lands.

Indirect Uses

The greatly increased productivity of modern agriculture is due in part to indirect energy uses, of which fertilizers and pesticides are the most important. During the decade 2001–2011, U.S. farms energy input (direct and indirect) totaled approximately 1.8×10^9 GJ/year (1700×10^{12} Btu/year). Fuel accounted for about 35% of this total, fertilizers about 30%, electricity about 25%, pesticides about 6%, and miscellaneous about 4%.[14]

Note: These relative quantities varied from year-to-year, largely in concert with the increase or decrease of the price of oil and natural gas, but overall the total usage was relatively constant at 1.8×10^9 GJ/year plus or minus 5% or so. It should be noted that natural gas is an important component in fertilizer production.

On-Farm Energy Production

Farms are slowly increasing their use of renewable energy sources, primarily solar energy and wind power. In the U.S., less than 2% of farms report onsite renewable energy generation.[15]

What has changed dramatically in the U.S. is the growing of corn for ethanol production and soybean oil for biodiesel production. Brazil was an early pioneer in the biofuels field producing ethanol from sugarcane as early as the 1930s, and then substantially expanding production during WWII.

Anaerobic digestion is another growing area of on-farm energy recovery and production. In livestock production, anaerobic digestion of animal residuals (manure) reduces nutrient levels and the overall manure volume to the point where the manure can be applied to land. The

[14] Beckman, J., Borchers, A., and Jones, C.A. (2013) p. 9. *Agriculture's Supply and Demand for Energy and Energy Products, EIB-112.* Washington, DC: U.S. Department of Agriculture, Economic Research Service.

[15] Ibid, p. iv.

digester gases resulting from the process are considered a biofuel (methane) and can be used directly for applications such as heating; alternatively, they can be used to generate electricity.

Food Processing

Farm products must be moved quickly and efficiently to processing plants or markets to prevent waste or spoilage. High value crops such as tropical fruits may be moved by air, sea, or in refrigerated trucks. Food processing encompasses slaughter houses, meat packing plants, canneries, grain milling, freezing, bakeries, and dairy products. Food processing is an extremely diverse sector of industry that makes use of process heat for cooking and preserving foods, refrigeration for freezing and cold storage, and many mechanical processes for cleaning, trimming, and packaging.

Many of the energy management options discussed elsewhere in this chapter are equally applicable to food processing operations.

Energy Storage for Process Industries

Energy storage is an old concept. An early application was pumped storage, where water was pumped from a low reservoir to a higher one at night, and allowed to flow back during the day to generate electricity during peak hours. Batteries are another well-known method for storing energy. Experiments have been done to store energy as high pressure air underground or using high speed fly wheels.

For process industries, energy storage provides an important tool for the energy manager in efforts to increase overall process efficiency. Much of the waste of energy that occurs in processes takes place in the form waste heat that is rejected to the atmosphere either directly from the process or indirectly via heat exchangers, cooling towers, stacks, and so on. There are a number of technologies available that permit energy storage at high and low temperatures. These processes will permit a better overall use of energy resources and can also help reduce the peak demand for certain energy forms, particularly electricity.

In the broadest sense, the quantity of energy stored in any media consists of four components: the sensible heat of the material in its solid and liquid phases, the latent heat of phase change, and any heat associated with chemical reactions.

Energy storage at high temperatures has been used for many years in industry. The metallurgical industries and their use of refractories is one example. The range of technologies that can be considered may be classified by operating temperature:

- **Sensible heat storage**
  Solids: Packed beds, fluidized solids (100–1500°C)
  Liquids: Hot water, organics, liquid metals, molten salts (200–800°C)
- **Latent heat**
  Solid/liquid phase change materials: Salts, metals (120–1400°C)
- **Chemical energy storage**
  Chemical or catalytic reactions (350–700°C)
  Thermal disassociation reactions (500°C).

Due to the wide variety of thermal energy use in processes, it is not possible to provide specific guidelines. However, in addition to the metallurgical industries, there have been many applications in food processing, clay and ceramics manufacturing, and industries using steam, such as paper, and textiles. The special case of building heating and cooling is discussed in Chapter 8. Chapter 12 describes other options for thermal and electrical energy storage in integrated building designs.

Process Control

Control of industrial processes is an important determinant of energy efficiency. We've already seen how control of boilers, pumps, fan, and compressed air systems has a critical impact on energy efficiency. The same is true of a variety of other process controls.

In the original concept, process controls were designed to accomplish one or more of the following:

- Maximize production throughput.
- Maintain product quality.
- Reduce human labor.
- Provide safety for equipment or personnel.
- Use raw materials efficiently.
- Control process parameters within specified limits.

Except for the most energy-intensive industries such as aluminum or cement production, the efficient use of energy for many years was a secondary concern, since its value was much less than the value of the final product, or of the labor that went into making the product. Today the situation is different.

Now the optimization of process controls to use energy efficiently is essential. Many new types of controls and methods have been developed to do this. Some examples include simple timers, sensors, feedback devices, meters, programmable logic controllers (PLC), microprocessors, large-scale numerically controlled equipment, distributed control systems (DCS), supervisory control and data acquisition (SCADA) systems, programmable automation controllers (PAC), and web-based facility automation systems. These technologies and control strategies often overlap. As the sophistication of process controls has increased, so have their costs. The increased cost of controls has generally been more than offset by savings of rising energy and labor costs.[16]

Improved process controls save energy by reducing downtime and speeding up production. Often this requires improved instrumentation to sense, monitor, and measure process parameters. With the rapid development and advances in microprocessors, "intelligent" controls are now widely found in industry. These advanced devices can sense changing conditions (such as tolerance variations as a tool wears) and adjust the process to accommodate the changed circumstances. Numerically controlled machines can optimize part layout and manufacturing, minimizing scrap and waste of materials.

Example. Optimal scheduling. Process heat is a major source of industrial energy use. Industrial heating operations such as baking, tempering, annealing, and drying can use more energy in start-up, preheating, and standby time than is actually used in making the product. The simple act of optimal scheduling—i.e., planning production so the equipment is used as long as possible once started up—can materially reduce energy costs.

CONCLUSIONS

Process energy—especially in industry—is a vital application of energy resources. Industrial processes are responsible for food preservation, clothing, pharmaceuticals, furniture, vehicles, and many other products that have improved the standard of living. Industrial processes also represent extremely diverse uses of energy, ranging from simple heating to the complex operations required to form, shape, and weld metals. Great improvements in industrial energy use efficiency have been made in the last three decades. Specific energy use—i.e., the energy required to produce a unit of any industrial product—has declined in every sector.

[16] Smith, C.B., Capehart, B.L., and Rohrer, W.M. Jr. (2007), Op. cit. pp. 14-34-36.

With rising energy costs, that means more goods can be produced with less energy expense. More significant, it also means that less CO_2 is produced per unit of production.

There are still areas for substantial improvements in process energy use efficiency. Studies of "best available technology" indicate there is a gap between the *average* energy use required to make specific products and the most efficient application. This gap represents an opportunity for improvement and further reduction of industrial energy use. New processes, better controls, and more efficient types of equipment are available to improve energy use efficiency.

CHAPTER 12

Integrated Building Systems

INTRODUCTION

Prior to the advent of low-cost and readily available fossil fuels, humans, in common with all other animals, satisfied their energy needs with their own energy and those renewable resources readily available to them.

The first human uses of energy are not known positively, but may be surmised to have been for warmth and security, followed by cooking and transportation. Early in human evolution the concept of shelter undoubtedly emerged, since we see this practice today by our mammalian relatives who are lower on the evolutionary ladder than we are.

Today the operation of buildings directly accounts for over 40% of the energy used by industrialized nations such as the U.S. and the European Community, and indirectly (in materials and construction) accounts for an additional amount. Of the energy use in buildings, over 40% is used for heating, over 10% is used for lighting, and the balance for air conditioning, hot water heating, refrigeration, computers, and other uses.[1]

Building energy use is influenced by a variety of parameters:

- Micro- and macro-climate.
- Site location.
- Building function.
- Occupancy and use.
- Building configuration.
- Building orientation.
- Building envelope design considerations.
- Building materials.

In previous chapters, we have seen how to achieve efficient energy use in HVAC systems, lighting, and processes. These are major components of buildings and have an important influence on building energy use. However, the importance of the building shell, or building envelope as we shall call it, must also be recognized. In this chapter, we shall discuss

---

[1] Refer to Table 2.3 for energy use by sector. To the residential and commercial uses listed, some portion of industrial energy use can be attributed to buildings, however, usually this is not differentiated from process use.

Energy Management Principles.
DOI: http://dx.doi.org/10.1016/B978-0-12-802506-2.00012-4

the importance of the building envelope, and then consider the integration of various subsystems (building envelope, passive systems, HVAC, lighting), in such a way as to maximize energy use efficiency. Unfortunately, the very process by which buildings are designed is often based on the work of separate disciplines; e.g., an electrical engineer (lighting), the mechanical engineer (HVAC), and the architect (building envelope). If their efforts are not carefully coordinated, the building suffers in overall efficiency. More often than not, this is actually what happens, since each designer is marching to his or her own drummer. To deal with this situation, a new discipline called Building Information Management (BIM) is becoming popular.

GENERAL PRINCIPLES OF ENERGY MANAGEMENT IN BUILDING SYSTEMS

Energy management in buildings can be approached by considering:
- The building site.
- Building envelope.
- Building systems.
   Table 12.1 summarizes general principles.

The choice of site determines the climatic conditions to which the building will be exposed. The building envelope determines how the site conditions influence the building occupants. It is literally a porous membrane that transmits energy, light, gases, and water vapor, back and forth between the building and its environment. (It is analogous to water entering the building through the walls, roof, and floors if it were immersed in water.)

The building systems supplement the naturally available light, heat, and cooling power of the environment. To the extent that the use of these supplementary systems is integrated with the capabilities of the building envelope and the site characteristics, energy use can be optimized. A building that is constructed without consideration of the site parameters, with the building envelope configuration determined solely by architectural considerations, and with the mechanical and lighting systems designed independently and not integrated, would be expected to use an excessive amount of energy.

In this chapter, we shall explore how each of these variables—site, building envelope, and building systems—can be manipulated to minimize energy use. Finally, when all three are carefully integrated, very efficient building concepts can be developed. Although not always the

Table 12.1 Energy management principles for integrated building systems

---

Site

- Take advantage of micro- and macroclimate conditions.
- Orient buildings for most favorable wind and sun conditions.
- Take advantage of shading provided by plants or topographical features.
- Use plants or topography for windbreaks.
- Use vegetation to influence microclimate (reduce heat absorption, provide evaporative cooling).
- Use bodies of water (natural or artificial) to influence microclimate.
- Select a site that minimizes transportation energy.
- Use contours by terracing for optimal energy use.

Building Envelope

- Provide shading to reduce solar heat gain and protect against wind losses.
- Optimize building volume, area, and layout for energy efficiency.
- Maximize use of daylighting.
- Tighten building envelope—minimize infiltration/exfiltration.
- Adequately insulate building envelope—minimize conduction losses.
- Improve glazing—reduce window heat gains and losses (low E-glass).
- Incorporate a *Trombe* wall and direct gain systems.
- Use high-performance facades.
- Provide thermal energy storage capacity—either passive or active.
- Incorporate renewable solar energy devices as structural elements in the building envelope (photovoltaics or hot water heating).
- Provide proper design of entrances (vestibules).

Systems

- Design lighting and heating systems to make use of daylight and solar heat gain.
- Optimize lighting and HVAC systems to deliver lighting and comfort conditions only to occupied areas; use unoccupied areas as buffers against unwanted heat gains and losses.
- Provide automatic controls for lighting and HVAC systems with local override capability, including the use of programmable electronic controllers.
- Optimize use of summer/winter outside air.
- Employ heat recovery from lights, equipment, and environment.
- Group systems so heat producing and using activities are adjacent to facilitate heat recovery.
- Use sensible zoning.
- Reduce peak demand.
- Employ cogeneration of heat and power.

---

case today, this approach is becoming increasingly common in building design as rising energy costs and an overarching desire to reduce greenhouse gas emissions create a greater imperative for more efficient energy use.

ENVIRONMENTAL CONFORMATION

A number of years ago, from a small cabin situated on the fringes of the great Mojave Desert, the authors carried out a series of studies regarding energy management practices of animals. It was here that we first observed what we have since come to call *environmental conformation*. This is the strategy employed by most animals that have to subsist on the sparse supplies of energy that nature makes available to them. This strategy stands in rather stark contrast to the common human approach, which we will term *environmental manipulation*.

Example. On a late fall or early spring morning, a desert side-blotched lizard (*Uta stansburiana stejnegeri*) may be observed to emerge slowly from its burrow and perch on the side of a rock that has been warmed by the morning sun. In doing so, it increases its body temperature to the point where it can actively forage for the insects that comprise its food supply.

This simple act is profound in its awareness of efficient energy use. The lizard has employed an understanding of microclimate (the exposed rock is warmer than its burrow), site orientation, and radiative heat transfer to assist its metabolism and, ultimately, its ability to capture food and escape predators.

Effects of Climate

Buildings are inextricably linked with climate. The relationship encompasses much more than just temperature. As noted in Chapter 8, human comfort depends on humidity and air velocity as well as temperature. The heat gains and losses in buildings depend on these factors, as well as on the intensity of solar energy, shading by trees or hills, wind velocity, and the duration of heating and cooling seasons. Olgyay was the first to observe that similar building forms have evolved around the world in response to similar climatic stresses.[2] In colder climates, houses of heavy timber construction are found. In the equatorial forests and tropical belts, the roof is more essential than the walls, which are often omitted to permit cooling. In hot arid regions, massive structures to provide shading and storage of "cool" are common.

Not only does the local environment affect the building but vice versa. The *urban heat island effect* (a phenomenon whereby the ambient air temperature above cities is warmer than the surroundings due to

[2] Olgyay, V. (1963) *Design with Climate*. Princeton, NJ: Princeton University Press.

additional energy absorption as well as heat given off by the city) is well known. However, the same effects occur on a smaller scale. A structure surrounded by nothing but concrete and asphalt will be hotter during the summer, since these materials absorb more heat, and the lack of vegetation means that evaporative cooling caused by plants (as well as the shading they provide) will be absent.

Microclimate Important

Weather maps and climate data represent average conditions portrayed on a large scale. Within a local region, there can be wide variations in the local climate. These *microclimate* conditions are important to recognize in the design of buildings.

For example, the desert cabin alluded to earlier in this chapter is on a north facing hillside in a narrow canyon at an elevation of approximately 1050 m (3450 ft). The canyon is on the east slope of one of the dominant mountains in the area. A stream flows from west to east in front of the cabin on years when there is adequate rainfall.

These factors combined to create weather conditions very different from the rest of the Mojave Desert; in fact, very different from areas only 3 or 4 km distant. We measured rainfall for a number of years in this location; it was frequently greater than Los Angeles. Snow falls several times a year, even though the summer days range above 40°C. Most striking though is the difference in temperature and vegetation between "our" side of the valley and the opposite, south-facing side. There, the vegetation is much scanter and less diverse, being limited to those species that can survive the intense sun to which they are exposed. Water has a higher specific heat than land so the effect of our nearby creek is to cause winter temperatures to be slightly warmer and the summer temperatures slightly cooler.

Similar microclimatic conditions can be found almost anywhere; the variations lead to certain building sites that are better than others. Topography is an important contributor to microclimate. For example, at our desert cabin location, during the hot summer months the prevailing breezes in the afternoon blow down our canyon and provide some relief from the heat. Olgyay describes how nighttime cooling creates currents of cold air. These flow through valleys in a manner analogous to water, causing local temperature minima. However, there are other locations such as valley slopes, which remain warmer unless exposed to wind.

The intensity of solar radiation is also greatest on the surface normal to the incoming radiation. By proper selection of a hillside site, one can maximize the winter heating and minimize summer heating.

BUILDING FUNCTION

Building function obviously plays an important role in energy use. Industrial buildings with heavy process energy requirements require different approaches compared to residential or commercial buildings.

OCCUPANCY AND USE

Occupancy also plays an important role in determining energy use. Is the building occupied 24 hours per day, as in a hospital? Are there special requirements for security or fire/life safety, as in banks and schools?

PASSIVE DESIGN CONSIDERATIONS

Passive design elements are those that require no external energy source to be effective, as opposed to *active design* elements. An example of the former is a window sunscreen, while example of the latter is a window air conditioner. Some typical passive design techniques are shading, site orientation, and building configuration.

Shading

Shading is an important design tool to reduce the effects of both sun and wind. For solar loads, the variation in the sun's elevation from summer to winter is a prime factor. On single story structures, horizontal overhangs effectively shield against the high summer sun, but allow the low winter sun to penetrate and warm the building. Vertical louvers are also useful.

Example. Native American cave dwellers. In ancient times, passive design techniques were widely employed. As Schoen points out in a case study, Native Americans in the Southwest used passive design techniques to accommodate a harsh environment.[3] Cave dwellers constructed buildings from sandstone and adobe. The heavy masonry walls had a large

[3] Schoen, R. (1978) "The Cliff Palace, a Pre-Columbian Cliff Dwelling as an Example of Designing for Efficient Energy Use", pp. 619–624 in *Efficient Electricity Use*, Craig B. Smith, ed., New York, NY: Pergamon Press.

thermal capacity and were able to moderate the large diurnal temperature swings experienced in the desert. The best caves faced southwest and were protected by the cliffs above. During winter, the sun would warm the structures, which reradiated thermal energy at night, helping maintain comfortable conditions. During the summer, the buildings were protected from direct sun by the overhang. At night, cooling breezes removed heat, and the structures were therefore cooler during the day (Figure 12.1).

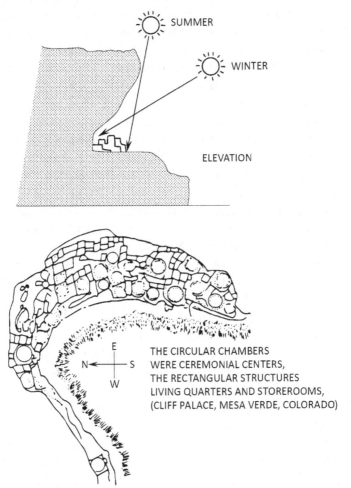

Figure 12.1 Passive design techniques used by Native American cave dwellers.[4]

[4] Smith, C.B. ed. (1978) pp. 620–621 *Efficient Electricity Use*, NY: Pergamon Press.

Building Configuration

The ratio of surface area to volume also influences energy use. Minimizing the ratio of wall area to enclosed volume minimizes heat loss (or gain). Tall buildings have reduced solar heat gains through the roof, but may be subjected to higher wind loads and subsequent losses due to infiltration. Buildings with large wall areas can use natural lighting more conveniently, but could also suffer greater heat gains and losses. Chimneys or tall structures can use natural or induced draft for heating or cooling. Each of these factors must be evaluated in terms of the specific use of the building and in terms of its specific microclimate.

Example. Passive cooling with a wind tower. Bahadori describes an interesting example from Iranian architecture.[5] The wind tower (called *Badgirs*) intercepts the prevailing summer winds, cools them, and uses cooled air to cool and ventilate the building. A wind tower operates on the same principles as a fireplace chimney. If air is heated, it expands and rises, causing an updraft. At night, heat stored in the tower walls is added to the air in the tower, causing a natural circulation and venting of the air in the building connected to the tower. During the day, the process is reversed (Figure 12.2). The walls of the tower have been cooled at night, so the warm incoming air at (A) is cooled by flowing down the tower to (B). As it is cooled, it becomes denser, going down through the tower and into the building. Wind increases this effect.

In addition to the sensible cooling described above, wind towers can provide additional cooling by virtue of evaporation. This can be done by trickling water down the walls of the tower, or by placing a water spray or pool at the bottom of the tower. Then the entering air is cooler and has increased moisture content (C). These towers have several disadvantages besides the initial cost; they must be closed up during the winter to prevent heat losses, and they need screens or other methods to keep birds, insects, and dust from entering the structure.

Many other interesting examples of passive design are reported by von Frisch, who has studied techniques used by a wide variety of animals to provide shelter for themselves.[6] He describes how paper wasps build nests that house their colonies. The nest is constructed of a paper-like material made from wood fibers. The temperature in the breeding combs is maintained at approximately 30°C by the activity of the special group of

[5] Bahadori, M.N. (1978) pp.144–154, v. 238, No. 2. "Passive Cooling Systems in Iranian Architecture," *Scientific American*, February 1978.
[6] Von Frisch, K. (1974) *Animal Architecture*, New York, NY: Harcourt Brace Jovanovich.

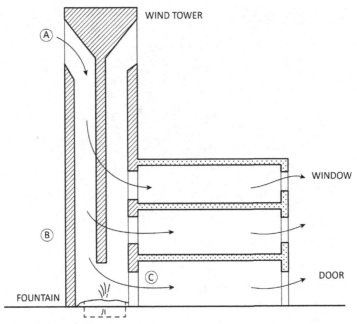

Figure 12.2 Use of a wind tower to cool buildings.

workers. Should the temperature get too high, the wasps bring in water to moisten the cells and cool them by evaporation.

Example. Nest configuration by desert rodents. These species employ passive design techniques to survive in a harsh environment. Their nests typically are found under a shrub such as a creosote bush or a clump of Ephedra. Their excavations produce a mound of soil that has several entrances and exits (Figure 12.3).

By tagging such animals with radioactive tags, it is possible to measure their depth as they rest in their burrows. At the same time, the soil temperature profile can be measured with a temperature probe. Studies made in this manner indicate that kangaroo rats (*Dipodomys merriami*) and other desert rodents adjust their position in their burrows to obtain the optimum temperature.

For example, Figure 12.4 shows typical spring and fall temperature profiles for the desert soil in the Owens Valley near Big Pine, CA. During the summer, measurements showed that tagged kangaroo rats were resting at depth of 20–30 cm during the day, where the temperature was 25–30°C. This temperature is at or near the thermal neutral zone for these animals, and therefore represents a minimum expenditure of metabolic energy.

Figure 12.3 Site of kangaroo rat burrow.

Similar measurements during the winter months indicated that the animals go to 0.5–1.75 m deep in the soil to escape the cold temperatures of the surface layers.

Excavations indicated that these desert rodents construct highly insulated nests. For example, San Diego pocket mice (*Perognathus fallax*) living in the vicinity of our desert cabin have nests at the bottom of burrows 1–2 m deep. These animals weigh on the average 20–30 g and live in nest 12–14 cm in diameter containing about 140 g of dried plant materials. The nests are spherical and therefore have the lowest possible ratio of surface area to enclosed volume (Figure 12.5).

In addition to the thermal capacity of the soil, the underground burrows have a high humidity. This is a decided advantage from the point of view of water retention for rodents attempting to survive in the desert.

Building Orientation

The orientation of a building on a site influences its exposure to sun and wind. The east and west faces receive the greatest amount of solar radiation during the summer and therefore should have their area minimized. As described above, south-facing areas with overhangs can be designed to remain cool in summer and warm in winter.

Site orientation should also take into consideration the prevailing direction of the winds. Windows and doors facing into winter wind should be minimized or at least weather-stripped and double- or triple-glazed in colder climates. At the same time, they should also be positioned to take

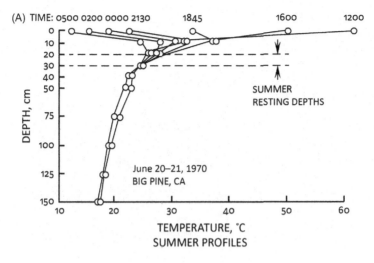

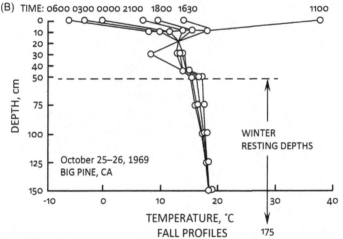

Figure 12.4 Kangaroo rats use of underground burrows to moderate seasonal temperatures.[7]

advantage of cool summer breezes. If this is not possible, consider using windbreaks of trees and shrubs as an alternate solution or vestibules.

Example. Building orientation by termites. This interesting example is provided by von Frisch. Most termites require warmth and high humidity. In tropical areas, they insulate the nest to provide proper

[7] Smith, C.B. (1981) p. 290 *Energy Management Principles*, NY: Pergamon Press. (Based on Kenagy, G.J. "Daily and Seasonal Patterns of Activity and Energetics in a Heteromyid Rodent Community," *Ecology*, 1973.)

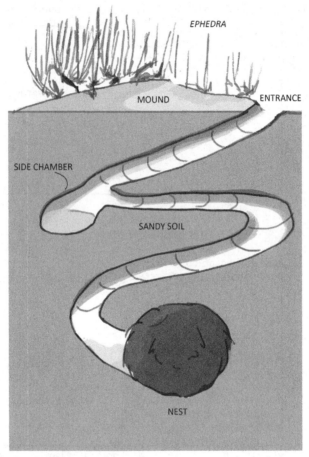

EPHEDRA

MOUND   ENTRANCE

SIDE CHAMBER

SANDY SOIL

NEST

Figure 12.5 Sketch of desert rodent nest.

temperatures. Compass termites (*Amitermes meridionalis*) in Australia orient the above-ground portion of their nests with the long axis north and south, so it is not overheated by the noonday sun. They congregate on the east and west sides in the morning and evening to get the benefit of the heat. Some nests reach a height of 3–4 m and require adequate ventilation and oxygen to support several million inhabitants.

Ventilation is achieved by chimney and duct systems that function much as the Iranian wind towers described previously. They bring in fresh air and exhaust waste gases. Some species construct the chimney walls of porous membranes that permit gas exchange but retain moisture inside the nest (Figure 12.6).

Obviously, few of us are prepared to live in caves, termite mounds, or burrows just to conserve energy. However, these examples illustrate

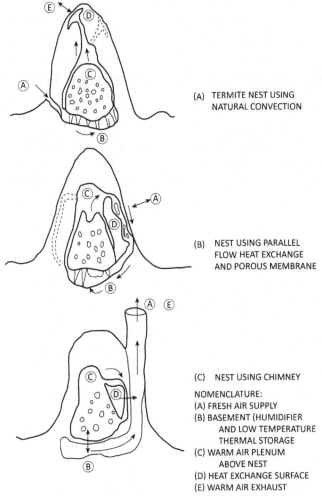

(A) TERMITE NEST USING
     NATURAL CONVECTION

(B) NEST USING PARALLEL
     FLOW HEAT EXCHANGE
     AND POROUS MEMBRANE

(C) NEST USING CHIMNEY

NOMENCLATURE:
(A) FRESH AIR SUPPLY
(B) BASEMENT (HUMIDIFIER
    AND LOW TEMPERATURE
    THERMAL STORAGE
(C) WARM AIR PLENUM
    ABOVE NEST
(D) HEAT EXCHANGE SURFACE
(E) WARM AIR EXHAUST

Figure 12.6 Termite air conditioning.[8]

concepts that are equally applicable to human dwellings, as we shall see subsequently.

To recapitulate, passive design techniques include the following:

- Building orientation on the site.
- Shading (by plants, topography, or structural elements).
- Thermal energy storage.
- Use of heat capacity and insulation properties of soil.
- Use of insulation.

[8] Smith (1981) Op. cit. p. 289. (Based on von Frisch, K. (1974) pp. 134–142. *Animal Architecture.* NY: Harcourt Brace Jovanovich.)

- Use of natural lighting (skylights).
- Natural convection heating and cooling.
- Use berms and windbreaks.

BUILDING ENVELOPE DESIGN CONSIDERATIONS

Five main elements of the building envelope determine the heat gains and losses that it experiences: walls, roof, floor, doors, and windows.

The basic equations that govern heat transfer through these elements were developed in Chapter 11 (see Equations 11.1−11.4).

Equations 11.1 and 11.2 can be rewritten in the following form:

$$\dot{Q} = UA\Delta T \quad W \qquad\qquad [12.1]$$

where

$\dot{Q}$ = heat flow, W
U = overall coefficient of heat transfer, $W/m^2 \cdot K$
A = area, m^2
ΔT = temperature difference, K
K   = temperature in degrees kelvin
where $T_{(K)} = T_{(^{\circ}C)} + 273.15$

The "U-value," or *overall heat transfer coefficient*, is useful for calculating heat gains and losses.

Equation 12.1 can also be written in this alternative form:

$$\dot{Q} = \Delta T/R \qquad\qquad [12.2]$$

where

R = thermal resistance, K/W
(*Note*: Some references express R in units of $m^2 \cdot K/W$.)

This equation is analogous to Ohm's law and is convenient in that thermal resistances can be added for layered media to give an overall thermal resistance:

$$R_{total} = R_1 + R_2 + R_3 \qquad\qquad [12.3]$$

Then the heat transfer coefficient can be found from

$$U_{total} = 1/R_{total}A \qquad\qquad [12.4]$$

The reciprocal of resistance is called the *thermal conductance* and is given by

$$C = 1/R \quad W/K \qquad\qquad [12.5]$$

Thermal resistances can be computed from the following relations:

$$For\ conduction \quad R = L/kA \quad K/W \qquad [12.6]$$

where

L = material thickness, m

k = thermal conductivity, $W/m \cdot K$

A = area, m^2

Table 12.2 lists values of thermal conductivity. These are representative values only. Actual material specification should be used for more accurate calculations. *Caution*: Verify thermal conductivity units when performing calculations!

$$For\ convection \quad R = 1/h_c A \quad K/W \qquad [12.7]$$

where

h_c = convective heat transfer coefficient, $W/m^2 \cdot K$

The convective (film) coefficient for still and moving air is also useful for these types of calculations. Table 12.3 shows typical values.

Table 12.2 Typical thermal conductivity values, k

Materials	Btu/h · ft · °F	W/m · K
Common brick	0.36	0.62
Concrete with stone aggregate	0.60	1.04
Window glass	0.42	0.73
Wood (fir)-against grain	0.067	0.12
Fiberglass (blankets/batts)	0.027	0.047
Steel	26	45
Ice	1.26	2.2
Water	0.34	0.59

Note: Thermal conductivity values vary widely depending on purity, temperature, and specific type of material. These values are indicative only and should not be used for design. The temperatures for which these data are valid are approximately 20–30°C. All values are rounded.

Table 12.3 Convective heat transfer coefficient, h_c

Air condition	Btu/h · ft² · °F	W/m² · K
Still air: horizontal surface (heat flow upwards)	1.63	9.25
Still air: vertical surface	1.46	8.25
Moving air: 24 km/h (15 mph) wind	6.00	34.0
Moving air: 12 km/h (7.5 mph) wind	4.00	22.7

Example. Consider a vertical concrete and stone aggregate wall 20 cm (8 in) thick. It is to be covered with 2.5 cm (1 in) of fir planks. What is the resulting U-value in SI units?

Outside surface (24 km/h wind):	$R = 1/h_c A = 0.029/A$	K/W
Fir planks:	$R = L/kA = 0.21/A$	K/W
Concrete wall:	$R = L/kA = 0.19/A$	K/W
Inside surface still air:	$R = 1/h_c A = 0.12/A$	K/W
Total resistance:	$R_{\text{total}} = 0.55/A$	K/W
Then	$UA = 1/R_{\text{total}} = 1.8A$	W/K

Note: In the example, these calculations are performed on a "per unit area" basis; e.g., 1 m^2. Therefore, $U = 1.8$ W/m$^2 \cdot$ K.

Glazing/Fenestration

Windows (glazing) are an important source of heat gain and loss. In some buildings, heat losses through windows can be equal to half of the heat loss through the walls. Likewise, during summer months, windows can be a source of heat gains. Window coatings, awnings, and movable sun shades can reduce heat gains.

Table 12.4 lists heat losses through various types of window units. Double and triple glazing substantially reduces the heat loss through windows; as the table indicates, the heat loss can be as little as one-third compared to a single pane of glass.

The type of window frame also contributes to heat lost; as Table 12.4 indicates, metal casements cause larger heat losses than wood. This is referred to as *thermal bridging*, where a high conductivity material acts as a "short circuit" for heat loss or gain. Aluminum window frames can account for 25% of total heat loss from the window, while wood frames are typically one-half this amount.

Example. Heat loss from a single story wood frame home constructed in 1940 in Southern California. This pre-WWII residence was originally uninsulated. When it was expanded, about half of the walls were insulated with 8.9 cm of fiberglass insulation, as was the entire attic. All the windows are single pane; most with aluminum casements. Table 12.5 shows the heat loss calculations. As a table indicates, windows account for 17% of the loss and cause greater heat loss than the walls. Clearly a major energy management opportunity is to insulate the floor where it is not slab-on-grade construction (about two-thirds of the house). This accounts for about 40% of the loss based on the present condition of the house. Note that the loss had already been reduced by about 50% compared to the original, as-built, uninsulated design.

Table 12.4 Heat losses through glazing systems, in W/m² · K

Glazing system type	Degree of exposure		
	Sheltered	Normal	Severe
Glazing Losses			
Single pane	5.0	5.6	6.7
Double pane			
Air space 3 mm wide	3.6	4.0	4.4
Air space 6 mm wide	3.2	3.4	3.8
Air space 12 mm wide	2.8	3.0	3.3
Air space 20 + mm wide	2.8	2.9	3.2
Triple pane			
Each air space 3 mm wide	2.8	3.0	3.3
Each air space 6 mm wide	2.3	2.5	2.6
Each air space 12 mm wide	2.0	2.1	2.2
Each air space 20+ mm wide	1.9	2.0	2.1
Casement Losses			
Single-glazed			
Metal casement (20%)[a]	5.0	5.6	6.7
Wood casement (30%)	3.8	4.3	4.9
Double-glazed[b]			
Metal horizontal sliding (20%)	3.0	3.2	3.5
Wood horizontal pivot (30%)	2.3	2.5	2.7

[a]Number in parentheses is frame percentage of total window area.
[b]With 20 mm air space.

Table 12.5 Heat loss calculation for a southern California residence, W/m² · K

Element	U-value W/m² · K	Area m²	Heat loss W/K	ΔT K	Total heat W	Percent %
Roof	0.4	149	59.6	15	894	9
Walls	0.8	125	100	15	1,500	14
Windows	4.5	26.7	120	15	1,800	17
Floor slab	6.0 W/m²	51	—	—	306	3
Subflooring	2.8	98	274	15	4,110	40
Infiltration (1 air change per hour)	See note[a]		120	15	1,800	17
Total					10,140	100

[a]363 m² × 0.33 W/m² · K.

Infiltration/Exfiltration

Another major source of heat gain and loss from buildings is infiltration. Infiltration is cold or hot air that enters through open doors or windows (unintentionally), through cracks, through the gaps around doors or windows that lack weather stripping, through paths that connect conditioned spaces with unconditioned spaces in the building (ducts, attic bypasses), or through penetrations for electrical conduits or plumbing. Infiltration is caused by a temperature or pressure difference within a building, for example, by the natural draft caused by a chimney or by air pressure created by wind. Surprisingly, about one-third of infiltration comes in through walls, ceiling, and floors; another one-third through ducts and fireplaces, and the balance through windows, doors, and plumbing penetrations. Exfiltration is air that leaves the building and occurs when the pressure in the building is higher than the surroundings.

Infiltration and exfiltration losses can be cut by

- Keeping doors and windows closed.
- Sealing cracks and other openings.
- Providing adequate weather stripping and insulation.
- Installing energy-efficient, high-performance windows.
- Using double doors or an enclosed foyer.
- Providing windbreaks or screens.

In reducing infiltration, it is important to still provide adequate outside air for safe ventilation. Inadequate ventilation can lead to health and safety problems if CO_2 or carbon monoxide were to accumulate in occupied spaces. In addition, paints and certain building materials containing formaldehydes can cause outgassing. Without adequate ventilation, occupants can become ill in such "sick" buildings (see Chapter 8 for more information on indoor air quality).

To estimate heat losses due to infiltration, common practice is to determine the total area (or length) of "cracks" (paths for infiltration) and then estimate the airflow through them using lookup tables of air leakage per unit area (or per length of crack) for different types of cracks at different pressure differentials. With knowledge of the volumetric airflow through all infiltration paths, the heat flow (loss or gain) can be estimated using Equation 12.8. This is called the *crack method*.

The second method is to estimate the number of "air changes" per hour, and then compute the total heat loss or gain. An air change refers to the amount of air infiltration required to replace the entire volume of the interior space. The number of air changes per hour will depend on the tightness of the building envelope. This is called the *air change method*.

The following standard data for air are useful for heat flow calculations (sea level, 760 Hg pressure, 20°C temperature):

	SI units	American/British units
Density, ρ	1.22 kg/m^3	0.075 pounds/ft^3
Specific heat, c_p	1.00 kJ/kg · K	0.24 Btu/lb · °F
Thermal capacity	1220 J/m^3 · K	0.018 Btu/ft^3 · °F

where we have defined thermal capacity as specific heat (c_p, J/kg · K) multiplied by density (ρ, kg/m^3).

The heat loss or gain can then be calculated as

$$\dot{Q} = (\dot{V})(\text{thermal capacity})\Delta T = \dot{V}c_p\rho\Delta T \qquad [12.8]$$

where

$\dot{V}$ = volumetric flow rate, m^3/s

With American/British units, $\dot{Q}$ will be in Btu/h or Btu/min depending on whether $\dot{V}$ is in m^3/h or feet^3/min.

For SI units it is convenient to express $\dot{V}$ in m^3/s or m^3/h, with the latter being preferred because standards are usually given in air changes per hour. For these two cases, the thermal capacity can be modified to read directly in Watts:

if $\dot{V}$ is m^3/s, the thermal capacity = 1220 W · s/m^3 · K

if $\dot{V}$ is m^3/h, then thermal capacity = 0.33 W · h/m^3 · K.

Example. Special shelter for a Mount Everest expedition. In 1981, a group of medical doctors climbed Mount Everest with the goal of making the first-ever measurements of arterial blood gases at altitudes in excess of 8000 m (26,250 ft).[9] They established a base camp laboratory at 5400 m (17,720 ft). This lightweight structure was 2.1 m high and 5.3 m long. It was fabricated from 7.5-cm-thick polyvinyl chloride foam with a fiberglass coating. The structure consisted of 12 sections (6 per side), each to be carried by a Sherpa. At the base camp, the sections were assembled into the complete structure. The design temperature for the laboratory was 40°C (140°F); temperature control was important for accurate measurements. Heat was to be supplied by a propane heater. The scientists needed to know the power required. The thermal conductivity of walls, roof, and floor was 0.028 W/m · K. Infiltration was to be held at 0.5 air changes per hour. The roof and floor areas were 11.1 m^2, while the wall areas totaled 31.1 m^2, and the shelter volume was 23.4 m^3.

---

[9] West, J.B. (1983) "American Medical Research Expedition to Everest, 1981," *The Himalayan Journal*, ed. Harish Kapadia, vol. 39.

Using these data, the heat load was determined to be 665 W. *Note*: To determine infiltration, the density of air has to be corrected for the altitude (at 5400 m, it is about 50% of the sea level value).

Blower door tests can be used to measure building air infiltration rates. In these tests, auditors mount a *blower door*—which is a large fan—into an exterior door frame and thoroughly seal the space around the fan. The blower door then pulls air from the inside to the outside of the building, creating a low pressure in the building. Because of the pressure differential, outside air flows into the building through openings, cracks, and gaps. Blower door tests help identify sources of infiltration and quantify the infiltration rate, before and after performing air sealing measures.

Reducing Building Heat Losses

Insulation and weather stripping can significantly reduce heat losses from existing buildings. Table 12.6 list some approaches that can be considered for existing buildings. Table 12.7 shows the benefit in terms of reduced U-value that can be obtained by increased insulation on existing buildings.

For example, Figure 12.7 shows R-13 fiberglass insulation batts added to an existing wall. This is often difficult and expensive for existing walls unless they are undergoing renovation, as was the case in this photograph.

For existing buildings, blown-in insulation will generally be less expensive. Figure 12.8 shows insulation blankets that have been added to the ceiling of a factory building. *Note*: The R-value refers to the thermal resistance of the insulation. The way we have defined thermal resistance on a per area basis (see Equation 12.2), the R-value is really an RA-value. For example, $R\text{-}19 = 19°F \cdot ft^2 \cdot h/Btu$ ($3.3\ K \cdot m^2/W$), which has units consistent with RA. Fiberglass and rock wool have moderate R-values ($15-26\ K \cdot m^2/W \cdot m$) per unit thickness, while polystyrene and polyurethane have high values of $28-60\ K \cdot m^2/W \cdot m$.

In new construction, the opportunities are much more extensive. It is possible to reduce energy use by as much as 60%, compared to conventional construction, for many types of more efficient buildings. In decades past, when energy was cheap, the energy savings did not justify the added capital costs. Today, with much higher costs, and the additional concerns for reducing greenhouse gases, the imperative to make buildings efficient should be obvious. One problem is that builders have little incentive for efficiency, unless forced to do so by incentives, codes, or regulation. This is easily understood, because it is the buyer who will be faced with paying the energy bill.

Table 12.6 Energy management opportunities in the building envelope

Improve Insulation of Exterior Walls with

- Loose fill blown into wall cavities.
- Spray on polyurethane insulation (observe fire codes).
- Slab and board insulation applied inside.
- Tongue and groove aluminum siding over fiber fill insulation applied outside.

Reduce Heat Losses Through Floors and Slab by

- Installing insulation under floor slabs where accessible.
- Insulating edges of concrete slabs.

Improve Insulation of Roofs and Ceilings with

- Installation of mineral fiber batts on suspended ceilings and under roofs.
- Sprayed polyurethane foam on exposed ceilings.
- Installation of suspended ceilings to reduce internal volume and provide an insulating attic space.
- Replacement of rigid preformed roof insulation material with material of higher thermal resistance during scheduled waterproofing replacement.

Reduction of Infiltration by

- Caulking cracks around doors and windows, installing weather-stripping and door sweeps, or using vestibules.
- Improving gaskets on doors and operable windows.
- Sealing plumbing penetrations and other gaps and openings such as in electrical outlets and ceiling fixtures.
- Closing construction gaps between precast concrete elements.

Reduction of Heat Loss Through Glazing by

- Replacing single pane glass with double (or triple) pane.
- Using windows with low-emissivity (low-e) coatings, including those that are spectrally selective and designed to either increase or decrease solar heat gain while maintaining good visible transmittance.
- Reducing glazed area when possible, unless the energy and aesthetic penalty of less natural lighting is larger than the energy benefit of reduced heat loss.
- Installing storm windows.
- Installing drapes over windows.

Example. A small office building was planned for Portland, OR. The size of the building was 102 m^2 (1100 ft^2) and design conditions called for 4000 heating hours per year. The local utility commissioned a study of three conditions:

- Base case: A conventional wood frame construction with single glazing and R-11 (walls) and R-19 (ceiling) insulation.

Table 12.7 Benefit of improved insulation

Component	Existing RA-value (h · ft² · °F/Btu)	Corrective insulation	New RA-value (h · ft² · °F/Btu)	New U-value[a] (Btu/ h · ft² · °F)
Cinderblock wall	2.7	1" polystyrene LD	4.17	0.15
Corrugated steel wall	0.85	1" polystyrene HD	5.0	0.17
Wood frame wall	4.3	3.5" blown fiberglass	12.0	0.06
Corrugated steel roof	0.78	4" fiberglass batt	16.0	0.06
Metal deck roof with rigid insulation	5.2	2" sprayed polyurethane	11.76	0.06
Concrete roof	2.0	2.5" sprayed urethane	14.7	0.06
Insulated concrete roof	7.1	1.5" sprayed urethane	8.82	0.06
Wood deck roof	2.7	3.5" foil-backed fiberglass	14.0	0.06
Insulated wood roof	7.1	2" foil-backed fiberglass	8.0	0.07
Shingle roof	4.7	3" fiberglass bats on ceiling	12.0	0.06
Single pane glazing	0.86	Acrylic storm window	1.8	0.40
Existing weather stripping	–	Improve weather stripping	–	–

[a]New U-value = (existing RA + proposed RA)$^{-1}$.

- "Energy efficient" construction, with increased insulation.
- "Highly insulated" construction.

  Table 12.8 summarizes the heat loss calculations.

  Based on heating costs of US\$0.11/kWh, the annual savings of the two alternatives were in the range of US\$2700–US\$3400 per year. The incremental costs of construction were estimated to be (as compared to the base case):

- Energy efficient structure: US\$1700 additional.
- Highly insulated structure: US\$3400 additional.

  Assuming base case construction costs of US\$200 per square foot (2014 costs), these energy improvements would add between 0.77% and 1.5% to the cost of construction. This is equivalent to a few dollars more per month on a typical 30-year mortgage. The total additional investment

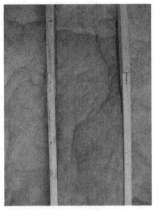

Figure 12.7 R-13 fiberglass insulation.[10]

Figure 12.8 Ceiling insulation blankets.[11]

would be recovered in the first year, without taking into consideration future escalation of electricity prices.

This example is from a few years ago. Today insulation standards are higher, especially in colder climates. The so-called "Super Insulated" buildings have U-values of $0.1-0.15\ \mathrm{W/m^2 \cdot K}$, corresponding to R-values of $38-57°\mathrm{F} \cdot \mathrm{h} \cdot \mathrm{ft^2/Btu}$.

The building envelope also needs to control moisture diffusion into the building. In new construction this is accomplished wrapping the entire

---

[10] Photo by authors, courtesy of Mr. Kim Koch, President, Farpointe Construction, La Habra, CA.

[11] Photo by authors, courtesy of Mr. Keith Garrison, General Manager, GBF Enterprises, Santa Ana, CA.

Table 12.8 Heat loss comparison for three building designs (heat loss in Watts)

Component	Base case construction	Energy efficient	Highly insulated
Ceiling Insulation			
6" R–19	935		
10" R–30		616	
12" R–38			484
Floor (Vented)			
No insulation	4,510		
Perimeter 1.5" R–6		1,023	
Under floor 6" R–19			836
Walls (Frame)			
Batts 3" R–11	1,301	1,301	
Batts 6" R–19			743
Windows			
Single pane	2,980		
Double		1,639	
Improved double			1,210
Doors			
Weather stripped	462	462	
Insulated			109
Infiltration			
Single glass	2,816		
Double glass + weather-stripped		1,848	1,848
Total Watts of heat loss	13,004	6,889	5,230
W/m^2	127.5	67.5	51.3
Approx. annual heating kWh (assume resistance heating)	52,016	27,556	20,920
Annual heating cost (2014)	$5722	$3031	$2301
Savings vs. base case	0	$2691	$3421

Note: *New U-value = (existing RA + proposed RA)$^{-1}$.

building with an impermeable material. In addition, some types of insulation have foil on one side. The foil acts as a heat reflector and moisture barrier. There are many online resources where it is possible to find information on insulation and reducing infiltration (e.g., the IECC, the U.S. DOE, the UK

Building Research Establishment, the North American Insulation Manufacturer's Association (NAIMA), and others). Some describe emerging technologies for the commercial and industrial sectors.[12]

Reduction of Heat Gains

Heat gains in a building are caused by the following:
- Solar heat load on walls and roof.
- Solar heat gains through windows.
- Heating from lighting.
- Heat from fan motors and other equipment in buildings.
- Heat from occupants.
- Infiltration.

Heat gains can be estimated using the same general procedures outlined above. The basic procedure consists of tabulating the solar heat fluxes incident on windows, walls, and roofs on a month-by-month basis, and then using the respective U-values to determine heat transmitted to the building. This process is easily performed using a computer simulation program. Many of the same measures that can be employed to reduce heat losses will also be effective in reducing heat gains.

Shading (by structures or plants) is an important passive heat control technique since it is one that can reduce solar heat gains while permitting solar heating during the winter. Shading can also involve reflective films on window glass to reduce solar heat loads. Some caution must be exercised, as it is possible to increase the glass temperature, causing thermal expansion, stresses, and cracking of glass. In addition, reflective films lower the visible transmission, in addition to reducing the solar heat gain coefficient. Another alternative (noted in Table 12.6 above) is the use of spectrally selective coatings on glazing to customize windows for either higher or lower solar heat gain coefficients, while simultaneously lowering the U-value and maintaining a good visible transmittance. Also consider using external louvers and drapes to reduce heat gains.

Many of the energy management techniques described in previous chapters have the effect of reducing heat loads in buildings (more efficient lighting, more efficient processes, reduced ventilation, etc.). Therefore, in conditioned buildings, these measures lead to the additional benefit of reducing energy expended for cooling.

[12] http://energy.gov/eere/buildings/downloads/new-generation-building-insulation-foaming-polymer-blend-materials-co2 (Accessed April 2015).

INTEGRATION OF BUILDING SYSTEMS

In the Portland building example described above, the heat losses ranged from 5.2 kW for the highly insulated structure to 13 kW for the conventional design. Small office buildings can be expected to have an installed lighting load of at least 10 W/m^2 or about 1 kW of lighting. Office equipment, including computers, printers, coffee machines, and possibly a refrigerator would add another 3 or 4 kW. Thus, in theory, the building has enough internal heat sources to meet part of its heating requirements, if all the equipment operated.

Of course, this is not practical since it is unlikely that the equipment (except lighting) would operate frequently enough, nor do they deposit heat in the right places at the right time.

Still, there are other sources of heat, including the occupants and solar heat gain.

Let's examine this last possibility:

For the building in Oregon, solar insolation of 300 langleys (mean annual value) is possible. This corresponds to about 12 MJ/m^2 per day, or, based on the roof area of 102 m^2 and 8 hours of daylight per day, an average power of

$$[(12 \times 10^6 \text{ J/m}^2)(100 \text{ m}^2)]/[(8 \text{ h})(3600 \text{ s/h})] = 41.7 \text{ kW}$$

is available. Thus, with a collection efficiency of only about 15%, the incident sunlight intercepted by the house could also provide all the necessary heating, so long as some means existed for storing energy to meet peaks and valleys in demand and to take care of overcast days.

So why aren't buildings designed to be energy self-sufficient? Historically, it was originally a matter of convenience. It was more convenient to be able to manipulate the environment at the flip of a switch or the turn of a valve. This approach was encouraged by the low cost and ready availability of energy during the last 100 years.

The lack of integration of energy activities within buildings also has been a consequence of the fragmented approach taken to design, wherein practitioners representing different disciplines are each responsible for one part of the project and have no incentive to seek optimal, overall, coordinated solutions. Clearly, it is possible and desirable to improve this situation.

Nonresidential buildings are more complicated. Energy self-sufficiency is generally not practical, simply because of the process energy use requirements. However, an integrated approach would go a long way toward reducing building energy use.

Energy use performance factors can be calculated for all types of building occupancies. Typically, these would relate annual energy use to building area, occupancy, or to industrial production. Data for 18 different building types ranging from office buildings to schools to hotels are tabulated.[13] These buildings are located in climates ranging from temperate to cold. Energy performance factors range from:

Per unit area	$MJ/m^2 \cdot yr$	$kBtu/ft^2 \cdot yr$
	494–2930	43.5–258

These numbers show the wide range of energy used in buildings. In the commercial buildings, energy use per unit area varies by as much as a factor of 6. If we broadened the sample to include oil refineries and steel mills, the ranges would be even greater.

In a 2009 study of residences[14] of different size and types throughout the U.S., these average values were found:

Per household	$GJ/home \cdot yr$	$MBtu/home \cdot yr$
	94.5	89.6
Per unit area	$MJ/m^2 \cdot yr$	$kBtu/ft^2 \cdot yr$
	517	45.5
Per occupant	$GJ/person \cdot yr$	$MBtu/personyr$
	37.2	35.3

The average family size was 2.54 persons, and the average residence size was $183 \ m^2$ (range was $79–230 \ m^2$). Average energy end use per household in 2009 is shown in Table 12.9.

Several changes are apparent in comparing residential energy use over the past several decades. In general, energy usage has decreased with the advent of more efficient lamps and appliances. However, offsetting this to some extent has been the growth of new electricity end uses. In the 2009 study cited above, over 75% of U.S. households reportedly had one or more computers. Over 70% had internet access. Supporting the computers were monitors, printers, scanners, fax machines, and other devices, many of them rechargeable.

---

[13] http://www.eia.gov/consumption/commercial/data/archive/cbecs/cbecs2003/detailed_tables_2003/2003set14/2003html/c3a.html (Accessed April 2015).
[14] http://www.eia.gov/consumption/residential/reports/2009/consumption-down.cfm#fig-4 (Accessed April 2015).

Table 12.9 Average energy end use per U.S. household (2009)

	MBtu/home · yr	GJ/home · yr	Percent
Space heating	37.2	39.2	41
Water heating	15.9	16.8	18
Air conditioning	5.6	5.9	6
Refrigerators	4.2	4.4	5
Lighting, appliances	26.7	28.2	30
Total	**89.6**	**94.5**	**100**

Source: U.S. Energy Information Administration: 2009 Residential Energy Consumption Survey, Table CE3.1: Final Energy Consumption and Expenditures Tables.

Residences are of interest because they fall near the lower end of the energy use scale for commercial buildings. The difference is most apparent in lighting; lower installed capacities are apparently acceptable in residences.

This wide range of energy use performance factors also occurs in buildings with similar functions, such as hospitals or office buildings, even when situated in similar climates. All this indicates that the way a building is designed and how it is used by the occupants has an important effect on overall energy use.

A growing number of designers are attempting now to develop integrated building designs that minimize use of nonrenewable energy forms while still providing an aesthetically pleasing and functionally satisfying building envelope.

The approach being taken goes something like this:

- Select a favorable site, and choose the most appropriate orientation.
- Plan the building to include passive design and energy storage capability.
- Include appropriate uses of solar energy for power and space and water heating.
- Design an efficient, tight, but adequately ventilated, well-insulated building envelope to minimize heat gains and losses.
- Select the most efficient of the new generation of EPA-approved building appliances such as refrigerators and water heaters.
- Plan for efficient task lighting systems, supplemented where feasible by daylighting.
- Use the most efficient new lighting technologies and controls, including compact fluorescent and LED systems.
- Today many localities have mandatory prescriptive or performance standards for optimizing energy use. For example, they may stipulate lighting energy budgets in W/m^2.

- Install efficient HVAC systems and provide control for operating them efficiently, when and where needed.
- Select the most efficient process equipment and recover heat where feasible.
- Create sensible building use patterns.

Using this approach, the heat losses and gains are smaller, so the mechanical systems can be smaller. Since some of the heating and lighting is being provided by passive systems, the electrical mechanical systems do not need to use as much energy. Savings on these systems (both capital and operating costs) is often more than enough to repay the added capital cost of the building.

Example. Proposed design of an industrial building in the state of Washington. Following a series of energy audits of U.S. government buildings in the state of Washington, the authors and their associates were tasked to develop the conceptual design of highly energy efficient buildings to replace older structures built during wartime. In this area of eastern Washington, the recorded temperature range is $-30°C$ to $+40°C$; the average wind speed is about 10 km/h, the average solar insolation is 370 langleys, and there are about 2900°C-days (5300°F-days) of heating per year.

One building was an 11,150 m^2 (120,000 ft^2) industrial shop building housing metal, wood, paint, sheet metal, welding, and electrical shops. The new building was proposed to replace a group of old (1940 era) wooden structures badly in need of repairs.

The new building was sited with the long axis east—west to permit utilization of solar energy. The building envelope design had double walls, an earthen berm, and double glazing. The walls and ceilings were designed for thermal resistances of $RA = 4.4 \text{ m}^2\text{K/W}$ ($RA = 25 \text{ ft}^2 \cdot \text{h} \cdot °\text{F/Btu}$). Infiltration was minimized through the use of air locks and vestibules for all personnel and loading doors. These measures reduced the size of the HVAC system needed.

The HVAC system was also designed to minimize energy use. During the summer, adequate cooling for this building could be provided by combination of precooling with spray pond air and by direct evaporative cooling. The summer cooling load was minimized by the insulative properties of the building, and by exhausting warm air from the ceiling overhead areas, thereby removing heat from lamps and skylights.

In winter, the design allowed the HVAC system to supply energy from a solar collection system and an off-peak, electrically heated hot water storage system. Heat would be supplied through ducts that recirculate the warm ceiling air, recovering heat from the lights and returning it to the HVAC system. The winter heating load would be minimized by recovering 80% of the heat exhausted from the building. Additionally, there would be some heat recovery from process equipment such as air compressors.

The lighting system design had an installed capacity of $10 \, \text{W/m}^2$ ($1 \, \text{W/ft}^2$). The design used efficient, high-pressure sodium lights on a photocell control for general area lighting. These were designed to deliver 32 Dlux (30 fc) of illumination at the floor plane. The photocell controls ensured that the lamps would be used only when the natural illumination fell below this level. They were switched on in banks to provide illumination only where required. The sodium lamp luminaires were mounted within skylight boxes that serve as return air plenums. In the design, these recovered the light fixture heat as well as a solar heat gain from the skylights.

The design was such that general illumination would normally be provided by a system of specially designed skylights. Glass panels were oriented to collect direct sunlight in uniform quantities both during the day and seasonally. Large diffusers distributed the light evenly on the work plane as well as on the ceiling for optimum brightness.

Task lighting was provided at each machine or workstation in the design. These were specified to be individually controlled by pressure-activated switches at the workstation, and automatically turned off (with a delay) if the area was not in use. Auxiliary areas were also illuminated with a combination of natural, fluorescent, and task lighting.

The shop contained machinery and power tools. The electrical system had a demand controller to optimize peak loads, and micro controls and time switches to provide effective overall control.

The original shop facilities used $2390 \, \text{MJ/m}^2 \cdot \text{yr}$ ($210 \, \text{kBtu/ft}^2 \cdot \text{yr}$). As designed, the facility was expected to use $570 \, \text{MJ/m}^2 \cdot \text{yr}$ ($50 \, \text{kBtu/ft}^2 \cdot \text{yr}$), a savings of over 75%.

Example. Retrofit of an industrial complex. The previous example concerned integrated design of a new facility. However, there are also benefits to be obtained from a similar approach applied to existing buildings.

The authors and colleagues carried out studies of several hundred buildings on a government reservation in Idaho. The team performed audits to establish where energy was being used, the qualities of steam, fuel, and electricity required annually, and the distribution between end uses such as HVAC, lighting, and processes.

Since many of the buildings were similar in design, approaches to more efficient, integrated use of energy could be somewhat standardized. The general approach involved tightening the building envelope to reduce infiltration and conduction losses, improving lighting system efficiency, and employing heat recovery where feasible. After these changes were evaluated, the HVAC systems were reviewed to improve their controls and performance.

The team prepared a report for each building. It summarized major actions in a matrix form (see Figure 12.9 for an example) to provide a quick visual reference to the changes recommended for each building.

Example. Integrated building retrofit with passive shading and active solar. This 186 m^2 (2000 ft^2) two-story single family residence located near the Pacific Ocean in Newport Beach, CA, had an open roof deck providing a view of the harbor and ocean. The roof deck was an attractive feature of the home but experience showed that it became uncomfortably hot during the summer months and damp with condensation during the cooler months, limiting its use. The owner contemplated installing a canvas shade but decided against this because the area was subject to occasional strong winds. After further investigation, the selected approach was to design a "solar canopy" that provided shade while at the same time generating electricity (Figure 12.10). This structure was designed to withstand both wind and seismic loads.

Measurements indicated that the residence used an average of 5000 kWh/yr, of which 75% fell in the local utility's lowest ("Baseline") or Tier 1 rate of US$0.13/kWh, and 25% was in Tier 2 at US$0.15/kWh. This load consisted primarily of lighting, an electric oven, a microwave oven, clothes washer and dryer, television, computers, and other small electronic devices. There was no air conditioning. Space heating and water heating were by natural gas.

The solar capacity was based on eliminating the Tier 2 usage (1250 kWh/yr) and about one-third (1250 kWh/yr) of the Tier 1 usage. This would yield the least capital cost and shortest payback, while providing some capacity for load growth to occur while remaining in the lowest rate tier.

The system has a nominal rating of 1.5 kW and uses 10 Sharp solar panels. Each panel is rated at 165 W peak and 150 W nominal, and produces 43.1 VDC open circuit voltage. An inverter rated at 2.5 kW

EMOs / BUILDING NO.	613	615	628	636	660	629	830
HVAC							
NIGHT SETBACK		●	●		●		
TEMPERATURE CONTROL			●			●	
ECONOMIZER/HEAT RECOVERY			●	●		●	
MINIMUM OUTSIDE AIR	●		●				●
BUILDING ENVELOPE							
DOUBLE GLAZING	●		●	●	●		
IMPROVE WEATHERSTRIPPING	●		●	●	●		●
INSULATION							
WALLS, 1" LD POLYSTYRENE	●		●				
WALLS, 1" HD POLYSTYRENE		●		●			
WALLS, BLOWN-IN FIBERGLASS FILL						●	●
ROOF, 4" FIBERGLASS BATTS		●	●	●			
ROOF, 3" FIBERGLASS BATTS						●	
ROOF, 1.5" SPRAY URETHANE FOAM	●	●					
ROOF, 2" SPRAY URETHANE FOAM							●
ROOF, 2.5" SPRAY URETHANE FOAM				●			
ROOF, 2" FOIL BACKED FIBERGLASS					●		
ROOF, 3.5" FOIL BACKED FIBERGLASS			●				
LIGHTING							
DELAMP		●	●				●
INCREASE FLUORESCENT EFFICIENCY	●						
REPLACE INCANDESCENT WITH FLUOR OR HPS			●	●			●
REPLACE INCANDESCENT AND MV WITH LPS						●	
INDIVIDUAL MANUAL SWITCHES	●	●		●			
CHANGE WALL COLORS						●	

Figure 12.9 Building Energy Management Opportunities.

converts 300–400 VDC to 240 VAC with an efficiency of 94%. The inverter includes a performance meter, which displays operating voltage, hours of operation, instantaneous power production (in units of W), energy (in units of kWh) generated on the current day, and cumulative energy generated since inception.

The system has operated for 12 years without failure, producing 27,673 kWh. Maintenance is minimal, amounting to occasional hosing

Figure 12.10 Solar canopy for shade and electricity generation.

off the panels during the summer months when there is no rainfall. During this period, the Tier 1 rate has increased by 24% and the Tier 2 rate by 39%, with further rate increases announced for the future.

When the system was installed in 2003, the California Energy Commission offered a rebate of US$4/net watt (net watt was after deducting inverter losses). In addition, there were State and Federal tax credits for installing renewable energy systems. At the time of installation, this system costs US$7/peak watt. Rebates and credits reduced this to approximately US$3/peak watt.

Since the system primarily generates power during the peak afternoon hours, it offsets electricity generated in natural gas-fired peaking plants, eliminating an estimated 1.5 tons of CO_2 for each year of operation. The increase in local electricity rates since the system was installed confirms it was a good investment.

Example. Computer data centers. Did you ever wonder how Google stores all that data you can access in microseconds, or where all those photos go that you put on Facebook? There are huge, warehouse-sized buildings scattered around the country that are full of racks of electronic storage media. A typical building might be 9290 m^2 (100,000 ft^2) in floor space. Inside, there are around 8000 to 9000 racks of electronic equipment, each a standard 19-in. rack. It takes about 20 kW per rack to power all of that electronic memory, plus an additional 10−30 kW per rack for cooling and all the auxiliary functions in the building. This equipment runs 24/7, so on average *each rack* requires about

Figure 12.11 Typical computer data center.[15]

350,000 kWh per year. Generating this amount of electricity from fossil fuels leads to dumping 241 tons of CO_2 per year into the atmosphere for each rack. If that number is hard to visualize, try this. The U.S. Department of Energy forecasts that by 2020, the carbon footprint of data centers will *exceed the airline industry*. With best practices, 20–50% of this energy use could be eliminated (Figure 12.11).[16]

This is a perfect example of the potential for an integrated approach to building design. More efficient cooling techniques are available, by distributing the racks in "hot and cold aisles" for better cold air distribution. Optimize lighting controls. (Computers work fine in the dark.) Use renewable energy (solar photovoltaic) and energy storage. Use cool storage and computer room air handlers rather than computer room air conditioners. Use liquid cooling (chilled water) directly in racks, rather than air. (Heat transfer is several thousand times more effective.)

PEAK DEMAND CONTROL

A utility must provide sufficient capacity to meet its customer's loads at all times, and still provide both spinning and cold reserve capacity in the case of equipment failure. Because of the high capital cost of new generating capacity, as well as the cost of transformers, substations, and

---

[15] Photo courtesy of Parsons Corporation.

[16] Anonymous (2013). "Quick Start Guide to Increase Data Center Energy Efficiency," Washington, DC: U.S. Department of Energy and General Services Administration. http://www.gsa.gov/graphics/pbs/data_center_quick_start_03_09_508_compliant.pdf (Accessed February 2015).

distribution systems, most utilities charge for the maximum demand (in kilowatts) made by their customers, regardless of the duration of this demand.

Therefore it is in the interest of the customer and the utility to keep the peak demand as low as practical. While this does not directly save energy, it definitely saves money. (See Chapter 5 for more information on demand changes and utility pricing structures and programs that encourage peak demand reduction.)

There are several techniques that can be used to reduce peak demand:

- Off-peak schedules.
- Demand limiting controls.
- Computer demand control systems.
- Energy storage.
- Onsite generation.

Off-peak scheduling applies to loads such as golf course irrigation, where pumps can be scheduled to operate at night or at other times of low demand. In a multishift factory operation, it may be possible to schedule certain energy intensive operations for off-peak hours, to reduce peak demand.

Demand limiting controls (electronic and digital controls) can be designed to provide an indication of the demand and can open relays, preventing a pre-established limit from being exceeded.

Computer demand controls can go a step further, projecting what the demand will be over the next demand interval, and if it appears that the demand limit will be exceeded, automatically drop certain nonessential loads from the line. Once the demand decreases, these loads restart automatically. If necessary they can be restored temporarily on a rotating basis, providing partial service. This latter approach is frequently used with air handlers, which may be shut off for periods of a few minutes to an hour without a noticeable effect on ventilation (depending on the type of occupancy).

Energy storage can also be considered to reduce on-peak demand. Air compressors can be operated off-peak and can store compressed air. Chillers can chill water off-peak and store it. Buildings can be precooled with air conditioning systems in the morning—or at night with outside air ventilation and by using the thermal mass of the building to store energy—to reduce daytime peak energy. Electric storage water heaters can heat water during off-peak hours for use during the utility's on-peak periods.

Having onsite generation capacity is another way to reduce peak demand charges, as well as to provide supplemental power and backup generation. Generation technologies could include gas or diesel engines, microturbines, solar photovoltaic systems, wind turbines, fuel cells, and combined heat and power systems that *cogenerate* both heat and power.

ENERGY STORAGE

Much of the energy waste that occurs does so simply because there is no convenient way to use it at the time it is available. It is also a matter of energy quality or available work. (See Chapter 7 for introduction to this concept.) Uses could be developed if the available energy could be stored. For example, if the energy being dissipated to the atmosphere from a cooling tower could be conveniently stored (at the proper temperature), it could be used for space heating.

A great part of energy use occurs as low temperature heat or ultimately ends up as low temperature heat. Much of this is used in heating and cooling buildings. As described in Chapter 8, space heating and air conditioning of buildings account for a large percentage of end-use energy throughout the world.

HVAC energy use is important because of its total share of national energy use and because of potential opportunities for fuel conservation through efficient energy use and energy management. It is also important in terms of load management, since heating and cooling systems frequently are key determinants of a utility's winter or summer peak load requirement.

Storage of energy, either in a 25−50°C (80−120°F) temperature range or a "cool" 5−20°C (20−70°F) temperature range, potentially has important advantages:

- Possibility to use off-peak generation.
- Possibility of capital cost savings.
- Possibility of improved energy use efficiency (e.g., by heat recovery).
- Possibility using alternative energy forms (e.g., solar).
- Possibility of fuel substitution (e.g., an available fuel for a scarce one).

Given these potential advantages, it is no surprise that heat/cool storage technologies are of interest to the energy manager, particularly in the design of new facilities where construction costs can be reduced (compared to retrofit installations).

Practical systems depend on overcoming a number of constraints, including cost, material availability and suitability, local weather and site

Table 12.10 Basic energy storage (heat/cool) system input parameters

Energy Source

Type: Electricity, heat recovery, fuel, alternate.
Availability: Continuous, noncontinuous, off-peak.
Backup required: Yes, no, maybe.

Site and Weather Data

Weather: Rain, snow, freezing, hail, wind velocity, insulation.
Temperature: Annual range, daily range, etc.

Projected Load

Utility experience: Heating/cooling, degree days, etc.
Customer load: Family size (or number of occupants), heated/cooled
   space volume.
Load characteristics: Hours of use, heat losses, infiltration, other heat/cool
   loads, etc.

System Concept

Design: Passive, active.
Storage media: Hot water, salts, brick or rock, cold water, ice, etc.
Equipment lifetime: Materials, number of cycles (if relevant).

Costs

Capital costs: Storage media, equipment, modifications (for retrofit projects only).
Operating costs: Utilities, maintenance, insurance, other.

conditions, utility rate schedules, and building codes and legal restrictions. Table 12.10 list parameters to be considered in an energy storage project.

The idea of energy storage is not a new one, having been practiced for centuries in human habitations. Today, heat storage is employed on a fairly wide scale in residences in Europe, and is beginning to be examined seriously in the U.S. Larger scale applications (commerce and industry) are still relatively rare, but are of increasing interest. They take on added importance for new facility designs incorporating alternative energy sources such as solar or wind. A brief summary of new technologies being studied in each of these areas is outlined below. Consult the literature for additional details concerning energy storage.

Cool Storage Techniques

Low temperature thermal energy storage, or "cool storage" as we shall refer to it here, has become increasingly prominent in recent years. This interest has arisen because of the shift to time-of-use metering by utilities, the use of night-radiation cooling, the increased efficiency of chillers with lower condenser temperatures, the ability to install smaller chilled water systems, and the rising cost of peaking power.

The various types of cool storage media that have been used include water, rocks, masonry, ice, clathrates, eutectic salts, metal hydrides, organic materials, and heat of solution materials. On a larger scale, one can also consider soil, lakes, or aquifers. The most widespread in commercial use are chilled water storage in large tanks, ice storage systems, and eutectic salt phase change systems.

The design of a particular storage system depends upon such diverse factors as the daily load curve, the type of heat sink, the storage volume available, the required temperature, and the period for which the cool temperatures must be maintained. The typical chilled water storage system for commercial buildings uses large insulated or underground tanks that can be located directly below the basement in a new design. In retrofit situations, the tanks are often placed below parking lots, tennis courts, etc.

The tank can be sized to serve different functions. For example, the storage can be used to condition a new building addition without increasing the capacity of the existing chillers. In a new building, the tanks can be sized to provide supplemental chilling so that chiller operation during the day does not exceed the building demand limit. Larger storage capacity can keep the chillers from contributing at all to the demand and allow the exclusive use of off-peak chilling. For a large load, say requiring a peak chiller capacity of two 300-ton chillers, with storage, two 200-ton chillers might prove adequate. (Typically two units would be provided so the load could be serviced when one unit is down for maintenance or repairs.)

In areas where time-of-use pricing for electricity occurs, as much as half of the electricity cost can be for demand charges. Energy storage can reduce this substantially, leading to savings that will pay for the system.

The configuration of tanks is significant in the system efficiency. It may be desirable to use two tanks to separate the chilled water 5.5°C (~ 42°F) from the return water at 15.5°C (~ 60°F). To minimize storage volume, this can be done in one tank with a piston or baffle system to separate the warm and cold liquids. A large tank with individual compartments can also be used. In the single tank mode, during on-peak hours, the chillers are off and the denser chilled water is drawn from the *bottom* of the tank and circulated through the building cooling coils. The warm return water is discharged through a diffuser into the *top* of the tank. During off-peak hours, the chillers are on and draw water from the *top* of the tank. Chilled water is distributed both to the bottom of the tank to recharge the system and to the building cooling coils to meet the minimal off-peak (nighttime) cooling load. With the proper tank arrangement, both heat and cool storage could be performed, increasing the versatility of the system.

In new installations, the relative costs of incorporating storage into the design goes down as the system capacity increases because of decreasing marginal cost for storage (large installations cost roughly US$2000 per ton of cooling capacity). This is the installed cost of chillers, ice storage tanks, pumps, and associated piping. In buildings over 100,000 m^2, the initial cost of the system with storage may be less than the cost of a large chiller system without storage.

Example. An office complex in Toronto, Canada. This facility used a storage capacity of 14 l/m^2 of floor space. The net cost increase for building the storage system and its interface was $100,000. Savings resulting from the system were estimated to be $30,000 per year in fuel and $10,000 per year in demand charges, resulting in a payback period of 2−3 years. (See Chapter 8 for an example that compares the electric load profiles of water cooled chiller systems with air-cooled chiller systems, with and without cool storage.)

In other parts of the world, particularly where water is less plentiful, rocks serve as heat storage media. Heat exchange is provided directly with the air, and the cooling source is often nocturnal evaporative cooling. A similar design uses a tank of water surrounded by rocks, so that the heat is transferred by convection to the rocks, by conduction to the water, and then by evaporation to the atmosphere.

Cool storage in ice might be appropriate in variable load HVAC systems such as might be found in a church. These operate at an evaporator temperature of less than 0°C (32°F), which is lower than what is actually needed for most cooling applications. As a result, chillers operate in a much less efficient region, causing significant increases in energy use. An improvement is possible with the use of clathrating agents that can raise the freezing temperature of water to the 10−15°C (50−60°F) range.

Combined cooling/heating of residences has been demonstrated in demonstration houses built by Oak Ridge National Laboratory using energy stored in ice in conjunction with a heat pump. This approach is referred to as an "annual cycle energy system" (ACES).

Eutectic salts are another media for storage. They offer the advantage of using heat of fusion of materials that is much higher than their normal heat capacity, resulting in higher volumetric efficiency. The problems with eutectic salts, such as poor nucleating that leads to super-cooling and settling that causes incomplete crystallization, need to be considered.

A typical eutectic salt that could be used for cool storage is sodium sulfate decahydrate mixed with sodium chloride and ammonium chloride. This mixture has a melting point of 13°C (55°F) that can be used to cool either water or air directly. Borax can be used as a nucleating agent, and inorganic thickening agents prevent settling. Mechanical stimulation may offer further improvement in preventing settling.

The salts can be contained in plastic trays that stack, leaving room for heat transfer between each level. The heat transfer medium in a large system would probably be water while a smaller (residential) design would be likely to use air, possibly with a solar heat source.

The heat of fusion of organic materials may also be used for cool storage. The paraffins C_{14} and C_{16} are examples with melting temperatures 3°C and 18°C, respectively.

The heat of fusion of water can also be used, in conjunction with the heat of solution of a dissolved salt to operate below 0°C. Other solutions of inorganic salts in water increase the volumetric heat capacity to utilize direct sensible cooling.

Problems with the various cool storage concepts discussed above include containment, corrosion, heat transfer equipment, control of vapors, effects of stratification, and costs.

Heat Storage Techniques

Heat storage is of interest due to the expansion of solar energy as a resource. To derive the full benefits of solar energy requires heat storage capability for night and cloudy day operation. Another major factor is the utilities' use of time-of-use metering and demand charges to stabilize their load curve.

The major heat storage media that have been deployed include water, rocks, ceramics, hydrated salts, and organic materials. Each of these has characteristics that make them applicable to particular types of installations. Consideration for any application should include the amount of heat necessary to be stored, the length of storage time, the volume available for storage, the heat transfer medium, the required temperatures, and the versatility to be used for cool storage.

Water storage gives the flexibility of use for both heating and cooling during the appropriate seasons. Hot water storage systems are currently available in the U.S. and Europe as package installations. These typically use storage water as the heat transfer and storage media. The ACES project described above is one example; another is electric water heating tanks heated during off-peak hours. The hot water is then circulated through radiators for space heating or hot water applications.

Rocks, ceramic bricks, or gravel (or any other bulk, low-cost, readily available, construction material) can be considered for thermal energy storage. A number of thermal energy systems using hot air as the heat transfer fluid and solar energy as the heat source have been demonstrated in residences. A large volume of material is required to provide adequate heat storage capacity.

The *Trombe* wall is another example. In a building, if a massive internal wall with large thermal capacity is exposed to sunlight during daytime hours, it will absorb solar energy. At night, as the interior space temperature drops, the wall will reradiate heat into the space.

Ceramics have been used in Europe for many years. These systems are basically ceramic bricks housed in an insulated cabinet. Electric resistance elements located between the bricks are energized during off-peak hours. During the day, space heating is accomplished when a small fan circulates air through the brick storage medium. Another similar approach makes use of existing radiant heating installations in concrete slabs. These can be scheduled to operate during off-peak hours.

The above systems employ sensible heat storage. Systems that employ latent heat storage are of interest because they permit smaller storage volume and weight. Materials with a solid—liquid phase change in the temperature range of interest are the principal candidates.

Hydrated salts are one example. Principal problems related to their widespread use include expansion and contraction during melting and freezing (this can affect heat transfer with fixed position heat exchangers), corrosion, toxicity and safety issues, and costs.

Finally organic materials such as paraffin P-116 offer phase change at a reasonable temperature. However, these typically have lower thermal conductivity than the hydrated salts. In addition, they are extremely flammable, which makes their use in buildings unattractive.

In summary, for storage of either heat or cool, there are three basic methods: sensible heat storage, phase change storage, or thermochemical storage, each with its own limitations and efficiencies. An example of thermochemical storage is the $CaO/Ca(OH)_2$ exothermic reaction. Adding water to CaO produces heat plus the hydroxide. The reaction is reversible if heat is added to drive off the water. In Europe, sensible heat storage costs range from €0.1/kWh to €10/kWh, depending on size and application. Phase change systems range from €10/kWh to €50/kWh. In Europe, it has been estimated that approximately 1.4 million GWh/yr could be saved in the building and industry sectors if heat and cold storage were widely adopted. This would also avoid the emission of 400 million metric tons of CO_2 per year.[17]

Electrical Energy Storage

Electrical energy storage is of interest both to the individual energy user and the supplying utility. From the user's viewpoint, storage would permit a more effective use of wind power and/or solar photovoltaic energy. This type of storage is currently practiced, but on a small scale, in limited (mostly remote) applications and at high costs.

Typical systems use conventional storage batteries, either nickel—cadmium for small systems or lead—acid for larger installations. In recent years, there has been considerable interest in developing new types of compact storage batteries and costs have fallen. The main impetus for this has been demand for longer range electric vehicles (see Chapter 10). Tesla Motors announced in 2015 that it is constructing a very large battery manufacturing plant in the state of Nevada. This facility will produce Tesla's lithium-ion batteries for its vehicles and also for commercial and residential electric power storage applications. High-speed flywheels are another means of energy storage with limited and special-purpose applications.

On a larger scale, electric utilities have done considerable research on large-scale battery storage facilities. There have also been research projects examining the feasibility of ground storage of compressed air or heat. For decades, utilities have used pumped hydro systems as a practical means of

[17] Hauer, A. (2012) "Thermal Energy Storage," an IEA-ETSAP and IRENA technology Policy Brief E17, January, 2012. Vienna: International Energy Agency. http://iea-etsap. org/web/Highlights%20PDF/E17IR_TES_Hauer_Jan2012_AH_GS_rev3%20HL.pdf (Accessed March 2015).

energy storage. In this system, water is pumped from a low reservoir to a high reservoir during off-peak hours, and then the flow is reversed during peak hours, generating electricity.

COGENERATION

Cogeneration—also referred to as combined heat and power—is the simultaneous or sequential production of mechanical energy (electricity) and thermal energy (process or space heat). As mentioned throughout this book, straight combustion of a fuel does not take advantage of its maximum potential to perform useful work. This situation can be improved by cogeneration. There are three main classes of cogeneration systems:

- Topping cycles.
- Bottoming cycles.
- Combined cycles.

Topping cycles produce electricity first and the thermal energy that results from the combustion process is used for process heat, space heat, or additional electricity production. Bottoming cycles produce high temperature thermal energy (typically steam) for process applications and then the lower temperature steam is recovered and used to generate electricity. Topping cycles are more common than bottoming cycles. Combined cycles are based on topping cycles, but the steam that is generated from the exhaust gas is directed into a steam turbine to produce additional electricity.

Since many industrial applications have a steady thermal (steam) requirement, but a variable electrical requirement with peaks and valleys, technologies that combine cogeneration with thermal energy storage are available to balance electrical and thermal loads better by "decoupling" the two when needed.[18] These systems have greater overall efficiency and take full advantage of any "excess" thermal energy produced during periods with higher demand for electricity for use during periods of lower electrical demand.

[18] Gellings, C., Parmenter, K.E., Hurtado, P. (2008). "Total Plant Energy Efficiency" in *Efficient Use and Conservation of Energy*, C.W. Gellings, ed., in *Encyclopedia of Life Support Systems*, developed under the Auspices of the UNESCO, EOLSS Publishers, Paris, France.

Typically cogeneration systems are feasible only in larger plants, are more expensive, and involve more complex equipment. However, they have proven to be cost-effective in many applications, including in the pulp and paper, chemical, steel, petroleum refining, and food processing industries, all of which have significant electrical and thermal energy requirements. If the energy manager encounters a situation where cogeneration might be applicable, there are several good sources of information.[19]

SUSTAINABLE DESIGN AND GREEN BUILDINGS
Sustainable Design

Sustainable design and "Green buildings" are important new concepts that have emerged in the last several decades and are finding growing acceptance among building designers, contractors, and building owners. The subject is broad enough to warrant an entire book in itself, and in fact such a book, which we highly recommend, has just been published.[20] Herein we provide a brief overview of this important new development.

The concept of sustainable design had its origins in 1987, when the United Nations Environmental Commission, chaired by G. H. Brundtland, first proposed sustainable development as:

... Development that meets the needs of the present without compromising the ability of future generations to meet their own needs.

While this statement could be applied to several different fields, it has been adopted by leading architects concerned with the built environment. Buildings have long lives—typically 40 years or more—so that impact on energy use (positive or negative) is long lived. Recognition of this fact, coupled with environmental concerns, has stimulated new approaches to building design.

Designers asked the question: could my building be entirely self-sufficient, without requiring external sources of energy? Or, carrying it a step further, could buildings be designed that would not need any utility services—not only no electricity or fuel, but no water, sewage, or waste disposal. Or, could realistic designs be developed that would at least minimize these requirements?

[19] Turner, D.W. (2007). "Cogeneration." Chapter 17 in *Handbook of Energy Efficiency and Renewable Energy*, Kreith, F. and Goswami. D.Y., eds. Boca Raton, FL: CRC Press.
[20] Iyengar, K. (2015). *Sustainable Architectural Design: An Overview.* London: Routledge.

Table 12.11 Six principles of sustainable design

- Use renewable energy and resources, as opposed to those that are finite.
- Design for recycling—plan for end-of-life reuse of materials.
- Insist on energy efficiency, not only in operation, but in fabrication and manufacture of materials.
- Eliminate or minimize waste by reuse, composting, or biodegradation.
- Substitute: Use local materials to minimize transportation energy and pollution; find alternate construction processes; use services with reduced impacts (rent vs. buy, etc.).
- Reduce impact: Select sustainably produced or recycled materials that are less energy intensive and nontoxic.

Sustainable design eliminates or minimizes negative environmental effects by using renewable materials and resources (e.g., solar energy vs. fossil fuels), by recycling (to avoid waste disposal), by material selection (avoiding energy intensive or finite minerals), and other means. Sustainable design also requires a more holistic view of design that considers sources and origins of materials, transportation, and extraction methods. Table 12.11 summarizes some basic principles of sustainable design.

Green Buildings and LEED Certification

In 1990, the UK Building Research Establishment published its Environmental Assessment Methodology, now known by the acronym *BREEAM*. It is reported to be the world's longest established and most widely used method of assessing, rating, and certifying building sustainability. More than 250,000 buildings have been BREEAM certified and over a million are registered for certification—many in the United Kingdom and in more than 50 countries worldwide. Another widely used certification system is *Green Globes*.

The "Green Building Initiative" got its start in the U.S. in the 1990s when the American Institute of Architects published an "Environmental Resource Guide." In 1993, the U.S. Green Building Council (USGBC) was established. The USGBC launched version 1 of its Leadership in Energy Efficient Design (LEED) building certification program in 1998. Today there are dozens of organizations promoting and certifying buildings that encompass an array of sustainability features. The World Green Building Council is an organization of GBCs from all over the world.[21]

[21] See www.worldgbc.org.

In the U.S., the Green Building Council's certification program for new construction and renovation of various types of buildings, including office buildings, schools, and others is now in its fourth version. Buildings are given a LEED award based on a numerical rating system.

Buildings are rated according to criteria in eight categories, namely:
- Location and transportation.
- Sustainable sites.
- Water efficiency.
- Energy and atmosphere.
- Materials and resources.
- Indoor environmental quality.
- Innovation.
- Regional priority.

Within each category, there are mandatory or required tests to be met, plus optional measures that are given a specified number of points. A total of 110 points is possible. For example, under the second category listed above, "Construction activity pollution prevention" is a mandatory requirement. "Site development—protect or restore habitat" is an optional requirement that yields two points.

Buildings are then certified on a scale:

Category	Points
Certified	40–49
Silver	50–59
Gold	60–79
Platinum	80–110

Buildings that are LEED certified encompass many if not all of the basic principles of sustainability described in Table 12.11. For example, "indoor and outdoor water use reduction" is a mandatory requirement for certification, as are "minimum energy performance," "storage and collection of recyclables," and "refrigerant management." Daylighting, renewable energy production, and rainwater management are examples of optional criteria that are heavily weighted.

In 2015, the U.S. Green Building Council reported that over 3.6 billion square feet of building space have been LEED certified.

Example. Punahou School's accomplishments in Green Buildings and Sustainability. We refer again to the energy audit conducted by the authors and their team at Punahou School that is described in Chapter 6. In the ensuing two decades since our initial energy audit was performed,

Punahou School has made great strides in creating a sustainable and green, energy-efficient campus.

Punahou has continued its drive for sustainability. The School's goals are to promote a sustainable community by educating and learning through curricular innovation, institutional action, and behavioral change.

The school is addressing energy usage on a broad front. All lighting on campus now uses energy-efficient lamps. Less paper (1.3 million fewer sheets per day) and 7% less water is now being used. There are more gardens on campus for growing healthy foods that are consumed on campus. Green waste and cafeteria fruit and vegetable peelings are recycled into compost on campus. There are water refill stations to reduce the use and disposal of plastic water bottles. Students at all grade levels are involved in energy and sustainability projects of one type or another, from working in the gardens to doing energy audits. One student project, for example, was to measure computer energy use and develop strategies so students do not have to charge their laptop computers during the daytime when Punahou's electricity rates are highest.

On a larger scale, recent building projects have more than fulfilled the efficiency goals proposed in our 1997 report. New construction includes the *Omidyar K-1 Neighborhood*, a new five-building indoor/outdoor section of campus opened in 2010 to serve kindergarten and first grade students. It was planned with sustainability in mind, not only for its own operation but also as a teaching example. It features photovoltaic solar power to provide 60% of its energy use, along with water conservation and many other sustainable features. It received LEED Platinum certification from the U.S. Green Building Council. Other features of the buildings include a ventilation system that is controlled by carbon dioxide sensors, occupancy sensors for lighting, water saving plumbing fixtures, and additional metering and monitoring on principle electric systems.

Another project that was awarded an LEED Gold certificate is the *Case Middle School*, a complex of nine buildings serving six through eighth graders. The buildings are designed to benefit from natural trade winds ventilation; interlocks prevent air conditioner operation when windows are open to the trade winds. There are three ice thermal storage plants that are operated at night and allow the building complex to avoid the high electricity rates charged during the peak hours.

Work is just beginning on a new project, a new learning environment for grades 2—5, where the School aspires to achieve the latest in sustainable building certification.

In 2008, Punahou entered into a power purchase agreement whereby a contractor installed a 468-kW photovoltaic solar energy system on the roof of the men's gym and four other buildings. The contractor financed, installed, and maintains the system, selling power to Punahou at a discounted rate compared to the electric utility. This system generates 750,000 kWh per year.

A unique feature of the K − 1 neighborhood is a building dashboard that can be viewed at http://buildingdashboard.com/clients/punahou.

Punahou School is an outstanding example of what can be accomplished by an integrated approach to efficient energy use.

CONCLUSIONS

More efficient building designs are possible. Current technology permits improvements, both in retrofitting existing buildings and in new designs. The savings available to the energy manager from either of these approaches is considerable. Energy-efficient buildings need not sacrifice comfort or aesthetics. In fact, the opposite may be true. They do, however, present a greater challenge to the designer's innovation and creativity.[22]

Some energy efficient buildings may have slightly higher capital costs. However, the potential energy savings compared to conventional buildings—often as much as 50%—will quickly return the added investment. In certain cases, an energy efficient building will actually be cheaper than the conventional design, owing to the reduced size of the mechanical and electrical systems. The savings can be expected to become even more significant in the future as energy costs escalate.

It is necessary to have close coordination between design disciplines to achieve a fully integrated (form, function, and use) efficient building. Rising costs will continue to create an economic incentive to encourage integrated building systems the future.

[22] There are hundreds of zero-net energy buildings designed in the last decade all over the world. See *The New Net Zero* (2014) by William Maclay, White River Junction, VT: Chelsea Green Publishing, and *The World's Greenest Buildings: Promise Versus Performance in Sustainable Design* (2013) by Jerry Yudelson and Ulf Meyer. Abingdon, UK: Routledge.

CHAPTER 13

The Economics of Efficient Energy Use

INTRODUCTION

Rationally evaluating the economic benefits and costs of proposed energy management projects requires a method for economic analysis that involves the various important factors. This chapter discusses some of the methods and considerations for performing economic analyses. We illustrate the most common methods with a simple example—given two motors with different costs, efficiencies, and lifetime, which yields the best return? We mention the limitations of each method.

In the real world, these decisions involve uncertainty. The longer the life of the investment, the greater the risk. Rates can change, equipment can break down, requiring unplanned maintenance expense, new taxes can be imposed, and so on. There are methods for evaluating large, long-term projects and their risk, but these are generally beyond the scope of this book.

GENERAL CONSIDERATIONS

An economic study may be defined as a comparison between alternatives that expresses differences between the alternatives so far as practicable in monetary terms. Where technical considerations are involved, such a comparison may be called an engineering economic study. Engineering economic studies stress the importance of clearly identifying the alternatives in terms of numerical data that can be compared on monetary basis. Many engineering economic studies finally reduce to the question: "Does it seem likely that a proposed investment will ultimately be recovered, plus a return that seems attractive considering prospective returns obtainable in investments of like risk?"[1]

---

[1] Grant, E.L. and Ireson, W.G. (1960) *Principles of Engineering Economy* (4th ed.), New York, NY: The Ronald Press.

Energy Management Principles.
DOI: http://dx.doi.org/10.1016/B978-0-12-802506-2.00013-6

Ultimately, the decision to go ahead with an energy management project will involve a number of considerations such as company policy, availability of funds, the priorities of contending projects, the availability of fuel, and predictions about future costs.

Although it is common to have procedures for the budgeting of investment funds, it often happens that there are no established criteria for comparing proposals. There are several different methods of performing economic analyses. Some have advantages over others depending on the type of problem being considered. Some are simple to apply, but may not give proper results for complex situations. The main types of economic analyses that are commonly used are as follows:

- Life-cycle cost.
- Break-even.
- Benefit/cost.
- Payback period (simple or discounted).
- Present worth (net present value).
- Equivalent annual cost.
- Capitalized cost.
- Internal rate of return.
- Advanced methods.

In most of these methods, it is necessary to project future cost to obtain accurate and meaningful results. Historically, the price of labor and materials has been increasing over the decades, subject to world economic conditions. Likewise fuel and energy prices, as we have seen in earlier chapters, have generally increased but have had periods where they declined. To make a meaningful analysis, one has to make some informed judgments about future escalation of materials, equipment, and construction costs, as well as future prices of electricity and fuels. Likewise, the cost of money has varied with global economic conditions, being at a low value at the present time. In past years, interest rates have been as high as 20% per year in the industrial economies and over 100% per year in many developing countries. Knowing the cost of money applicable at the time of implementing a project is clearly an important variable. It will also depend on whether the project is implemented by a government entity or by private industry. Table 13.1 provides a suggested checklist of information that is important for conducting energy management economic studies.

Table 13.1 Checklist of data for energy management economic studies

- Investment costs
- Expected economic life in years (or capital recovery specified by management to be used for energy projects)
- Estimated salvage value at end-of-life
- Annual cost of taxes
- Annual cost of insurance
- Annual cost of materials
- Annual cost of direct labor
- Annual cost of indirect labor
- Annual cost of maintenance and repairs
- Annual cost of power and fuels
- Annual cost of supplies and lubricants
- Annual costs associated with space occupied by equipment
- Other annual indirect costs
- Desired return on investment
- Energy cost escalation factors
- Cost of money (interest rate)

BASIC CONCEPTS OF ENERGY STUDIES

At this point, it is appropriate to mention some of the main concepts that form the basis for energy management economic studies:

- The study is made from the viewpoint of the owner(s) (or investor) of an enterprise or facility; i.e., the ultimate payer and benefactor of the savings.
- The study is a comparison of energy management alternatives and deals with prospective differences between the alternatives.
- The effects of the decision are in the future and begin at the time of the decision.
- Insofar as possible, the differences between alternatives should be reduced to differences in money receipts and disbursements.
- If there are no economic differences between alternatives, then intangible and subjective differences may be relied upon to choose between the competing alternatives.

Each economic analysis method can be used to compare the relative benefits of several energy management alternatives. However, each method has advantages and disadvantages. Some methods may yield different results for different conditions, so one should be familiar with the limitations of each approach as well as the valid conditions of applicability.

EXAMPLES OF ECONOMIC ANALYSIS METHODS

To illustrate the various economic analysis techniques, we will analyze a typical energy management comparison problem using each of the economic analysis methods listed above. The question is to determine which of two 7.5 kW (10 hp) 3-phase AC electric motors should be selected based on the given economic data and operating efficiencies. One motor is NEMA premium rated and the second is an inexpensive "standard" motor found online (Figure 13.1).

The motor properties are as follows:

Motor properties	Motor A	Motor B
Output rating	7.5 kW (10 hp)	7.5 kW (10 hp)
Efficiency	85%	91.7%
Initial cost	US$450	US$993
Replacement life	8 years (16,896 h)	20 years (42,240 h)
Salvage value	US$150	US$330
Annual maintenance	US$100	US$100
Electricity cost	US$0.10/kWh	US$0.10/kWh

Based on the data given for the two motors, the annual operating cost of each motor can be calculated. The yearly operating time of the motors is

$$(8\ h/day)(22\ days/month)(12\ months/year) = 2112\ h/year$$

Electricity used by each motor assuming constant-rated load during 1 year is

$$Motor\ A:(7.5\ kW/85.0\%)(2112\ h/year) = 18,635\ kWh/year$$
$$Motor\ B:(7.5\ kW/91.7\%)(2112\ h/year) = 17,274\ kWh/year$$

Figure 13.1 7.5 kW (10 hp) Electric motor.

The *annual* operating cost of each motor is therefore given by the following:

	Motor A	Motor B
Annual maintenance and overhead costs	US$100	US$100
Electricity costs (at US$0.10/kWh)	US$1864	US$1727
Total	US$1964/year	US$1827/year

Thus, the annual cost savings (excluding replacement costs) of purchasing the initially more expensive motor B is US$137. Note also that motor B uses 1361 kWh less per year for the same useful work output. Therefore, besides yielding a direct cost savings to the operator of the electric motors, the fuel otherwise required to generate 1361 kWh/year of electricity has been conserved. This converts to approximately 2.5 barrels/year of crude oil in equivalent energy considering conversion efficiencies. It is also equivalent to approximately 1 ton of CO_2 *not emitted* for each year of operation.

The following sections describe the main feature of each of economic analysis type and give an example of the application of each method based on evaluating the two electric motors.

Life-Cycle Costing

As a first approximation to a detailed economic analysis, life-cycle cost can be used to assist in making a decision between competing energy management options. Life-cycle costing is based on a consideration of all costs associated with an alternative during its entire lifetime. If an energy manager was to make a decision on motor A versus motor B based on the initial costs only (which is often done by most individuals purchasing competing merchandise if the "functions" are essentially the same), then he or she would select motor A. This type of decision gives no thought to future costs and possible future savings of one alternative over another.

Life-cycle costing is said to be a "first approximation" since it gives no consideration to cost of money (interest). It is a better method than only considering initial cost as will be shown. Future costs to be considered are annual maintenance costs (labor and material), operating costs (electricity), and replacement costs (labor, equipment and salvage value, if any).

Table 13.2 shows the example of calculating the life-cycle cost for an assumed desired life of 20 years. The life-cycle costing analysis method indicates that motor B will save US$2827 over the lifetime of the

Table 13.2 Life-cycle costing example (in US$)

Item description	Notes	Motor A Per year	Motor A Total	Motor B Per year	Motor B Total
Annual maintenance costs		100	2,000	100	2,000
Operating costs @US$0.10/kWh	(1)	1,864	37,280	1,727	34,540
Replacement costs	(1,2)	56.25	1125	49.65	993
Salvage values		18.75	−375	16.50	−330
Total life–cycle costs	**(3,4)**		**40,030**		**37,203**

Notes:
(1) Assumes no escalation in electricity price.
(2) Assumes replacement motor price stays the same and ignores the cost of money. Motor A has 2.5 replacements.
(3) Life-cycle is for 20 years.
(4) Life-cycle advantage of Motor B over Motor A is US$2827.

investment compared to motor A. Thus, based on life-cycle cost analysis, the energy manager would select motor B.

As noted in Table 13.2, this example assumes energy and equipment costs remain constant over a 20-year period. While this is unrealistic, it can be justified in this case on the basis that either motor would be similarly affected and the outcome would be unchanged. For greater accuracy, the future costs and savings would be discounted, using methods discussed later in this chapter.

Break-Even Analysis

It is often necessary to choose between two alternatives, such as motors A and B, where one alternative may be more economical in a situation and the other may be more cost-effective for another set of conditions. By holding all but one of the points of difference between the two alternatives constant and allowing the value of one to vary, it is possible to determine the value of the variable that results in the two alternatives being equally economical. That particular value is defined as the "break–even point" for that particular variable. This terminology is derived from business ventures when the investors want to define the critical variable in profit and loss functions for the business. For example, one hears hotel managers remarking that their break-even point is 80% occupancy, or steel mill operators speaking of a break-even point at 67% of capacity.

These are the critical points above which these operations will just start to make a profit.

For the motor example, there are several ways that break-even analysis might be employed. We might want to know: (i) how many hours of full-load operation are necessary before motor B is as economical as motor A or (ii) what future cost of electricity will result in motor B being more economical than motor A for 3 years equivalence (operating hours) of full-load operation? The following examples illustrate the method.

Example. How many hours of full-load operation are necessary before motor B is as economical as motor A?

Solution. Let X represent the number of hours of full-load operation for the two motors to economically break even. The total cost may be equated as follows (neglecting replacement and interest costs):

Total costs = initial cost + annual maintenance cost + energy cost

Equating the total cost of motors A and B and then solving for X gives

US\$450 + (\$100/year)(1 year/8760 h)(X) + (7.5 kW/0.85)(US\$0.10/kWh)(X) = US\$993 + (US\$100/year)(1 year/8760 h)(X)+ (7.5 kW/0.917)(US\$0.10/kWh)(X)

Solving for X yields

$$0.88235X - 0.81788X = 543; \quad X = 8423 \text{ h}$$

Thus, in 4 years motor B is as economical as motor A, and after that, saves money. Again, remember that this analysis ignores interest and escalation. It is also of interest to show the break-even calculation graphically, as in Figure 13.2.

Example. At what future cost of electricity will motor B be more economical than motor A for 3 years equivalence (operating hours) of full-load operation.

Solution. Let Y represents the cost of electricity where the two motors economically break even. The equation from the first example may be rewritten as follows:

Equating the total cost of motors A and B and then solving for Y gives

US\$450 + (US\$100/year)(3 years) + (7.5 kW/0.85)(3 years)(2,112 h/year) (Y US\$/kWh) = US\$993 + (US\$100/year)(3 years) + (7.5 kW/0.917) (3 years)(2112 h/year)(Y US\$/kWh)

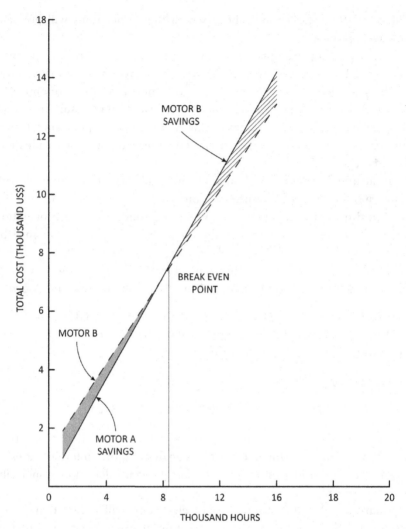

Figure 13.2 Break-even analysis example.

Solving for Y yields

{Motor A}   {Motor B}
$750 + 55,906Y = 1293 + 51,821Y$; or $4.085Y = 543$ and $Y = US\$0.133$

These results show that the full-load operation break-even point is about 8423 h, and electricity cost break-even point (for 3 years of operation) is US\$0.133 per kilowatt hour. These break-even values are interpreted to mean that: (i) for more than 4 years of use, motor B will be more economical to operate, otherwise the lower initial cost motor A should be purchased. (ii) If electricity costs were less than US$0.133 per kilowatt hour over the next 3 years, then motor A should be purchased and operated.

Break-even analysis is a useful tool for isolating a particular variable to study differences between two alternatives so that the decision may be made relative to the anticipated value of the variable and with respect to the break-even point. Most purposes do not require the exact value of the break-even point to be known, but only whether or not the estimated actual value will be above or below the break-even point.

COST OF MONEY

Up to this point, the simple economic analyses discussed (life-cycle costing and break-even analysis) have not considered the cost of money (the interest one must pay for the use of money for capitalizing or financing energy management projects). For significant capital investments, or for situations where the funds for financing an energy project must compete with demand for funds for other investments (which is almost always the case), the cost of using the money and the ultimate savings and return on the investment must be considered.

The economic analysis methods described in the following sections all consider the cost of money. Up to this point, we gave little attention to the rigorous formulation of basic economic theory. However, to put the following discussion and examples on a firm basis, we will utilize the accepted formulae for the respective analyses. For the remainder of this chapter, we will use the nomenclature and formulae shown in Table 13.3 to facilitate the understanding of the various analyses. The respective examples presented with each type of economic analysis will serve to illustrate the various economic formulations and terms as they are used.

It is not within the scope of this chapter to develop economic theories from first principles nor to illustrate the many uses for economic studies; rather the goal is the illustrated application of some specific types of economic analyses to assist an energy manager in making rational economic decisions. The reader is encouraged to read the references cited in this chapter for further clarification of the formulae and to obtain a better understanding of the use of economic studies in making business decisions.

Table 13.3 Formulae and nomenclature for economic analyses

Formulae	
Single payment:	*Payback period*
Compound amount factor	Given annual return R, find n to repay P:
Given P, to find S:	$n = \log[R/(R - ip)]/\log(1 + i)$
$S = (1+i)^n(P)$	
Present worth (value) factor	*Nomenclature*
Given S, to find P:	A_s = annual energy savings (\$)
	e_o = annual energy cost escalation (fraction)
$P = S/(1+i)^n$	e_f = annual escalation of fixed charges (fraction)
	ITC = investment tax credit (fraction of investment)
Uniform annual series:	
Sinking fund factor	i = interest or discount rate (fraction)
Given S, to find R:	L = useful lifetime of project (years)
$R = Si/((1+i)^n - 1)$	M = maintenance costs (fraction of investment)
	n = payback period, also year for evaluating discount factors (years)
Capital recovery factor	
Given P, to find R:	O_m = annual operating and maintenance costs (\$)
$R = Pi(1+i)^n/((1+i)^n - 1)$	P = an initial or present investment (\$)
	PW = present worth (\$)
Compound amount factor	PV = present value (same as PW)
Given R, to find S:	R = an annual cost to recover a present or future investment (\$)
$S = R((1+i)^n - 1)/i$	
	ROI = return on investment (percent)
Present worth factor	S = any future sum or investment (\$)
Given R, to find P:	S_v = salvage value (\$)
$P = R((1+i)^n - 1)/i(1+i)^n$	tr = state and federal tax rate (fraction)

Benefit/Cost Analysis

The calculation of the benefit/cost ratio for competing alternatives is useful for (i) comparing more than one alternatives and (ii) when contemplating replacement of an existing system. The purpose of benefit/cost analysis is to determine if the additional initial costs of a more expensive system are merited in light of long-term cost factors. The simple criterion governing the decision is that if the benefit/cost ratio is greater than 1.0, then the additional initial expense will result in long-term savings.

As recalled from the enumeration of the basic concepts of economic analysis, it is the differences between alternatives that are important. Thus, the benefit/cost ratio to be computed is the ratio of the annual savings (benefit of the one alternative relative to the other) to the yearly

Table 13.4 Benefit/cost example

First cost of motor A: US$450
First cost of motor B: US$993
Motor B exceeds motor A by US$543
Annual operating, maintenance, and energy costs:
  Motor A: US$1964
  Motor B: US$1827
Motor A exceeds motor B by (neglecting replacement costs)
  $1964-1827 = US\$137$
Capital recovery factor (CRF) = 0.231 (for 5 years at 5%)
Amortization cost for additional capital investment (motor B compared to
  motor A) is
(0.231) $(543) = US\$125$
The benefit/cost ratio is
Annual savings/amortized cost = $137/125 = 1.096 > 1.0$
If the replacement cost of motor A at the end of 5 years is added into the annual
  cost of motor A by using the equivalent annual cost for a payment made in
  year "n" (5 years) at interest i (5%), the sinking fund factor required is 0.181.
  Multiplying the sinking fund factor times the difference between the 5-year
  replacement cost and salvage value for motor A gives
Additional equivalent annual replacement cost = $(0.181)(450-150) = US\$54$
The new benefit/cost ratio is then $(137 + 54)/125 = 1.53 > 1.0$

Notes:
1. The capital recovery factor is that factor, when multiplied by a present debt
   (which from the point of view of the lender is a present investment), gives
   the uniform end-of-year payment necessary to repay the debt (the lender's
   investment) in n years and interest rate I, in this case, 5 years at 5% interest.
2. This analysis assumes that the cost of Motor A does not increase in 5 years.

amortized cost of the initial investment difference for the two alternatives.
The example in Table 13.4 shows that the benefit/cost ratio for motors A
and B is 1.096 neglecting the replacement cost of motor A and 1.53 if
the equivalent annual replacement cost of motor A is included in the
annual savings. Therefore, the additional US$543 cost of motor B will
result in long-term savings.

If motor B was being considered to *replace* an existing motor A, then
the initial cost difference would be the full US$993 cost of Motor B since
the capital cost of motor A was spent in the past and has no bearing on
the present and future costs being considered (this is referred to as a
"sunk cost").

Payback Period Analysis

The purpose of payback analysis is to determine the time to repay an investment. The payback period is used to compare various alternatives, with the energy management alternative having a shorter payback being the preferred investment.

The payback period is defined as a number of years required for the net cash flow to equal zero. Simple payback may be calculated by not considering the interest rate. A more refined method is also available for calculating the payback period considering the effective cost of money (interest).

The examples in Table 13.5 show the methods for calculating the crude and more exact payback periods for the electric motor example. The simple payback period for installing motor B to *replace* an existing Motor A is about 7.25 years and the payback considering 5% interest is 8.53 years. As long as a consistent approach is applied to the calculation of payback period, either method can be used for a preliminary comparison of energy management alternatives. A second example might be to compare whether to purchase Motor A versus purchasing Motor B. In this case, the incremental *cost difference* of the two motors would be used and the simple payback would be 3.96 years or 4.53 years with the interest rate included.

Table 13.5 Payback period example

Motor B initial investment: US$993
Annual operation savings: US$137per year
1. Simple payback period:

$$Payback = investment/savings\,per\,year = 993/137 = 7.25\,years$$

2. Payback period formula:

$$n = \log[A/(A - iP)]/\log(1 + i)$$

where
 $A =$ annual savings ($137)
 $P =$ initial investment ($993)
 $i\ =$ interest rate (5%)
 $n\ =$ payback period, years
Therefore

$$n = \log[137/(137 - (0.05)(993)]/\log(1.05) = \log(1.516)/\log(1.05)$$
$$= 0.180699/0.021189 = 8.53\,years$$

Except for the special case where money is so limited that no outlay of capital funds can be made unless the investment can be recovered in a very short time (e.g., 6 months to a year), the payback period is not a preferred method of ranking the relative merits of a set of energy management alternatives. The primary objection to the payback period analysis is that the payback period fails to account for the results of investments after the time that payback has been achieved. For example, one energy management alternative may have a relatively short payback period, say 2 years, but essentially no useful return after that period. Another energy management proposal may have longer payback period, say 3 years, but sufficient continued savings for many years to give an overall rate of return of 25—30%. This latter energy management alternative would be preferred from an economic point of view.

Present Worth Analysis and Net Present Value

The purpose of present worth (also called present value) analysis is to determine the worth, or value, of a future disbursement or receipt (or series of uniform disbursements or receipts). Present worth analysis is used for (i) comparison of alternative series of estimated money disbursements and (ii) placing a valuation on perspective money receipts, and (iii) trial-and-error calculations to determine unknown rates of return.

When doing a present worth comparison, the same period of service should be considered for the two alternatives. Thus, in the case of the two motors, we must consider enough renewals of each motor to serve for a number of years that is the least common multiple of the estimated lives (8 years for motor A and 20 years for motor B), in this case 40 years. Therefore as shown in the example of present worth analysis in Table 13.6, Motor A must be renewed at the end of every eighth year at a cost of US$450 and motor B replaced the end of 20 years at a cost of US$993. Note that the sixth 8-year replacement of Motor A and the third 20-year replacement of Motor B were not required in the present worth analysis since the equivalent equal time of service can be just prior to the end of the service life of the two motors.

The result of the present worth analysis example indicates that motor B has a lower present value of US$2054 as the net cost of 40 years' service. Thus, motor B is preferred over motor A when compared by their present worth (Table 13.6).

Table 13.6 Present worth analysis

Motor A	
Motor A initial investment	US$450
Single payment PW of cost of 8-year renewal: (450−150) (0.6768)	203
Single payment PW of cost of 16-year renewal: (450−150) (0.4581)	137
Single payment PW of cost of 24-year renewal: (450−150) (0.3101)	93
Single payment PW of cost of 32-year renewal: (450−150) (0.2099)	63
Uniform annual series PW of annual expenditures for 40 years (1,964) (17.159)	33,700
Total PW of net cost of 40-year service	US$34,646
Motor B	
Motor B initial investment:	US$993
Single payment PW of cost of 20-year renewal: (993−330) (0.3769)	250
Uniform annual series PW of annual expenditures for 40 years (1827) (17.159)	31,349
Total PW of net cost of 40-year service	US$32,592

Therefore, motor B shows a prospective savings of US$2054 in the present worth of the net cost of 40 years of service with interest at 5%

Notes:
At 5% interest, 8 years, single payment PW factor is 0.6768
At 5% interest, 16 years, single payment PW factor is 0.4581, etc.
At 5% interest, 20 years, single payment PW factor is 0.3769
At 5% interest, 40 years, uniform annual Series PW factor is 17.159.

Another application of the present worth method is to determine the *net present value*, defined as the sum of all income (or savings) over the specified time period, less the sum of all costs (expenses or outgoing cash flows) over the same period of time. In the motor example, we have the present worth of expenses. The present worth of the annual savings of motor B compared to motor A, US$137, is given by

$$(US\$137)(17.159) = US\$2351$$

To achieve this savings requires expenditure of $543 (the difference in the cost of motor B vs. motor A). This is a present cost and does not need to be discounted. Thus the net present value of motor B, compared to motor A, over 40 years with interest assumed to be 5% is

$$\text{Net present value} = US\$2351 - 543 = US\$1808$$

The present value is positive, so the investment will save money.

Equivalent Annual Cost Analysis

To compare nonuniform series of money disbursements and for alternatives having different useful lives, the equivalent annual cost method of analysis is useful. Reducing each nonuniform series of disbursements (and different life alternative) to an equivalent uniform annual series of payments allows for a consistent basis for comparison. The calculated equivalent annual costs provide an accepted method of comparing the long-term economy of two energy management alternatives for specified rate of interest. Obviously, the option that has the lowest equivalent annual cost will be the preferred investment.

The equivalent annual cost is calculated by using the previously discussed capital recovery factor (see Table 13.4 for the definition) and the present worth factor (refer to Table 13.6). The capital recovery factor spreads a present cost over the n years of lifetime of the investment, while the present worth factor converts a future disbursement (or receipt) back to a present value so that it, in turn, may be spread over the investment lifetime to obtain the equivalent annual cost (or savings) using the capital recovery factor at the specified interest rate.

Table 13.7 demonstrates the procedure of applying the equivalent annual cost analysis method for comparing the two electric motors. The results of the analysis again point out that Motor B is a better investment than Motor A for the long term. The example analysis shows that Motor B has an equivalent annual cost of US$1907, which is US$127 less than Motor A.

Capitalized Cost Analysis

For very long lifetime investments, or for present worth analyses in which the least common multiple of the lives of the two alternatives is greater than 25 or 30 years, the economic study can be for a perpetual period with valid results. Present worths for an assumed perpetual period of service are referred to as *capitalized costs*.

The calculation of the present worth of an infinite series of renewals (i.e., continual replacement of the investment at the end of its useful life) begins by converting the periodic renewal cost into an equivalent perpetual uniform annual series using the sinking fund factor for the appropriate

Table 13.7 Equivalent annual cost analysis

Motor A	
Motor A initial investment equivalent annual cost:	US$70
(450) (0.1547)	
Actual annual cost	1964
Total annual cost	US$2034

Motor B	
Motor B initial investment equivalent annual cost:	US$80
(993) (0.0802)	
Actual annual cost	1827
Total annual cost	US$1907

Savings in annual cost using motor B is 2034−1907 = 127
Note that if there were a nonuniform annual cost for the motors, the procedure
  would be to first determine the present value of the net cost (or savings) for
  each year during the investment lifetime, and then determine the equivalent
  annual costs by employing the capital recovery factor at work as was done for
  the initial cost. In the example above, the salvage value of the two motors has
  been ignored.
Note also the uniform annual series capital recovery factor at 5% interest for
  8 years is 0.1547, and for 20 years is 0.0802.

number of years and interest rate. The net effective renewal cost is the
initial cost minus the salvage value. The capitalized costs are simply the
annual costs divided by the interest rate.

The example in Table 13.8 illustrates the method of capitalized cost
analysis for the two electric motors.

To calculate capitalized costs, it is necessary to make the assumption
that the expenditure series already estimated for 20 years will repeat
themselves in succeeding 20-year periods. Thus, this approach assumes
that selecting motor A involves spending US$450 immediately and every
eighth year thereafter and also spending US$1964 every year. Similar
assumptions are made for motor B. The results of the capitalized cost
analysis show that motor B has a prospective savings of US$2609 over
motor A in capitalized cost for an assumed 5% interest rate.

INTERNAL RATE OF RETURN

In most instances, the primary criteria for judging the relative merits of
proposed investments should be internal rate of return (IRR). The rate of

Table 13.8 Capitalized cost analysis

Motor A	
Motor A initial investment:	US$450
Present worth of an infinite series of renewals (every 8 years) (450) (0.10472)/0.05 = 942	942
Present worth of perpetual annual expenditures of 1964 1964/0.05 = 39,280	39,280
Total capitalized cost, motor A	US$40,672
Motor B	
Motor B initial investment:	US$993
Present worth of an infinite series of renewals (every 20 years) (993) (0.03024)/0.05 = 601	601
Present worth of perpetual annual expenditures of 1827 1827/0.05 = 36,540	36,540
Total capitalized cost, motor B	US$38,134

Thus motor B shows a prospective saving of $2538 in capitalized cost with interest at 5%.

Note: The uniform annual series sinking fund factor for 8 years at 5% interest is 0.10742, and for 20 years is 0.03024.

return criterion can be applied using several different methods of computation and should use compound interest methods. The IRR method of analysis has the advantage that it is more universally used and therefore comparison of energy management uses of available funds can be compared directly with competing uses of funds such as increased production, improved company operating efficiency, or manufacturing cost reduction schemes. No matter how intrinsically satisfying it may be to save energy costs by reducing energy use, corporate decision makers must choose between competing uses of funds based on maximizing return on the investments of available company monies.

The previous types of economic analysis assumed an interest rate or minimum attractive rate of return. It may be more appropriate to compute a prospective rate of return on investment and compare this with the previously assumed interest rate or some other company standard.

The calculation can be done with a computer program or a spreadsheet program, or by a trial-and-error method to find the interest rate at which the present worth of the net cash flow is zero. Specifically, the process involves assuming two or more interest rates, calculating the present worth (or equivalent annual cost), and using interpolation or a graphical

Table 13.9 IRR analysis

For this example, we assume both motors have an economic life of 20 years and ignore escalation, salvage value and depreciation. IRR is that rate that makes the Net Present Value equal to zero

Motor A

| Motor A initial investment: | US$450 |
| Annual expenditures for maintenance and electric power | 1964 |

Motor B

| Motor B initial investment | US$993 |
| Annual expenditures for maintenance and electric power | 1827 |

Annual savings (earnings) of motor B over motor A: 1964−1827 = US$137.
Present worth of expenses is differential cost of motor B over motor A, US$993-450.
Net present value = Present worth of earnings − present worth of expenses
Zero = (137) (UASPWF, $n = 20$, i)−(993-450)
Rearranging
(UASPWF, $n = 20$, i) = 543/137 = 3.964
The equation for the (UASPWF, $n = 20$, i) is $[(1 + i)^{20} − 1]/[i (1 + i)^{20}]$ (from the Table).
This can be solved directly for the value of i (if you enjoy tedious calculations involving logarithms). An easier method is to look up several values in a table (or calculate with a spreadsheet), and then interpolate, as shown here for $n = 20$ years:

Interest rate I	(UASPWF, $n = 20$, i)
10%	8.514
20%	4.870
30%	3.316

With these three results, we know the IRR is between 20% and 30%. Try 25%:

| 25% | 3.954 (very close; try 24.9%) |
| 24.9% | 3.969 (Bingo) |

Thus the IRR in this example is 24.9%.

Note: UASPWF refers to uniform annual series present worth factor.

method to find the rate of return. The example in Table 13.9 shows the procedure for two electric motors with equal lives. Note that we first calculate the net cash flow for each year of the investment (minus signifies net cost, plus signifies net savings). Then we calculate the present worth

Table 13.10 Nonuniform annual cost example (US$)

Yr	Item	Motor A	Motor B	Difference	PW (0%)	PW (10%)	PW (5%)
0	Cost	−450	−993	−543	−543	−543	−543
1	Main.	−100	−100	0	0	0	0
	Elect.	−1864	−1727	+137	+137	+124	+130
2	Main.	−100	−100	0	0	0	0
	Elect.	−1957	−1814	+143	+ 143	118	+130
3	Main.	−100	−100	0	0	0	0
	Elect.	−2055	1905	+150	+150	+113	+130
4	Main.	−100	−100	0	0	0	0
	Elect.	−2158	−2000	+158	+158	+108	+130
Net present value:					+45	−80	−23

based on various estimated interest rates. The present worth factors are found by looking them up in handbooks or an online reference, or by calculating them for the particular interest rate and period using the following expression: Uniform Annual Series Present Worth Factor $= [(1 + i)^{20} − 1]/[i(1 + i)^{20}]$.

This IRR analysis example shows that purchasing motor B rather than A will result in a 24.9% return on investment. From a corporate standpoint, the decision to proceed with this energy management option must be weighed against the returns of other options and other possible investments that are currently available to the firm.

In this case, we simplified the analysis by assuming the annual costs were uniform. For a nonuniform series of expenses, the analysis would be as shown in Table 13.10.

The calculation has been truncated at 4 years for convenience. Normally it would be continued for the life of the project. Note each project outflow is listed as a negative (net expense) and any savings as a positive. Then each line item is converted to a present worth using a present worth factor and assumed interest rate for that year. Then the total gives the net present value. At 0% interest, the net present value is positive. Trying another rate, 10%, the net present value turns negative, so we know the IRR for the first 4 years of this investment is between 0% and 10%. Trying 5%, the result is still negative. With one final iteration (not shown), we find that with an IRR of 3%, the net present value is zero. If the series was extended to year 10 and beyond, the savings would grow and the IRR would increase.

SIMPLIFICATION OF ANALYSIS

Once a minimum attractive rate of return is stipulated or prospective rate of return is calculated, there may be an advantage in using the annual cost method or present worth method to make economic studies at the plant operating level or design level. The minimum attractive rate of return becomes the standard by using it as the interest rate in compound interest conversions for comparisons based on equivalent uniform annual cost or present worth analyses. The advantage of this approach is that the annual cost and present worth methods are somewhat easier to apply because of the absence of the trial-and-error calculations needed in computing rates of return. Therefore, operating plant personnel and designers may be less resistant to making economic studies based on annual cost than they would to rate of return.

Because of greater simplicity, it may also be desirable to base many energy management economic studies at the design level on a minimum attractive rate of return *before* income taxes. In general, someone who understands the income tax aspects of the subject should make a sufficient review of such studies to ensure that they are made in circumstances where the before-tax and after-tax analyses will lead to the same energy management decisions.

ADVANCED ECONOMIC ANALYSES

For large capital expenditures requiring decisions at the top of the corporate organization, it may be advisable to employ more advanced types of economic analyses. Some other considerations that may have an important impact on corporate wide energy management decisions, if included in the economic analyses, are as follows:
- State and federal tax rates.
- Provisions for risk and uncertainty.
- Energy cost escalation.
- Fixed cost escalation.
- Labor cost escalation.
- Depreciation schedules.
- Investment tax credits.
- Fixed asset taxes.
- Rental space escalation.
- Property insurance.
- Retraining of personnel.

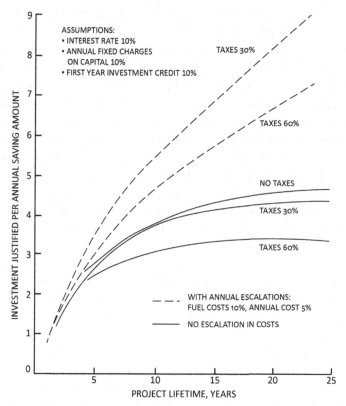

Figure 13.3 Investment per annual dollar saved.

As an example of how two of these parameters can affect an energy management decision, Figure 13.3 shows the maximum justified capital expenditures that can be invested per annual dollars saved for various tax rates and cost escalation rates for project lifetimes of 1–25 years.[2] Zero cost escalation and no tax consideration cases are shown for comparison. The curves illustrate the significant impact that cost escalation can have on investment decisions, particularly for project lifetimes greater than 5 years. For example, for a 15-year investment lifetime, a 30% tax rate, 10% interest rate, and a 10% first-year investment credit, and no escalation, an investment of about US$4.25 per annual dollar saved would be justified. In comparison, with an energy cost escalation of 10% per year and 5% for other annual costs, an investment of about US$7.10 would be

[2] Smith, C.B. ed. (1978). p. 39, *Efficient Electricity Use,* 2nd ed., New York, NY: Pergamon Press.

justified. Thus, about a two-thirds increase in capital investment can be justified for reasonable annual escalation rates for energy and other costs.

Risk needs to be considered in any investment decision, although it is often neglected because it is difficult to quantify. Risk becomes even more important the longer the lifetime of the project or investment.[3]

For example, suppose management decides to install a more efficient but costly color film processing machine in a printing plant. The machine works perfectly and is expected to pay for itself in 5 years. It is installed about the same time digital cameras take off in the marketplace. Due to lack of film processing orders, the plant is closed 4 years later. Thus, because of an unforeseen market risk, what should have been a money-saving investment turns into a loser. With the advent of the new digital technology, no one will buy this perfectly good used machine, so even the salvage value is lost.

Plant closures are one risk. New environmental regulations, changing market conditions, high maintenance costs, inflation, technological innovation, or equipment failure to perform as expected are other typical risks that can change the project outcome.

One way to handle risk is to increase or decrease the discount rate. Thus if the company expects a 10% rate of return on a "normal" project, it may use 12% to account for a higher degree of risk. Another approach would be to use a sensitivity analysis. The return for project might be calculated with three different sets of annual costs (or three sets of annual incomes). If even in the worst case scenario, the rate of return is acceptable, the project could then proceed.

For complex situations, consider using Monte Carlo or other simulation tools to study different conditions. Details are beyond the scope of this book.[4]

COMPARISON OF METHODS

Table 13.11 outlines the advantages and disadvantages of each of the methods described above and provides our recommendations. For quick analyses on simple projects with short lifetimes, use the payback method.

---

[3] See Thuesen, G.J. and Fabrycky, W.J. (2000), *Engineering Economy*, 9th ed. New York, NY: Prentice Hall, for additional details on advanced topics in engineering economy.

[4] Kreith, F. and Goswami, D.Y., eds. (2015). Chapter 11, "Economics Methods," by Short, Walter and Ruegg, Rosalie, in: *Handbook of Energy efficiency and Renewable Energy Technologies*, London: CRC Press/Taylor Francis Group.

Table 13.11 Comparison of methods

Method	Table/figure number of example	Advantages	Disadvantages	Recommendations
Life-cycle costs	13.2	Includes operating costs, replacement costs, life effects	Normally does not include escalation and interest	Use for small projects with short lives
Break-even	Figure 13.3	Simple comparison of two alternatives	Does not include cost of money, escalation, or effects beyond break-even point	Do not use except in special cases
Benefit/cost	13.4	Simple criterion for acceptance	Difficult to relate to management financial criteria; does not consider the effect of different project lives	Do not use
Payback period	13.5	Easily understood by management	Does not consider variable project lifetimes, future events	Recommended for small project with lives less than 5 years
Present worth	13.6	Takes into account escalation, cost of money, variable project lives	More calculation effort involved	Recommended as best overall method for all types of projects
Annual cost	13.7	Similar to present worth	More complicated computationally	Do not use
Capitalized costs	13.8	Useful for long-lived projects	For long-lived projects should include escalation	Do not use
Internal rate of return	13.9	Similar to present worth; results easily understood by management	Needs iterative calculations	Recommended for use

For more complicated project, use the net present value or rate of return method. Computer programs or spreadsheets can be used to facilitate all types of economic analyses.

EFFECT OF TAXES AND ESCALATION

A more involved example will illustrate important effect of taxes and escalation.

Example. A system to recover compressor waste heat and use it for heating domestic water. An industrial building had an electric domestic water heater with a 760-L (200 gallons) capacity. The average demand for water was about 700 L/h at 50°C. Measurements indicated that the water heater used approximately 27.5 kW during the summer and 40.7 kW during winter. Water input temperature ranged from 4°C to 18°C.

In the same building, there was a reciprocating chiller unit rated at 40 tons (in terms of equivalent heat, 140 kW). This unit had a compressor motor rated at 37 kW (50 hp). Given that the average power required for water heating was 34 kW, it was determined that even if the chiller operated at half load, it rejected three times as much heat as was required for water heating. The energy manager in this case considered several alternatives and decided to add a second hot water tank, a heat exchanger, and a recirculating pump. The recirculating pump used 1 kW. The total system cost, including design, equipment, and controls was US$100,000. The annual savings without interest costs or escalation is

$$(34 - 1 \text{ kW})(8760 \ h/\text{year})(\text{US}\$0.10/\text{kWh}) = \text{US}\$28,908 \text{ per year,}$$
$$\text{suggesting a 3.5-year simple payback.}$$

The following additional information applies:
- Power used by pump motor, 8760 kWh/year
- Electricity price escalation, 5% per year $= e_o$
- Maintenance and operating costs, $1000 per year $= O_m$
- Tax rate, 35% = tr
- Investment tax credit, 10% = ITC
- Project life, seven years = L
- Salvage value, $0 = S_v$
- Insurance, $250 per year
- Interest, 5% $= i$

First calculate the annual savings in energy costs using the escalated energy price. Note that

$$e_o = (1+i)^n$$

The initial energy savings is US$28,908 as determined above. The energy cost for operating the heat recovery system is US$876 per year. These numbers escalate as follows:

Year	1	2	3	4	5	6	7
Energy savings (US$)	28,908	30,353	31,871	33,465	35,138	36,895	38,739
Energy cost (US$)	876	920	966	1,014	1,065	1,118	1,174

The operating costs for the system are summarized in Table 13.12. Note that the investment is recovered using the capital recovery factor for 5% interest and 7-year project lifetime, or CRF = 0.1728.

The significance of the CRF can be seen as follows. It can be approximated by straight-line depreciation and average interest defined as

$$i_{ave} = \tfrac{1}{2}Pi[(L+1)/L] \tag{13.1}$$

where

i_{ave} = average interest, %
P = present investment, $
L = Project life, yr
i = interest rate, %

Thus the capital recovery factor can be viewed as a way of depreciating the item while meanwhile earning a return on the investment, or

CRF = (approximately)depreciation + interest.

For $L = 7$ years, $i = 5\%$, and $P = 1.0$, we obtain

$$CRF = 1.0/7 + (0.05/2)[(7+1)/7] = 0.171$$

This result is very close to the value shown above (0.1728).

Thus the columns of costs under item A in Table 13.12 include all the expenses involved in operating the system as well as recovering the investment. A simple payback of 3.5 years was determined previously, based on the annual savings of US$28,908 per year. Now a more accurate value can be obtained by summing the net savings year by year until the investment is repaid. This exercise yields the following paybacks:

Before taxes: 85 months (7.1 years)
After taxes: 113 months (9.5 years)

Table 13.12 Summary of calculations for heat recovery example (in US$)

Description/year	0	1	2	3	4	5	6	7	Totals
A. System Costs									
Install system	100,000								
Capital recovery (5%, 7 yrs)		17,280	17,280	17,280	17,280	17,280	17,280	17,280	
Operation and maintenance costs		1,000	1,000	1,000	1,000	1,000	1,000	1,000	
Energy costs (escalated)		876	920	966	1,014	1,065	1,118	1,174	
Insurance costs		250	250	250	250	250	250	250	
Salvage value		0	0	0	0	0	0	0	
Total costs		19,406	19,450	19,496	19,544	19,595	19,648	19,704	136,842
B. Savings due to Heat Recovery System									
Energy savings	0	28,908	30,353	31,871	33,465	35,138	36,895	38,739	235,369
C. Profit and Taxes									
Gross profit (B − A)		9502	10,904	12,375	13,921	15,543	17,247	19,036	98,527
Taxes (35% of B − A)		−3,326	−3,816	−4,331	−4,872	−5,440	−6,036	−6,662	−34,484
ITC (10% of $100,000)		10,000	–	–	–	–	–	–	10,000
Net profit		16,176	7087	8044	9048	10,103	11,210	12,373	74,042
Annual return on investment, % (after taxes)	0	16.18%	7.09%	8.04%	9.05%	10.10%	11.21%	12.37%	

Notes:
crf for 5% and 7 years is 0.1728.
Repay original investment in:
Before taxes: 85 months, 7.1 years
After taxes: 113 months, 9.5 years

Similarly the return on investment for the entire 7-year period is found as

ROI before taxes equal $98,527/100,000 = 98.5\%$, 14.1% per year average.

ROI after taxes $= 74,042/100,000 = 74.0\%$; 10.6% per year average.

Note that the after-tax calculations include the investment tax credit benefit but do not include depreciation. If depreciation was added, the before-tax profit would be lower, taxes would be reduced, and the ROI would be higher.

The bottom line of Table 13.12 shows the annual return on investment after taxes. Note that this return is (i) after recovering the original investment and (ii) after earning 5% interest for 7 years on the original investment.

This example, although it is more complicated than the examples involving motors A and B, illustrates a number of important points for economic analyses:

- Escalation can be a critical factor.
- Annual costs are almost never uniform.
- Taxes must be considered in most decisions.
- Investment tax credit can be an important consideration.

Note also that once Table 13.12 has been constructed (e.g., in a spreadsheet), all of the various types of economic analysis can readily be performed.

FINANCING ENERGY MANAGEMENT PROJECTS

This chapter has focused on economic issues and so far has not considered the financial aspects of energy management projects. In government or municipal organizations, funding at low-interest rates may be available for financing projects. In the commercial or industrial sector, this is generally not the case.

Energy management projects in business must compete with other projects—such as expanding the plant or modernizing production facilities—for scarce capital. Sometimes it is the availability of capital—not the expected return on investment—that is a determining factor. Thus, if the firm is capital short, it may not be inclined to invest in energy management projects no matter what the return on the investment is.

General experience with industry indicates that paybacks of 1–2 years at most will be considered. Longer term pay off projects will be deferred in favor of other investments.

Access to capital varies from business to business and from country to country. During the early part of the twenty-first century, most businesses in the U.S. and Europe could obtain capital from banks or investors at interest rates in the range of 5–10% per year. Following the global recession in 2008, world economic conditions had pushed interest rates even lower, down to a few percent.

In the developing countries, capital is much more difficult to obtain and interest rates typically have been higher in the range of 10–20% per year.

Sometimes special funding is available for certain projects. In the U.S., Asia, and Europe, government subsidies and low-interest loans encourage renewable energy projects, improved insulation, and more efficient equipment. Other subsidies include investment tax credits typically, 10–30% of the equipment cost in the year in which it is installed, accelerated depreciation, property tax exemptions or deductions, or cost sharing. In the developing countries, many governments are subsidizing investments that reduce dependence on expensive imported fuels.

In summary, the energy manager should be aware of the prospects for financing proposed projects before submitting them to top management.

CONCLUSIONS

Several different types of economic analysis procedures exist that can be used for deciding on the best investment from the available energy management alternatives. An energy manager should be familiar with at least two or three of the analysis methods and be able to apply them for making investment decisions and making recommendations to upper management. The limitations of the analysis being used should be understood, with more exact analysis methods used for larger investments or where differences between options appear to be small. The implementation of all energy management opportunities must be based on a sound, rational economic basis, rather than on a subjective approach or emotional desire to save energy. For investments with economic lives greater than a year or two, it is important to include escalation of costs in the analysis.

CHAPTER 14

Implementation and Continuous Assessment

INTRODUCTION

After initiating and planning an energy management program, carrying out audits and identifying projects, and analyzing potential energy savings and project costs, it is time to enter the third phase of the energy management program cycle. The third phase is where the energy manager really begins to enjoy the fruit of the labor. However, before actually realizing the program's benefits, a little more planning must be done. First, it is necessary to firm-up specific goals, determine how best to prioritize projects, and then plan for project implementation and subsequent performance assessment. Once those details are taken care of, the energy savings and process improvements begin. Along the way, it is also critical to inform, train, and motivate personnel. Experience has shown that energy waste is as much influenced by the attitude and lack of awareness of personnel as it is by the inefficiency of equipment and systems.

As projects are completed, it is important to assess the program to evaluate its benefits and to determine if changes are required. Assessment is useful because it permits experience gained in one project to be applied to others. Also, corporate management support for the energy management program depends on its record of performance. We dedicate a large portion of this chapter to methods for measuring, verifying, and reporting energy usage, savings, and investments.

Planning and assessment activities should not be limited to the startup of a program, but should be carried out continuously to ensure the program's ongoing success and improvement. An element of this ongoing effort should be to anticipate future expansions and new facilities and plan for energy efficiency in those new spaces.

GENERAL PRINCIPLES FOR IMPLEMENTING AND ASSESSING ENERGY MANAGEMENT PROGRAMS

Table 14.1 outlines the general principles for implementing and assessing an energy management program. The main steps include setting program

Energy Management Principles.
DOI: http://dx.doi.org/10.1016/B978-0-12-802506-2.00014-8

Table 14.1 Principles for implementing and assessing an energy management program

---

Set goals

---

a. Define energy management metrics (energy indices and other key performance indicators).
b. Establish general goals and specific targets for success.

Prioritize and implement projects

---

a. Develop approach and criteria to prioritize projects.
b. Establish schedules, budgets, and resources for implementing projects.
c. Define criteria for evaluating project performance.
d. Design and implement projects.

Inform, train, and motivate personnel

Measure, verify, and report

---

a. Develop an M&V plan.
b. Track project information, recording actual project implementation dates, measure details, and costs.
c. Document any implementation problems (installation, operation, or maintenance).
d. Measure and analyze energy savings achieved.
e. Assess project economics, including estimating cost avoidance.
f. Conduct an energy management assessment to score the cultural and organizational aspects of the program.
g. Compare and document actual performance versus goals.
h. Evaluate ancillary benefits or drawbacks.
i. Develop corrective actions to remedy problems encountered.
j. Make recommendations for future projects.
k. Disseminate reports.

Establish the basis for an ongoing program

---

a. Continuously assess the program to ensure that the savings are still occurring.
b. Plan new facilities for energy efficiency.

---

goals; prioritizing and implementing projects; informing, training, and motivating personnel; measuring, verifying, and reporting program performance; and establishing the basis for an ongoing program. We use the remainder of the chapter to discuss each of these steps in more detail.

ESTABLISHING GOALS

Energy management program goals can take many forms. Some goals may reflect corporate or government mandates and others might be defined by the energy manager or energy management committee. They may include specific targets such as "reduce energy costs by 10% in the next 5 years;" or they might be more qualitative like "increase employee awareness." They may apply to one system or area of a facility, or they might be corporate wide. The most important thing is that there must be goals! Progress toward meeting the goals must also be measureable in some way so that the program can be evaluated and improved on a regular basis. Examples of metrics for measuring progress toward goal attainment include the following:

- Annual energy costs.
- Annual energy use.
- Peak demand.
- Greenhouse gas emissions.
- Return on investment.
- Cost avoidance.
- Production rate.
- Product quality.
- Occupant comfort, health, or safety.
- Energy indices (J/unit of production, J/m^2 of floor space, J/person, etc.).
- Projects identified.
- Projects completed.
- Number of employees trained in energy management practices, etc.

In some cases, some or all of the goals may be set prior to initiating an energy management program. In fact, one of the goals might be to start an energy management program. In other cases, the goals may begin to materialize during the program initiation and planning phase and then they might solidify during the audit and analysis phase. As important as it is to have concrete goals, it is equally important to revisit those goals and adjust as necessary as the program matures.

PROJECT PRIORITIZATION AND IMPLEMENTATION

Once a program is up and running and goals have been set, how should an energy manager prioritize projects? Prioritization will depend on the goals and on decision criteria important to the organization. Economic considerations are often on the top of the list, but a variety of other factors may also influence project prioritization.

One effective approach for the energy manager to employ with investment-wary top management is to suggest making all operations and maintenance improvements first. These types of improvements generally require little or no capital investment. The limited resources required can usually be squeezed out of operating funds in one way or another. Some operation and maintenance improvements also have the advantage of being easy to implement quickly, so that savings start to accumulate right away. Demonstrating low-cost savings early on is an excellent way to convince any skeptics of the value of the program. In this scenario, the next step would be to get management to agree to "set aside" the savings resulting from the first implementation phase for use in financing subsequent phases in the plan, which might include more capital intensive projects. As the possibilities for saving money and energy become more fully understood by all parties, then it will become easier to take greater risks for greater gains.

There will always be criteria for determining which investments are suitable. What is acceptable for one organization may be very different for another. Industries with limited access to investment capital may seek a payback period of 1 year or less. Larger firms with access to capital may settle for 2—3 years. Certain government agencies may find 3—10 years acceptable. Other times the present worth or internal rate of return will be the determining factor (see Chapter 13). Alternatively, the organization may choose to work with an energy service company (ESCO) who can offer them an energy performance contract. In that arrangement, the ESCO guarantees savings and the organization can finance the project and pay off their debt with energy cost savings realized through the project. Regardless, the energy manager needs to establish what criteria are acceptable to his or her management early in the process.

Example. Project prioritization. You are the energy manager of the XYZ Company. You have recently completed energy audits of several buildings and have identified some projects. Preliminary engineering analyses have confirmed the technical feasibility and an economic analysis indicates the projects are economically viable. Four projects have been identified so far (Table 14.2).

How should you proceed?

These projects may be ranked several different ways (Table 14.3).

The best answer for project prioritization depends on the criteria selected. If money for new projects is limited, the strategy might be to implement projects that require little or no capital cost, beginning with project number 3. If fuel supplies were short or onsite emissions limited, project 2 might be selected first. If electricity was in short supply, or subject to curtailment, project 1 might be implemented first.

Table 14.2 Costs and energy savings for XYZ company's energy management opportunities

Project	Initial cost (US$)	Energy savings (units/yr)	Energy cost savings (US$/yr)
1. Modify lighting controls	2,000	25,000 kWh	2,250
2. Install heat recovery system	30,000	1,250 GJ	5,875
3. Setback temperature at night	0	200 GJ	940
4. Insulate building attics	7,500	150 GJ	705

Note: In the above data, 25,000 kWh × 3.6 MJ/kWh = 90 GJ.

Table 14.3 Ranking of XYZ company's potential projects

Ranking criterion	Project ranking (best→worst)
• Least capital cost	3, 1, 4, 2
• Greatest energy savings	2, 3, 4, 1
• Greatest money savings	2, 1, 3, 4
• Shortest simple payback	3, 1, 2, 4
• Greatest energy savings per dollar invested	3, 1, 2, 4
• Projects that reduce electricity use	1, 2, 3, 4

Table 14.4 Codes and regulations to consider for energy projects

• Building code
• Electrical code
• Mechanical code (heating, cooling, ventilation, etc.)
• Plumbing code
• Fire prevention code
• Health codes
• Occupational safety and health administration (OSHA) regulations
• Public utility regulations (electricity, steam, gas, water, sewer, telephone)
• Air and water pollution regulations

So far we have only considered economic criteria. Other factors may enter into the decision. For example, it could be that the buildings associated with project 4 are occupied under terms of a 4-year lease. Since investment would not be recovered during the lease, this project might be eliminated or given a lower priority. Conformance with local building codes may come into play with projects 1, 2, and 3 (Table 14.4).

Personnel may need to be trained to operate or maintain the heat recovery system. Will there be human aspects to consider, such as employee relations or information needs? Will a night shift walk off the job if the night setback plan is implemented? Will some employees feel "deprived" if their lighting controls are the ones modified? All relevant factors need to be considered when prioritizing projects for implementation.

After project prioritization, the next step is to develop and carry out a detailed implementation plan. Depending on the degree of rigor of the initial energy and economic analyses conducted when identifying and prioritizing projects, development of the plan may require more or less additional analysis. At a minimum it will include working with contractors as needed to design the project and specify equipment, contacting utility representatives for program information, exploring financing options, and getting firm quotes from contractors. This is also the time to plan how project performance will be measured and verified (see below for information on developing measurement and verification (M&V) plans). Once management, the energy committee, and any other stakeholders are satisfied with the implementation plan, it is time to start construction!

INFORM, TRAIN, AND MOTIVATE PERSONNEL

Technology and economics can only go so far in implementing an energy management program. Ultimately, the success of the program depends on the informed and willing participation of employees and management. It requires the participation of individuals ranging from the custodians and maintenance personnel to the building manager and ultimately to top management. It is important to understand that without support of operating personnel, no energy management program will succeed. It is not sufficient to have improved lighting systems, more efficient air conditioning and heating systems, and better designed buildings, if operating personnel do not understand *how* to use them, do not know *why* to use, and are not *motivated* to use them properly.

An important consideration for any energy management program is the training of all operating and technical staff personnel. One approach is to develop a series of training programs that all employees must take. We recommend that such programs be carried out at all levels within the company and at each facility operated by the organization. For nontechnical personnel whose activities do not directly involve energy use, training can consist of briefings that describe the importance of energy management not only at the site but also in employees' homes and communities. For managers and technical personnel whose decisions have a great bearing on energy use, more detailed programs are justified. At a minimum, the training should include engineers, building managers, maintenance personnel, power plant operators, vehicle drivers, office personnel, and production personnel. Consider including the following elements in the training program:

- A review of current energy use in the nation, its relationship to the economy, to the environment, to employment, and to industrial productivity. This helps put the importance of the facility's energy management program into perspective.
- The types of energy use at the facility and the relative importance. What uses the most energy? Is it HVAC, process equipment, lighting? What are the costs?
- General techniques for energy management and efficient use. Emphasize that being cognizant of low- or no-cost efficient operation and maintenance practices can have a significant impact on energy use, and those types of actions can involve nearly everyone in an organization. Also describe other large-scale projects.
- Specific actions that can be undertaken at the facility to improve energy use efficiency and reduce cost. This element of the training should provide staff with a list of items they can all begin considering immediately (e.g., notify the energy manager if you see lights on in unoccupied areas or misuse of compressed air), as well as some longer term goals for personnel directly involved in energy systems. During the discussion, encourage staff to make suggestions for energy management projects.

Motivation is another essential feature of an energy management program. Many companies with successful energy management programs offer recognition and a nominal prize for personnel who provide suggestions for improving energy use in the facility. This type of acknowledgment is very effective in increasing awareness of possible energy management opportunities. If possible, consider implementing an incentive program that goes one step further by rewarding building managers and other personnel for achieving major reductions in energy use in their facilities. The incentive should be monetary and linked to the level of savings obtained. Also, comparing energy use across facilities within an organization can be extremely helpful in identifying underperformers and helping motivate them to improve. Friendly competition is also a great tool.

The labor force at any facility—whether it be a university, an office building, or a manufacturing plant—represents an important factor in the organization's operating budgets. No energy management measures should be undertaken that would materially reduce performance or productivity. Operating personnel should be considered both as a source of ideas and as partners in efforts to effect more efficient energy use. Consultation with operating staff members will often lead to recommendations for improved efficiency. If there are plans to modify equipment, change lighting systems, and so on, it is imperative to inform the personnel in the facility being modified of the changes, the reasons for making them, and what steps are being taken to reduce inconvenience. Having advance knowledge gives them an opportunity to ask questions and make any necessary adjustments in their own activities. If care is taken in informing operating personnel of the needs for an energy management program, and if their support and participation are solicited, the program is much less likely to meet with objections once it is implemented.

Example. Engaging Hospital Staff in an Energy Management Program. During an energy management program involving hospitals, the audit team spoke to many of the hospital personnel, including nurses, secretaries, kitchen employees, and administrators. Most were quite friendly and curious about what an energy management program involved. Some noted particular energy-related problems they faced such as rooms that were constantly too cold or improperly ventilated. Some, however, did have difficulty relating to energy management, particularly since, unlike buying a new fuel-efficient automobile, they felt that the cost/benefits to the hospital would have little direct impact on their own working conditions and lifestyles. They showed more interest in the program upon learning of the relationship between energy and jobs.

A key to creating a supporting, energy conscious staff within a large organization such as a hospital, therefore, is making individuals aware of the economic benefits of energy use that will impact their own lifestyles. They must also understand that energy management measures do not necessitate a decrease in the comfort of their working environment; in fact, the energy management program, in many cases, will yield a more comfortable, pleasant working environment.

Table 14.5 summarizes measures that can be considered for informing and motivating employees.

Table 14.5 Motivating personnel for energy management

1. Post signs near light switches, televisions, and other energy-using appliances and equipment that can be turned off when not in use.
2. Develop a webpage, posters, newsletter articles, and pamphlets with ideas for energy management in the facility and in the community.
3. Develop seminars, workshops, briefings, or videos that illustrate energy management possibilities and projects.
4. Start energy management suggestion boxes in various areas of the plant and inform all personnel of their existence and purpose; offer awards for suggestions.
5. Create energy management goals for the plant and announce these goals to all staff with a specific timetable. Let staff know that the savings will be of *direct benefit* to them.
6. Have groups of employees meet periodically to review energy use in the facility and to share ideas concerning methods for increasing efficiency.
7. Keep staff informed by issuing, on a regular basis, reports of progress and cost savings.
8. Sponsor contests, award programs, or incentives to employees, departments, or divisions that show the most successful energy management efforts.
9. Set aside 1 week during October of each year to be "energy management week," and provide booths, displays, speakers, and utility representatives as part of an in-house program or as a community service function.
10. Publish case studies or project reports that describe successful projects in sufficient detail so they can be tried in other divisions or facilities.

MEASUREMENT, VERIFICATION, AND REPORTING

An essential element of the implementation and continuous assessment phase of an energy management program is to take measurements, monitor equipment and facility energy use, interview employees, and perform analyses to verify that systems are operating as expected, and energy use and performance targets are being met. In addition to evaluating energy savings and other benefits from equipment measures (e.g., maintenance, advanced controls, retrofits, new systems or processes), this step includes assessing goals that relate to *behavioral* aspects of personnel and management (i.e., the organization's overall approach to managing energy). It also includes reporting how results compare with goals and disseminating information to the energy management committee and other stakeholders. As noted in Chapter 4, these activities reflect the fundamental management concept that people are only able to operate effectively if they know what they are supposed to accomplish and they receive feedback that tells them how well they are doing.

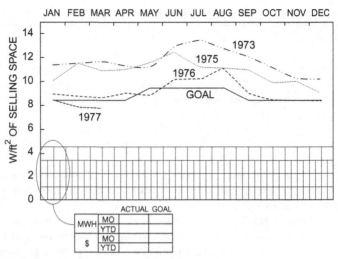

Figure 14.1 Energy tracking in a large department store.

Example. Energy tracking in a large department store. Figure 14.1 shows how a large Seattle department store energy manager tracked energy use in the stores under his supervision. He provided each store manager with a goal, expressed in electrical watts per square foot of selling space. The goal reflected seasonal variations: higher usage in winter and lower in summer. The energy manager reported back to each store manager his or her usage for a base year, actual usage for the last 2 years, and usage for the current year. He graphed monthly usage against the goal. He used the same chart to report monthly and year-to-date kilowatt hours and cost. Walking through a large department store with the energy manager on one occasion, he observed a highly illuminated empty display space. Turning to the store manager, he was overheard to remark "That level of illumination is unnecessary. We're not trying to sell the shelf."

That example illustrates a very clear energy performance goal, detailed measurements of actual usage, and effective reporting comparing the goal with actual energy performance (not to mention direct communication to "inform, train, and motivate" the store manager to not over-illuminate the shelf!).

There are numerous potential measurement, verification, and reporting techniques. The best approach for a given energy management project will depend on the scale of the energy management project and overall program, the type and complexity of project, resources available (personnel and equipment), and the audience receiving the reported results. Costly projects may demand a higher degree of rigor in analyzing impacts because of the higher risk associated with them; the analysis may also be more complex. Conversely, inexpensive maintenance measures may not justify comprehensive analysis, even though they may be just as complex to analyze accurately. The degree of rigor may also be mandated by internal management or by a third party, such as a regulator or a utility who is providing an incentive based on verified energy savings. Moreover, in the case of an energy performance contract where there is a performance guarantee with an ESCO, it is very important to have robust savings estimates since the arrangement for the facility's repayment of the debt hinges on savings actually realized.

The subsections below describe aspects of measurement, verification, and reporting in more detail.

M&V Plan

The first step is to develop an M&V plan that takes into account the considerations mentioned above such as project type and scale and stakeholders involved. Examples of performance evaluation techniques used in the M&V plans are as follows:

- Spot measurements of key parameters affecting energy use (temperature, pressure, flow, amperage, voltage, power factor, light levels, power draw, etc.).
- Temporary or continuous metering of energy use or energy use indicators.
- Engineering calculations, usually coupled with some measurements of key parameters.
- Simulation models, often calibrated with utility or submetered data.
- Statistical techniques such as regression analysis using utility or submetered data.

It is best to develop the M&V plan prior to project implementation so that the costs for M&V can be factored into the implementation cost and overall project economics. Early planning and accounting for M&V costs is especially important for large projects with higher financial risks. Establishing the plan prior to implementation also allows for more straightforward characterization of baseline (preimplementation) conditions for subsequent comparison with postimplementation conditions. It may be more challenging to estimate savings after the fact if the baseline has not been well established with appropriate measurements.

At a minimum, a well-conceived M&V plan for a given energy management project will include the information in Table 14.6.

Table 14.6 Information to include in an M&V plan

Summary of the project (overview)

- Current state of equipment or processes and associated issues.
- Proposed modifications and expected outcome, including justification for expectations.

Available baseline data

- Detailed description of all baseline equipment, such as age, model number, rated capacity, and efficiency.
- Operation and maintenance practices, including type of controls and control settings, load factor, maintenance condition, and operating schedule.
- Other variables affecting energy use, such as weather, occupancy, building characteristics, production volume, product type, season, and shift.
- Interactive effects with other systems (lighting and space cooling interactions, heat recovery for use in other process, etc.).
- Historical utility data, submetered data, weather, occupancy or production data, and spot measurements.

Specifications for proposed modifications

- Detailed description of proposed equipment (for retrofit and replacement projects).
- Modifications to operation and maintenance practices.
- Expected changes in other variables such as building modifications, occupancy changes, production increases, and new products.
- New interactive effects to consider.

M&V approach to evaluate performance

- Specification of measurement equipment and measurement accuracy.
- Type and duration of measurements needed, including any additional measurements required to characterize the baseline and all postimplementation measurements.
- Analysis approach, such as engineering algorithms, simulation modeling, or statistical techniques.
- Scope of analysis (whole facility or subsystem).

Analysis period and conditions

- Duration for measurements and savings calculations (weeks, months, years).
- Conditions for savings estimates (weather, season, occupancy, production level).
M&V budget
Reporting guidelines

M&V plans may also include documentation of project costs and pre- and postimplementation energy costs to assist in verifying energy cost savings and project economics.

M&V Standards

The Efficiency Valuation Organization (EVO) develops and promotes standards for evaluating the performance of energy efficiency, water efficiency, and renewable energy projects. The standards are documented in various International Performance Measurement and Verification Protocol (IPMVP®) reports.[1,2] The standards provide guidance for developing and carrying out M&V plans to quantify savings actually achieved as a result of project implementation. IPMVP documents are great resources for facility energy managers to use when internally evaluating energy management activities; they are also widely used by service providers and independent evaluators to verify savings claims. The IPMVP savings verification process includes the following:

- Specification of the measurement boundary (e.g., entire facility or just one system).
- Appropriate baseline period for estimating preproject energy use.
- Appropriate reporting period for estimating postproject energy use and savings.
- Any adjustments for weather, production, occupancy, etc.

In some cases, an energy manager may want to determine the savings actually achieved during the conditions in the reporting period relative to the conditions in the baseline period. In that type of analysis, factors such as changes in weather, production, occupancy, or even economic conditions may influence savings.

In other cases, it may be more desirable to predict savings for a fixed set of conditions, such as a typical weather year or a projected future period when production has doubled. This is where normalization comes in. Normalizing savings by one or more parameters allows for more accurate comparison of energy use between pre- and postconditions for a fixed set of conditions.

There are four IPMVP options for estimating savings.

- **Option A. Retrofit isolation: Key parameter measurement—** Incorporates short- or long-term measurement of one or more key parameters affecting energy use (e.g., spot measurements of lighting

---

[1] EVO. (2014). *International Performance Measurement and Verification Protocol: Core Concepts.* Washington, DC: Efficiency Valuation Organization.

[2] EVO. (2012). *International Performance Measurement and Verification Protocol: Concepts and Options for Determining Energy and Water Savings.* Volume 1. Washington, DC: Efficiency Valuation Organization.

power draw), coupled with engineering estimates for other parameters (e.g., operating hours).
- **Option B. Retrofit isolation: All parameter measurement**—Relies
  on short- or long-term measurements of system energy use (e.g., metering of motor power draw) during baseline and reporting periods.
- **Option C. Whole facility**—Uses whole facility utility data for baseline and reporting period.
- **Option D. Calibrated simulation**—Employs building or system
  simulation modeling—sometimes calibrated with utility or submetered
  data—to simulate baseline and reporting period energy use.

It is beyond the scope of this book to describe these options in detail,
but the protocol is available for free download in several languages from
EVO's website (www.evo-world.org). For applications where adherence
to the standards of IPMVP is it important or mandated by your organization or an independent reviewer, refer to the specific guidelines set forth
for developing M&V plans and reports.

Uniform Methods Project

The U.S. Department of Energy's Uniform Methods Project (http://
energy.gov/eere/about-us/ump-home) is another resource for proven
approaches to evaluate energy savings from energy management projects.
This effort is currently underway, with about a dozen protocols already
developed for common commercial and industrial energy efficiency measures such as energy-efficient lighting and controls, chillers, unitary and
split AC and heat pump systems, HVAC controls, retro-commissioning,
variable frequency drives, compressed air, and whole building retrofit
projects. The measure protocols are based on IPMVP options but provide more detail than IPMVP for applying the options to specific
measures.

Energy Management Assessments

An energy management assessment is a process to evaluate an organization's progress toward developing and improving an energy management
program. Generally energy management assessments use some sort of
scoring system to rate how far along the spectrum an organization is in a
variety of categories, with particular focus on the cultural and organizational aspects of an energy management program. For example, there is
an ENERGY STAR® energy management assessment tool available for

free download (it is called the Energy Program Assessment Matrix).[3] The tool is a spreadsheet-based matrix that consists of a series of energy management categories and three possible ratings to score an organization's status for each category: "little or no evidence," "some elements," and "fully implemented." The bullet points below summarize the energy management categories in the tool and how they tie to the three phases of an energy management program we present in this book.

- **Commitment to continuous improvement**—This relates to aspects of the initiation and planning phase discussed in Chapter 4, including importance of management commitment, naming an energy manager, and establishing an energy management committee. It also addresses the organization's progress toward formalizing an energy policy.
- **Assess performance and opportunities**—This is akin to the audit and analysis phase we introduce in Chapter 4 and address in detail in Chapters 6–12, including establishing baseline conditions, understanding parameters affecting energy use, and identifying opportunities for better energy management.
- **Set performance goals, create action plan, implement action plan, evaluate progress, and recognize achievements**—All of these categories fall into the implementation and continuous assessment phase described in this chapter.

The ENERGY STAR tool is just one example. Other similar tools are also available from utilities and energy management professionals.

The purpose of doing an energy management assessment is to see what is working well and what parts of the program could use improvement. It can also be used to set goals. Performing assessments on a periodic basis helps the energy management committee or corporate management track progress, or lack thereof. The results from these types of assessments tend to be somewhat qualitative, but nevertheless very informative.

Example. Energy management assessment comparison. Figure 14.2 shows a hypothetical example of how the results from the ENERGY STAR Energy Program Assessment Matrix (or a similar assessment tool) can be converted to a radar chart to compare the performance of two facilities in an organization. The chart shows relative performance in six energy management categories. Higher values in the radial direction

---

[3] ENERGY STAR Energy Program Assessment Matrix. Available from: http://www.energystar.gov/buildings/tools-and-resources/energy-program-assessment-matrix-excel.

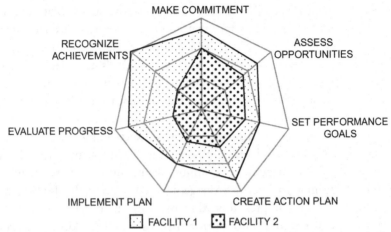

Figure 14.2 Comparison of energy management assessment results for two facilities.

represent higher performance using a rating of 1–3, where 1 is "little or no evidence" and 3 is "fully implemented." Facility 1 is farther along in all categories, but could still use improvement in areas such as setting performance goals and implementing the action plan. Facility 2 is in the earlier stages of energy management—they have some elements of a commitment, have started to assess opportunities and set goals, but have not progressed very far in terms of creating and implementing an action plan, or evaluating and recognizing achievements. This type of comparison serves two purposes: besides identifying areas for improvement for an individual facility, it also shows how one facility in the same organization has progressed more quickly. Benchmarking comparisons such as these may help motivate underperforming groups.

Reporting

Some form of tracking or documentation is essential to record performance of the implemented projects as well as to report on the overall progress of the energy management program to the energy management committee, corporate management, and other staff. Reporting can take many forms. Typically, databases or spreadsheets are used to track project level data, energy savings, and other key performance metrics. Energy managers can then analyze the data and use it to create tables and graphs for periodic reporting. Several companies offer software tools for displaying historical and current facility or system level energy and cost data, analyzing savings, benchmarking facilities with other similar facilities in an organization, diagnosing problems such as spikes in usage, and tracking usage versus targets. These types of tools are nice because they can be accessed online and may present facility data in real time, allowing for ongoing monitoring and prompt feedback.

When it comes to written reports or presentations, the format, length, and content will depend on specific requirements of the organization, but any type of reporting should present data clearly in tables and figures that compare specific performance targets with program achievements (energy savings, demand reductions, cost savings, greenhouse gas reductions, etc.). Reporting should also address attainment of more general program goals such as greater staff awareness of energy efficiency, improved comfort, and greater productivity. All formal reporting should end with next steps and recommendations for improving the program based on lessons learned from work carried out to date.

There are numerous ways to present data effectively when reporting on an energy management program. We have already provided a few examples of the types of graphs to use. Figure 4.4 (see Chapter 4) and Figure 14.1 (above) show energy use over time to illustrate trends and show reductions in usage due to energy management actions. Both of those figures normalize energy use by a key metric for the building—in one case it is energy use per capita, in the other it is energy use per square foot of floor area. Figure 14.1 is especially informative since it compares usage with goals "at a glance." Figure 14.2 shows an entirely different type of energy management data. The radar graph format depicts relative areas of strength and weakness for different energy management categories. This format is useful to present more qualitative information. Other common graphs for displaying *and analyzing* data are regression plots and CUSUM (cumulative sum) control charts.

Example. Regression plot of air conditioner data. In a hypothetical example, an energy manager performs maintenance measures to improve the performance of a three-ton unitary air conditioning system. The resulting energy savings vary with weather, with greater savings on hot days when the system runs longer. Figure 14.3 compares pre- and postimplementation average daily energy use (kWh) as a function of cooling degree days (CDD). The data points reflect submetered energy data collected over the course of several weeks using data logging. Performing a simple linear regression analysis on both the pre- and postdata illustrates the dependency of energy use on weather (measured in terms of CDD)

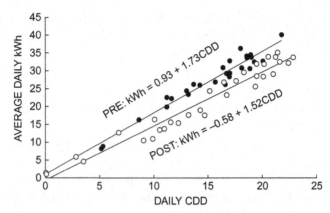

Figure 14.3 Comparison of pre- and postimplementation data using regression analysis.

and also shows savings trends. The equations for the regression lines take the standard form of a line: $y = mx + b$, where y is energy use (kWh), m is the slope of the line (kWh/CDD), x is CDD, and b is the y-intercept. This information can be used to estimate cooling season energy savings for different weather scenarios.

It is important to note that there are often multiple independent variables that affect energy use and savings for a given system. In the above example, we have assumed CDD is the key variable and have neglected other factors. However, other variables such as humidity, business operating hours (are they different on different days?), time-based electricity rates (do higher weekday rates result in reduced operation relative to weekends?), occupancy (how does it affect load?), and production (in a manufacturing plant, does greater production affect air conditioning load?) may also affect energy use. Therefore, to get the most accurate estimates, regression analysis should consider all potential variables and include those that significantly influence usage.

Example. CUSUM control chart to illustrate natural gas savings. A grain drying operation made process improvements to reduce the water required for cleaning the product. Less water equated to less natural gas use for drying. Figure 14.4 is a CUSUM chart that shows how preimplementation (baseline) natural gas use during months 1–11 compares with postimplementation use during months 12–24. We have also added a hypothetical goal to illustrate how an energy manager might compare actual results to a performance target.

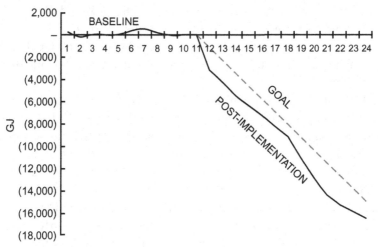

Figure 14.4 A CUSUM chart shows accumulation of energy savings over time.

Developing the CUSUM chart in Figure 14.4 was a three-step process. The first step was to use regression analysis of *actual* measured baseline natural gas usage as a function of heating degree days (HDD) to come up with a *predicted* baseline. This process resulted in a linear regression equation to represent the *predicted* baseline, which we refer to as the preimplementation reference case. (The approach was very similar to the method illustrated in Figure 14.3, but in this case the dependent variable is monthly natural gas use in GJ and the independent variable is monthly HDD.) The second step was to subtract actual measured energy use during the baseline period (months 1−11) and during the postimplementation period (months 12−24) from monthly gas use predicted for the reference case to determine the difference in usage for each month. The third step was to sum differences in energy use for each month using a cumulative sum approach and then plot the results.

The results depicted in the CUSUM chart are as follows:

- Data for the actual baseline period is roughly a horizontal line—A horizontal line in a CUSUM chart means no savings relative to the reference case. This means that the reference case prediction did a good job representing actual baseline usage.
- Data for the postimplementation period has a downward slope— Downward slopes in CUSUM charts reflect energy savings. Steeper downward slopes mean savings are accumulating at a higher rate than less steep downward slopes.

- The postimplementation curve is below the goal—This means that the project savings exceed the hypothetical goal we set.
- Total accumulated savings for months 11–24 are 16,460 GJ. This level of savings is equivalent to the avoidance of roughly 830 mt of CO_2 emissions.

CUSUM charts are useful for reporting savings trends and deviations from trends. A sudden change in slope in this type of chart indicates something happened to change the rate that energy savings accumulate. Upward slopes means energy use is increasing, flat slopes means energy use is staying constant, and downward slopes mean energy use is decreasing.

In addition to reporting energy use and energy savings data, it is often useful to report economic information. After all, a particularly important parameter for businesses is the impact on corporate profits. One metric is *cost avoidance*. Cost avoidance is the difference between actual energy cost savings achieved and projected escalated energy costs if the project had not been implemented.

Example. Reporting cost avoidance. In the XYZ Company, Project 1 (modify lighting controls) saves 25,000 kWh per year relative to the base year. However, during the second year, right after project implementation, the plant adds a new wing to the building to accommodate increased production. This wing causes additional lighting electricity use of 20,000 kWh. Rate escalation has been constant at 5% per year. What are the cumulative project savings and what costs were avoided after 3 years, compared to the base year? Base year lighting energy was 250,000 kWh (Table 14.7).

Total cost avoidance = 74,799−67,707 = US$7092.

In this example, the project saved money the first year. During the second and third years, escalation increased the price of electricity and increased production caused more electricity to be used, resulting in an increase in the electricity bill relative to the base year. *Without* the energy management program, however, costs would have been US$7092 higher than they actually were. This is a case where adjustment of the performance goal makes sense to reflect a new baseline condition.

Monitoring, Targeting, and Reporting

Energy management professionals sometimes refer to the process of establishing goals and measuring, verifying, and reporting results as

Table 14.7 Cost avoidance report for XYZ company's lighting controls project

Year	Energy used (kWh/yr)	Energy saved (kWh/yr)	Net energy (kWh/yr)	Energy cost ($/kWh)	Annual energy cost ($/yr)	Annual energy cost savings ($/yr)
Base	250,000	0	250,000	0.0900	22,500	0
1	250,000	25,000	225,000	0.0900	20,250	2250
2	270,000	25,000	245,000	0.0945	23,153	<653>
3	270,000	25,000	245,000	0.0992	24,304	<1804>
Three-year total cost and savings:		75,000	715,000	–	67,707	<207>

Without the project, the costs would have been

1	250,000	0	250,000	0.0900	22,500	0
2	270,000	0	270,000	0.0945	25,515	0
3	270,000	0	270,000	0.0992	26,784	0
Three-year totals		0			74,799	0

Monitoring, Targeting, and Reporting (MT&R). MT&R combines many of the same principles we have introduced here to set performance targets, measure and analyze data, monitor energy performance relative to targets, report results, and use findings to improve the program.

ESTABLISH THE BASIS FOR AN ONGOING PROGRAM

For long-term success in current facilities and for future success in new facilities, it is essential to establish the basis for an ongoing energy management program that includes continuous program assessment and addresses energy management in facility expansion plans.

Continuous Program Assessment

Energy management efforts should be ongoing. It is not sufficient to have an active program for 1 year, then stop in the mistaken belief that "everything that can be done has been done." Priorities will change, staff will turnover, building occupancies will vary, production will expand or decrease, maintenance will slip, old habits may return—there are a thousand reasons why the program should receive continuing review. It is true that as time passes, the easiest things will have been done, and it will become tougher to find major new savings. The program should then shift to a series of cost-effective "fine-tuning" operations. Because it is quite unlikely that energy costs and associated environmental issues will ever decrease to the point that effective energy management is no longer needed, there will be continuing challenges for the professional energy manager to address. Table 14.8 lists items to consider in maintaining an ongoing program.

Planning New Facilities for Energy Efficiency

Corporations frequently face the need to expand their facilities. Effective energy management programs will have a provision for optimizing the efficiency of future designs that goes beyond local building code requirements. Table 14.9 outlines energy management considerations involved in planning new facilities at existing sites and at new locations.

Existing site. At an existing site, there are two options: modify existing buildings (retrofit) or construct new facilities. In either case, the site is fixed, and this largely determines the available energy resources, transportation options, and design weather conditions.

When retrofitting or refurbishing existing facilities, the building orientation is established, and planning must accommodate the existing circumstances. In these cases, it is important to address building envelope improvements, at least in the parts of the building being modified. Improvements include additional insulation, better glazing, weather stripping, and a number of other measures that may not be cost-effective if undertaken separately, but which become so when undertaken as part of a retrofit or expansion project. Energy management plans should also stipulate minimum, cost-effective, energy efficiency requirements for lighting, HVAC, and process equipment, since these are major energy users. Designs should include better methods (daylighting, heat recovery, advanced processes), improved controls (occupancy sensors, economizers, variable frequency drives), and more efficient equipment (LEDs, heat pumps, premium efficiency motors).

New construction on an existing site offers other additional opportunities beyond those for retrofits of existing facilities. It may be possible to select the orientation of the new facilities on the site. Having that flexibility helps optimize energy use. For example, in a cold climate, the building orientation should provide shelter against prevailing winds to

Table 14.8 Develop continuing energy management efforts

Continue to measure results by collecting, tabulating, analyzing, and charting data

---

- Track, compare, and analyze utility bills.
- Measure, monitor, and analyze system or equipment level data.
- Calculate, monitor, and analyze key energy indices, such as energy use per unit of production, taking into consideration effects of complicating variables such as weather, operating hours, production schedules, product mix, and building occupancy.
- Assess indices for individual departments or areas and for the whole plant or facility.
- Compare energy indices with past performance and theoretical maximum performance, if feasible.
- Assess trends in other important metrics such as product quality, operation and maintenance costs, greenhouse gas emissions, and employee engagement.

Continue energy management committee activities

- Hold meetings on a regular basis.
- Have each committee member be a communication link between the committee and the department supervisors represented.
- Periodically update energy management project lists.
- Plan and participate in further building and site energy surveys.
- Communicate energy management techniques through interchange of ideas and documentation of new methods.
- Study the competition.
- Use outside consultants.
- Plan and conduct a continuing program of activities and communication to keep up interest in energy management within the company or organization.
- Develop cooperation with community organizations (clubs, professional organizations, utilities, technical societies, etc.) in promoting energy management.

Continue to involve employees

- Maintain the efforts to motivate personnel listed in Table 14.5.
- Continue periodic energy management training inside and outside company, perhaps with support from utility or other organizations.
- Send occasional emails on energy management tips to company employees.
- Encourage service on the energy management committee; perhaps rotate committee members so more people have a chance to participate.
- Maintain a library of handbooks and other resources on energy management.
- Have facility engineers hold periodic technical talks on lighting, insulation, steam traps, and other subjects.
- Extend efforts to the community through presentations to local organizations (social, professional, technical groups, etc.).
- Regularly recognize people responsible for energy management achievements, including publicizing in company newsletters, public news outlets, or social media.

(Continued)

Table 14.8 (Continued)
Continuously evaluate program

- Review progress in energy savings targets and other program goals by utility bill monitoring, equipment submetering, staff interviews, data analysis, and reporting.
- Evaluate original goals for reasonableness, percent completion, etc.
- Consider program modifications such as revising priorities or change selection criteria.
- Revise goals as necessary based on past results and new priorities.

Table 14.9 Planning new facilities: energy considerations

Expansion of an existing site (retrofit)

---

1. Building envelope changes (insulation, glazing, reflecting).
2. Lighting and HVAC improvements (controls, more efficient equipment).
3. New processes (more efficient equipment and methods).

Expansion on an existing site (new construction) Items 1–3 (above) and

---

4. Building orientation considerations (reduce heating and cooling loads).
5. Provisions for alternative energy sources.
6. Incorporation of passive design techniques.

Relocation to a new site (existing building) Items 1–3 (above) and

---

7. Site weather conditions.
8. Transportation patterns (distances, public transport availability, etc.).
9. Energy sources (future availability and costs).

Relocation to a new site (new building) Items 1–9 (above) and

---

10. Site layout and planning.
11. Energy storage capability (link to alternate fuels prospects).
12. Building energy performance specifications.
13. Incorporation of waste heat recovery.
14. Incorporation of water management and recovery.
15. Incorporation of appropriate metering and submetering.
16. Inclusion of a SCADA system or energy management and control system.

---

minimize heat losses and infiltration. Orientation and shading techniques can also be used to maximize the heating benefit from the winter sun, while decreasing the solar heat load during the summer. It may also be feasible to include alternative energy sources (solar water heating, ground-source heat pumps, solar photovoltaic systems) in new construction or at least make provisions for them to be installed at a later date. In addition, new construction may allow utilization of passive design techniques (skylights, berms, energy storage concepts).

New site. Many of the same opportunities exist when relocating to a new site. However, with a new site, it is possible to be selective about the location. Obviously, many factors influence the selection of the site, especially its costs, zoning limitations, proximity to suppliers and customers, and location relative to the workforce. A number of parameters also influence energy use; since they are partially determined by the site choice, they can be manipulated to reduce energy costs.

Site weather conditions are one example. In selecting a site, do not overemphasize published *macroclimate* weather data. Local conditions or *microclimate* data (a sheltered valley, for example) may be helpful in choosing between two alternative sites. Site selection will have an important effect on employee transportation energy use, and could ultimately affect recruitment and employee turnover. Distance, accessibility, and location relative to public transport should be considered.

A fuel and utility study will provide information concerning costs and availability of energy for the facility. Today it is no longer possible to take these essential services for granted; it may be necessary to obtain special permits or licenses for certain utilities; others may not be available at any cost.

New construction on a new site provides a planner with the greatest number of options. In addition to the site selection opportunities discussed above, it allows for completely new facility and system designs. As discussed in Chapter 12, the new discipline of sustainable design and green buildings will not only save energy, it will reduce the long-term harmful effects of greenhouse gas emissions and needs strong consideration. With increasing energy costs, spending more on efficient designs is often economically justified, especially considering the 30—40 year life of a typical building. The same is true for spending more on efficient equipment. In most cases, an outside architect-engineering firm will design new facilities. Consider providing them with performance specifications that stipulate the energy performance you expect from the building, especially in terms of installed lighting capacity and annual heating and cooling use.

Many of the alternative energy concepts under consideration today are expensive or uneconomic because they require new energy storage capacity, expensive modifications to existing buildings, and expensive new equipment. These disadvantages become advantages in new design. Energy-efficient buildings with tighter envelopes require smaller systems for heating and cooling. Energy storage capacity can be designed into the construction often at little or no additional cost.

New facilities can be oriented on the site not only to minimize the heating and air conditioning impacts of the local weather but also to facilitate the flow of people and materials within the building and between buildings.

A new design can often more cost-effectively incorporate ingenious methods for heat recovery and water use management. As noted in previous chapters, heat recovery can come from such diverse sources as waste hot water, lighting, air conditioning chillers, building exhaust air, air compressors, or process equipment. In many areas, water is also scarce resource. Wastewater may have discharge limitations or special costs. Water management should be considered to reduce water use and losses and to reclaim, treat, and reuse it where feasible.

It is also useful to provide additional metering, monitoring, and control capacity in new facilities. These types of additions are very inexpensive if done at the time of original construction. Options to consider include separate electrical meters for each major building wing, steam and water meters, relays or sensors on major items of equipment, SCADA systems, and energy management and control systems. Additional instrumentation provides the energy manager with information necessary to continuously assess how well the building and systems are performing and to ensure they are performing according to design.

Finally, local and national tax regulation should be reviewed. In many areas, there are significant tax credits for energy-efficient buildings. These credits apply to both passive design concepts and to the inclusion of alternative energy sources. They basically take the form of special investment tax credits. There are also, in some areas, loans and incentives offered by the local utilities to encourage energy-efficient buildings.

ISO Standard 50001

The International Organization for Standardization (ISO) has developed a standard for energy management called ISO 50001:2011—Energy Management Systems.[4] The standard addresses many of the principles we

---

[4] ISO. (2011). *ISO 50001:2011—Energy Management Systems*. Geneva, Switzerland: International Organization for Standardization.

have introduced here, including establishing, implementing, maintaining, and improving energy management systems. The purpose of the standard is to help organizations achieve continual improvement in energy performance. It provides a framework that includes guidance for setting targets and measuring and reporting results. Because it uses a similar model as other standards in use in industry (e.g., ISO 9001:2008—Quality Management Systems and ISO 14001: 2004—Environmental Management Systems), organizations familiar with those standards may find it relatively easy to implement. It is also possible to get certified to it.

CONCLUSIONS

Need for Assessment

There are several key points we wish emphasize in this chapter and by extension in this book. After initiating an energy management program using the principles we have described, it is important to assess and continuously monitor the program to assure that its goals are met and continue to be realized. Energy management requires more than technology. It depends on the motivation and training of personnel. Economics are always critical. Regulations and codes can provide barriers to or incentives for energy management programs. These are among the most obvious hurdles to be overcome.

If we shift our viewpoint from that of an individual, to an organization, to a community, or to a state, the situation becomes more complex.

The Tragedy of the Commons

Nearly 50 years ago, economist Garrett Hardin wrote an article with this title in the journal *Science*.[5] Hardin's title stems from a pamphlet written more than 100 years earlier describing a custom whereby villagers shared a common field where their stock (sheep or cattle) could graze. The point was made that if unregulated, it would be to each villager's benefit to put as many animals on the common field as possible, with the result that overgrazing would destroy the field, to the detriment of all. (This was the tragedy.)

Let us think of our planet's atmosphere as the commons. The highly industrialized nations of the world have freely tapped this universal

[5] Hardin, Garrett. (1968). "The Tragedy of the Commons," Science, vol. 162, no. 3859 pp. 1243–1248, Dec. 1968.

resource as a sink for waste gases. Now other nations want to increase their use of this resource for the same purpose.

One can imagine a hypothetical back-room conversation in the United Nations between an energy minister from India or China and one representing the U.S. or the OECD.

India/China: "What? You are saying that after you used 30% of the world's energy for the last 50 years, we can't do the same because it will cause global warming? We have as much right to the atmosphere as you."

There is no easy solution to this conundrum, and barring some vastly different political awareness and political compromise, there is only one proven way to give time to find a solution.

Energy Management a Stewardship

We live in an uncertain world. This was mentioned in the first chapter, where the prospect of wars being fought over energy resources was considered. Our energy options are definitely uncertain, inasmuch as no one can foretell with certainty what future energy costs will be, what energy forms will be available and when and if practical alternative sources will arise.

If past history is a guide, we can hope that the world's critical problems associated with the use of carbonaceous fuels will stimulate efforts directed at new energy technology innovation. Whether these innovations will actually occur and whether they will occur in a timely manner is of course uncertain.

Given these doubts, it is comforting to know that we can be confident about our abilities to use existing energy resources more efficiently and in ways that will reduce the burden placed on our common atmosphere. This is in our grasp today and is economically viable.

Energy management programs have the further advantage that they are implemented readily (without government intervention) and quickly (weeks to months). While such efforts probably will not in themselves solve the world's energy problems, they help buy time for us to correct the grosser inadequacies of our energy use patterns, and give us an opportunity to develop nonpolluting approaches.

Let us end on a positive, certain note. There is much that can be done, if we will but get on with the job. We will benefit; the children of the world benefit; perhaps we can improve the impression future historians will have when they look back at our generation and consider our stewardship of the earth's finite and precious resources.

APPENDIX A: ABBREVIATIONS, SYMBOLS, AND UNITS

A	area	m^2
A	(unit)	amperes
B	magnetic field intensity	weber/m^2
B_{cm}	nonflow availability	J
B_{cv}	flow availability	J
Btu	(unit)	British thermal unit
bbl	(unit)	oil barrel
C	capacitance	F
C	contrast	dimensionless
C	thermal conductance	W/K
Cap	rated capacity	ton
Ccf	(unit)	100 ft^3
CDD	(unit)	cooling degree days
CF	coincidence factor	dimensionless
CFM	(unit)	cubic ft/min
C_L	connected load	W
COP	coefficient of performance	dimensionless
CRF	capital recovery factor	dimensionless
CRI	color rendering index	dimensionless
CUSUM	cumulative sum (chart)	acronym
C_t	current transformer ratio	dimensionless
cal	(unit)	calorie
cd	(unit)	candela
cm	(unit)	centimeter
c_p	specific heat	J/kg · K
D	diameter	m
D	demand (electrical)	kW
D	distance	miles
De	equivalent diameter	m
Dlux	(unit)	10 lx
E	energy, work, heat	J or kWh
E	specific energy	J/kg
EER	energy efficiency ratio	Btu/Wh
Ef	emission factor	g CO_2/gallon
$EFLH$	equivalent full load hours	h
EMO	energy management opportunity	acronym
$EUPF$	energy use performance factor	varies
e_o	electricity price escalation	dimensionless
F	force	N

(*Continued*)

(Continued)

F	(unit)	farad
f	friction factor	dimensionless
fc	(unit)	foot-candle
ft	(unit)	foot
f	frequency	Hz
G	1 billion—10^9	giga
g	acceleration of gravity	m/s^2
g	(unit)	gram
H	(unit)	henry
H	pump head	m
H	enthalpy	J
H	annual operating hours	h
HDD	(unit)	heating degree days
$HSPF$	heating seasonal performance factor	Btu/Wh
$HUOD$	hours use of demand	h
HVAC	heating, ventilating, and air conditioning	acronym
Hz	(unit)	hertz
h	specific enthalpy	J/kg
h	(unit)	hour
ha	hectare	$10^4\ m^2$
h_c	convective heat transfer coefficient	$W/m^2 \cdot K$
hp	(unit)	horsepower
I	illumination (illuminance)	lm/m^2
I	interest rate	dimensionless
IE	interactive effects factor	dimensionless
IPMVP®	International Performance Measurement and Verification Protocol	acronym
IRR	internal rate of return	dimensionless
ITC	investment tax credit	dimensionless
i	electrical current	A
in	(unit)	inch
J	(unit)	joule
j	$\sqrt{-1}$	dimensionless
K	(unit)	kelvin
K_h	meter constant	kWh/revolution
k_h	material constant	dimensionless
k	thermal conductivity	$W/m \cdot K$
k	one thousand—10^3	kilo
L or l	length, thickness	m
L	inductance	H
L	(unit)	liter
L	luminance	cd/m^2
L	project life	yr

(Continued)

(Continued)

LI	luminous intensity	cd
LF	load factor	dimensionless
l	load fraction	%
lb	(unit)	pound
lm	(unit)	lumen
lx	(unit)	lux
M	$1,000,000$—10^6	mega
M	molar mass ratio	dimensionless
MBtu	(unit)	10^6 Btu
M&V	measurement and verification	acronym
$\dot{m}$	mass flow rate	kg/s
m	mass	kg
m	(unit)	meter
min	(unit)	minute
mo	(unit)	month
mpg	(unit)	miles/gallon
mt	(unit)	metric ton
N	motor speed	Hz or RPM
N	(unit)	newton
N or n	number (quantity) of something	dimensionless
NPV	net present value	$
O_m	maintenance and operating costs	$/yr
P	pressure	N/m^2 or Pa
P	present investment	$
Pa	(unit)	pascal
P_t	potential transformer ratio	dimensionless
PW	present worth	$
PWF	present worth factor	dimensionless
ΔP	pressure drop or differential	N/m^2
p	period of time	s or h
pf	power factor	dimensionless
pp	pumping power	W
Q	heat transfer	J
$\dot{Q}$	rate of heat transfer	W
Q	specific power	W/m^2
quad	(unit)	10^{15} Btu
R	resistance	ohms
R	thermal resistance	K/W
R	universal gas constant	$8316.6 \, J/kg \cdot K$
ROI	return on investment	dimensionless
RPM	(unit)	revolutions per minute
r	radius	m

(Continued)

S	apparent power	kVA
S	entropy	J/K
s	(unit)	second
S_v	salvage value	$
SEER	seasonal energy efficiency ratio	Btu/Wh
T	temperature	°C, K, °F, or °R
T	torque	$N \cdot m$
T or t	time	s
tr	tax rate	dimensionless
U	overall heat transfer coefficient	$W/m^2 \cdot K$
U	internal energy	J
V	volume	m^3
VARs	(unit)	volt–ampere-reactive
VAR	reactive component of power	VARs
v	velocity	m/s
v or V	voltage	V
V	(unit)	volt
$\dot{V}$	volumetric flow rate	m^3/s
W	(unit)	W
W	work done	J
$\dot{W}$	rate of work, power	W
w	weight fraction	%
X	reactance	ohms
x	excess air fraction	%
yr	(unit)	year
Z	impedance	ohms
z	elevation	m

Greek letters

Δ	incremental change	dimensionless
ϵ	effectiveness	dimensionless
ϵ	emissivity	dimensionless
η	efficiency	dimensionless
θ	phase angle	degrees
μ	specific internal energy	J/kg
μ_o	potential energy	J/mol
ρ	density	kg/m^3
ρ	line resistivity	ohms/m
σ	Stefan–Boltzmann constant	5.67×10^{-8} $W/m^2 \cdot K^4$
Φ	phase	dimensionless
Φ_L	luminous flux	lumen
ω	angular velocity	radiant/second
Ω	(unit)	ohms

APPENDIX B:
UNITS AND CONVERSION FACTORS

INTRODUCTION

This book has been prepared using the international systems of units (SI) that has been or is being adopted by all nations of the world. In SI practice, the approved units for energy and power are the joule and the watt:

Energy, heat, work: joule (J) = 1 newton meter = 1 watt second.

Power: watt (W) = 1 joule per second = 1 newton meter per second.

When SI units are used for calculations, they are frequently preceded by a prefix that is a multiple of 10. Then the same basic unit can be used to measure a very large or a very small quantity. Thus, a millimeter is slightly more than 1/16 of an inch, while a kilometer is slightly more than ½ mile. The prefixes, symbols, and multipliers are:

tera (T) 10^{12}
giga (G) 10^9
mega (M) 10^6
kilo (k) 10^3
hecto (h) 10^2
deca (D) 10
deci (d) 10^{-1}
centi (c) 10^{-2}
milli (m) 10^{-3}
micro (μ) 10^{-6}
nano (n) 10^{-9}
pico (p) 10^{-12}.

The SI system of units has several important advantages: (i) it is universal, with a single unit for each quantity (e.g., m for length); (ii) it uses decimal arithmetic, facilitating changes, calculations, and conversions; and (iii) it is coherent, meaning that when two units are multiplied or divided, the product or quotient has the units of the resultant quantity (e.g., m and ha are coherent while ft and acre are not).

The common units of the SI system are the kilogram (mass), the meter (length), the second (time), the newton (force), the watt (power), and the joule (energy). Table B-1 lists common conversion factors for energy and power. Table B-2 includes other useful SI conversion factors.

Table B-1 Conversion factors for energy and power
Energy, heat, work

Multiply	By	To obtain
Btu	2.931×10^{-4}	kWh
Btu (mean)	1.055×10^{3}	J
cal	4.190	J
kcal	4.190×10^{3}	J
ft · lbf	1.356	J
N · m	1.000	J
kgf · m (kilopond · m)	9.807	J
erg (1 dyne · cm)	1.000×10^{-7}	J
electron volt	1.602×10^{-19}	J
kWh	3.6×10^{6}	J
therm (10^{5} Btu)	1.055×10^{8}	J
Btu/lbm	2.323×10^{-3}	MJ/kg
Btu/short ton	1.162×10^{-6}	MJ/kg
Btu/gal (US)	2.79×10^{-4}	GJ/m^3
Btu/ft^3	3.72×10^{-5}	GJ/m^3

Power

Multiply	By	To obtain
Btu/s	1.055×10^{3}	W
Btu/min	17.58	W
Btu/h	0.2931	W
cal/s	4.184	W
ft · lbf/s	1.356	W
ft · lbf/s	1.818×10^{-3}	hp
ft · lbf/min	2.260×10^{-2}	W
hp	7.46×10^{2}	W
kgf · m/s	9.807	W
erg/s	1.000×10^{-7}	W

Table B-2 Other SI conversion factors

Multiply	By	To get
Length		
in	2.54	cm
ft	12	in
ft	0.305	m
km	0.621	Statute miles
m	3.281	ft
m	39.37	in
mile	1.609	km
mile	5280	ft
mile	1609	m
yard	0.9144	m
Area		
in^2	6.452	cm^2
ft^2	0.0929	m^2
acre	0.4047	ha
m^2	1.0×10^{-4}	ha
Volume		
ft^3	0.0283	m^3
gal (US liquid)	3.785	liter
gal (US liquid)	3.785×10^{-3}	m^3
gal (UK liquid)	4.546	liter
gal (UK liquid)	4.546×10^{-3}	m^3
barrel (42 US gal)	0.159	m^3
m^3	1.0×10^3	liter
Pressure and force		
lb/in^2 (psi)	0.06895	bar
psi	6.895×10^3	Pa
lb/ft^2	47.9	Pa
lb (force)	4.448	N
N/m^2	1.000	Pa
Mass and density		
ounce (avoirdupois)	28.35	g
lb (avoirdupois)	0.4536	kg
mt	2205	lbs
short ton (2000 lbs)	0.9072	mt
long ton (2240 lbs)	1.016	mt
kg	1.0×10^{-3}	mt

(Continued)

Table B-2 (Continued)

Multiply	By	To get
g/cm^3	1.0	mt/m^3
lb/ft^3	16.02	kg/m^3
lb/in^3	27.68	g/cm^3

Speed

mph	0.447	m/s
km/h	0.6214	mph

Thermal conductivity

$Btu/h \cdot ft \cdot {}^\circ F$	1.73	$W/m \cdot K$

Light

illumination (fc)	10.76	lx
illumination (lx)	0.0929	fc
luminance (fl)	3.43	cd/m^2
luminance (cd/m^2)	0.292	fl
luminous flux (lm)	0.001496	light watt
lumen $\cdot$ h	5.386	J

APPENDIX C: ENERGY MANAGEMENT DATA

Table C-1 Energy content of fuels and equivalencies

Fuel	Btu/gal, lb, ft^3	MJ/kg	GJ/m^3 or MJ/l
Hydrogen[a]	61,000 Btu/lb	120	0.011
Propane	91,500 Btu/gal	50	26
	21,500 Btu/lb		
Butane	94,670 Btu/gal	45	26
	19,520 Btu/lb		
LPG	90,000−105,000 Btu/gal	50	25−29
	(average 21,500 Btu/lb)		
Natural gas[a]	960−1550 Btu/ft^3	47−56	0.036−0.058
	(average 1000 Btu/ft^3)		
Manufactured gas[a]	460−650 Btu/ft^3	29−41	0.017−0.024
	(average 565 Btu/ft^3)		
Methane (organic digester)[a]	500−700 Btu/ft^3	28−38	0.019−0.026
Crude oil (1 bbl = 42 gal)	5.8×10^6 Btu	−	38.5
Oil no. 1	136,000 Btu/gal	46	38
	19,800 Btu/lb		
Oil no. 2	138,500 Btu/gal	45	39
	19,400 Btu/lb		
Diesel	130,300 Btu/gal	43	36
	18,400 Btu/lb		
Gasoline	127,600 Btu/gal	48	36
	20,750 Btu/lb		
Kerosene	135,000 Btu/gal	46	38
	19,810 Btu/lb		
Anthracite coal	12,000−13,000 Btu/lb	28−30	45−48
	24−26 MBtu/ton		
Bituminous soft coal	10,000−15,000 Btu/lb	23−35	32−47
	20−30 MBtu/ton		
Wood (12% moisture)[b]	8000−10,000 Btu/lb	19−23	8−10

[a]Data apply at standard temperature and pressure.
[b]Typical information for other woods, ranging from pines to live oak: specific weight, 28−61 lb/ft^3; energy content, 18−36.6 MBtu/cord or 225−475 kBtu/ft^3.

Table C-2 Green house gas equivalencies[a]

Energy use form	Multiply by	To get
U.S. non-baseload electricity	6.896×10^{-4}/kWh	metric tons of CO_2
Natural gas consumption	5.302×10^{-3}/therm	metric tons of CO_2
Natural gas consumption	5.302×10^{-5}/ft^3	metric tons of CO_2
Crude oil consumption	0.43/bbl	metric tons of CO_2
Coal consumption	1.862/short ton	metric tons of CO_2
U.S. passenger vehicle gasoline	8.887×10^{-3}/gal gasoline	metric tons of CO_2
U.S. home electricity use[b]	7.270/home	metric tons of CO_2
U.S. home total energy use[c]	10.97/home	metric tons of CO_2

[a]Does not include any greenhouse gases other than CO_2.
[b]Based on average of 114 million U.S. homes, 2012 data, electricity only.
[c]Includes electricity, natural gas, LPG, and fuel oil used by 114 million U.S. homes.
Source: U.S. Environmental Protection Agency (9/9/2014). Clean Energy: Greenhouse Gas Equivalencies Calculator http://www.epa.gov/cleanenergy/energy-resources/refs.html. Accessed April 8, 2015.

Table C-3 Miscellaneous data and conversion factors

- Speed of sound in dry air at standard pressure $\approx 331.4 + 0.6 \times T(^{\circ}C)$ m/s. So, at 20°C, it is approximately 343 m/s (1127 ft/s).
- Speed of light = 2.998×10^8 m/s
- Charge on the electron = 1.602×10^{-19} coulomb
- 1 coulomb/second = 1 ampere
- Siemens = unit of electrical conductance = amperes/volt
- 1 ton of refrigeration = 12 kBtu/h ~ 3.5 W cooling (to produce 1 ton requires ~ 1.0 kW of air conditioning)
- 1 lbm of steam ~ 1000 Btu ~ 1 MJ
- Specific heat of water = 4.179 kJ/kg $\cdot$ K = 1.0 Btu/lb $\cdot$ °F
- 1 cord = 128 ft^3 of stacked wood 8 ft $\times$ 4 ft $\times$ 4 ft or 80 ft^3 of solid wood

INDEX

Note: Page numbers followed by "*f*" and "*t*" refer to figures and tables, respectively.

Hot water and water pumping, 247–251
Hours use of demand (HUOD), 120–121
Human comfort and health, requirements
    for, 130–133
Hydrated salts, 323
Hydrogen, 22
Hydropower, 15, 23
Hysteresis, 105*t*, 261–262

I

Illuminance, 192–194
Illuminating Engineering Society of North
    America (IESNA), 193–194
Illumination, 190, 192–193
    clean luminaires to increase, 212
    pragmatic aspects of, 191–192
    task-oriented, 209–210
Impeller, 145
Implementation and continuous
        assessment, 56–58, 359, 360*t*
    establishing the basis for an ongoing
        program, 379–385
    continuous program assessment, 58,
        379–380
    energy efficiency, planning new
        facilities for, 380–384
    ISO Standard 50001, 384–385
    general principles for, 359–360
    goals, establishing, 56, 360–361
    informing, training, and motivating
        personnel, 57, 364–366, 367*t*
    measurement, verification, and
        reporting, 57–58, 367–379
    energy management assessments,
        372–374
    M&V plan, 369–370
    M&V standards, 371–372
    monitoring, targeting, and reporting,
        378–379
    reporting, 374–378
    Uniform Methods Project, 372
    need for assessment, 385
    prioritizing and implementing projects,
        57, 361–364
    tragedy of the commons, 385–386
Incandescent lamps, 195, 197–198
    60W incandescent A-19 lamp, 203

India
    electric locomotives in, 219–220
    energy use in, 229–230
    industrial energy use in, 229–230
Induction heating, 256
Induction motors, 262–263
Industrial Assessment Center (IAC), 114
Industrial energy use, 229–230
Inefficiency
    aggregate impact of, 109*t*
    causes of, 105*t*
    factors contributing to, 104–109
Inexhaustible energy resource, 14
Infiltration, 300
Infrared heating, 258
Initiation and planning phase, 47–50
    energy champions, 47–49
    institutional barriers, addressing,
        49–50
    management commitment, importance
        of, 47
Insolation, 308, 311
Institutional barriers, addressing, 49–50
Insulation, 128, 238, 253, 302–307, 303*t*,
        304*t*, 305*f*, 306*t*
    R-13 fiberglass insulation, 305*f*
Integrated building systems, 283
    building envelope design considerations,
        296–307
        glazing/fenestration, 298–299
        heat gains, reduction of, 307
        infiltration/exfiltration, 300–302
        reduction of building heat losses,
            302–307
    building function, 288
    cogeneration, 325–326
    energy management principles for,
        284–285, 285*t*
    energy storage, 318–325
        cool storage, 319–322
        electrical energy storage, 324–325
        heat storage, 322–324
    environmental conformation, 286–288
        effects of climate, 286–287
        microclimate important, 287–288
    Green buildings and LEED certification,
        327–330

Printed in the United States
By Bookmasters